Contemporary Futurist Thought

Science Fiction, Future Studies, and Theories and Visions of the Future in the Last Century

by
Thomas Lombardo, Ph.D.

Bloomington, IN

Milton Keynes, UK

AuthorHouse™
1663 Liberty Drive, Suite 200
Bloomington, IN 47403
www.authorhouse.com
Phone: 1-800-839-8640

AuthorHouse™ UK Ltd.
500 Avebury Boulevard
Central Milton Keynes, MK9 2BE
www.authorhouse.co.uk
Phone: 08001974150

First published by AuthorHouse 6/20/2006

ISBN: 1-4259-4577-5 (sc)

Library of Congress Control Number: 2006905558

Printed in the United States of America
Bloomington, Indiana

This book is printed on acid-free paper.

To Seven Young Souls on their Journeys in the Future

Bryan, Kristin, Tom, Daniel, Matthew, Emily, and Daniel

Acknowledgements

The beginning of this book and its companion volume, *The Evolution of Future Consciousness*, first sprang into existence in my mind in 1992 while I was standing in a check-out line in a Safeway grocery store. I had just moved to Arizona and aside from being hired as the new chair of the psychology and philosophy departments at Rio Salado College I was also put in charge of "Integrative Studies" - the capstone course for the Associate degree offered at Rio Salado. Integrative studies could focus on whatever theme I chose to select, as long as the theme somehow pulled together the breadth of undergraduate courses, especially the sciences and humanities that students took in completing their Associate degrees. I had been thinking about different possible topics for the course but that day in Safeway I still hadn't decided on anything yet. As I was waiting in line in Safeway I started browsing through the paperback bookstand in check-out and noticed that Alvin Toffler, author of the highly popular book *Future Shock*, which I had read back in the 1970's, had a new book out titled *Powershift: Knowledge, Wealth, and Violence at the Edge of the 21st Century*. In a flash - it hit me - why not do the Integrative Studies course on the 21st Century - in fact, on the future? Most, if not all of the main areas of study in an undergraduate college program could be addressed and synthesized in the context of the future. Having had a long standing fascination with and love for science fiction since childhood, as well as studying evolution and the nature of time in college and post-graduate school, I felt I had some background and expertise on the topic of the future. Creating a course on the future, it seemed to me, would be an interesting and challenging endeavor, and I felt that such a course would be of great value for students. Shouldn't we all try to think about and understand the future? So I charged into the topic and in the following months read whatever books I could find on the future beginning with Toffler's *Powershift,* his earlier book *The Third Wave,* and John Naisbitt's *Megatrends* series, and constructed a course on the future.

Once I started teaching the course I realized that although there were many good books that dealt with selected aspects of the future, no one book (that I was aware of) covered all of its main dimensions. I was reading books on the future that dealt with science and technology, human society and culture, and economics and politics, but there were always major gaps in any individual reading. Hence, I started to write short articles for my students to supplement and integrate what they were reading. In fact, the first paper I wrote explained the value of thinking about the future - a short two-page piece titled "The Nature and Value of Thinking about the Future." This paper was the first reading I assigned in the class; to me it made sense to begin a course by asking what the value of studying the main topic of the course was. This beginning paper, over the course of a dozen years, grew into chapter one of *The Evolution of Future Consciousness.*

There were multiple sections of Integrative Studies offered at Rio Salado and early on I found a group of adjunct faculty who were interested in teaching sections of the course. Two of these faculty members, Robert Brem and Dr. Matt Bildhauer, were highly enthusiastic about the course and became very good friends in the coming years. We spent many hours in restaurants and coffee shops discussing and debating the future and related topics in history, philosophy, social studies, and science. I am grateful for their friendship, collegiality, and unwavering support as the course evolved in those early years. Matt, with a Ph.D. in philosophy, stimulated me into writing a supplemental paper for students on how thinking about the future developed in religion and philosophy through history, and this early paper grew into chapters three and four of *The Evolution of Future Consciousness*.

Eventually, the various readings for students grew and coalesced into an entire textbook on the future. In those first years I immersed myself in the literature on the future and encountered the works of Kevin Kelly, Walter Truett Anderson, Riane Eisler, Barbara Marx Hubbard, Hazel Henderson, and Francis Fukuyama, among others, and became a devout reader of *The Futurist* magazine. Pulling all this material together, I wrote *The Odyssey of the Future*, which covered all the main areas of future studies from science, technology, and space travel to culture, society and religion. After finishing the book, I gave a copy of it to the Chancellor of Maricopa Community Colleges, Dr. Paul Elsner. (Rio Salado is part of the Maricopa college system.) Dr. Elsner read the book and was extremely positive in his comments on it. In fact, in the following couple of years he supported the establishment of a Futures Institute at Rio Salado and funded the development of a website which I created on the study of the future. Through the subsequent years, Paul has become another good friend who emerged as a consequence of my work in the study of the future, and has remained highly encouraging in my ongoing writing every since.

Yet to really test the waters regarding the value and substance of my book, I sent a copy to Dr. Wendell Bell, former chair of sociology at Yale University and one of the best known academic futurists in the country. To my great satisfaction, Wendell was also very complimentary in his reaction to the book and encouraged me to publish it. Events in my life though took some unexpected twists and turns and the book was never published, but many of the elements and themes of my continuing work on the future were begun in that book. Much of *Contemporary Futurist Thought* began in the writing of *The Odyssey of the Future*. In those early years, Wendell and Paul did much to validate my confidence in what I was doing since I very much respected the intelligence, wisdom, and scholarship of both of them. Additionally, during this time, I began a correspondence with Dr. Richard Slaughter, former President of the World Futures Studies Federation, and Richard accepted for publication my first article on the future. I thank Richard for validating my work, as well as providing a host of interesting writings of his own on the future that I have read over the years.

Also during this period I shared my writings, ideas, and enthusiasm on teaching the future with various colleagues at Rio Salado College. I want to especially thank Vernon Smith, Beatriz Cohen, and Larry Celaya, fellow faculty members, and my academic supervisor Vice-President Karen Mills, all of whom read sections of my early book and collaborated with me on work I was doing on futures education.

By the late 1990s, I began to attend the annual meetings of the World Future Society (WFS) and started to give presentations. I met Dr. Peter Bishop, faculty member in the Master's Program in Future Studies at the University of Houston at Clearlake, and Peter was also supportive of the academic work I was doing. At one of the earliest meetings of the WFS that I attended, I met a very energetic, intelligent, and extremely warm and friendly person on the shuttle from the airport to the convention hotel. His name was Jonathon Richter and over the last half a dozen years we have become great friends and professional collaborators, writing papers and giving presentations together. Jonathon and I continue to work on developing educational approaches to teaching the future and enhancing future consciousness. I also met Rick Smyre of Communities of the Future at one of the convention meetings and he has become another friend and colleague who I met because of my futurist work. Rick has also been very supportive of my writing and educational efforts.

Eventually Paul Elsner retired as Chancellor, the Futures Institute ended, the Integrative Studies course was dropped as a requirement in Maricopa, and I took down my website, but I continued to read and write on the future. I discovered new non-fiction writers such as Ray Kurzweil, Robert Nisbet, Lee Smolin, Sally Goerner, Peter Watson, and dove back into contemporary science fiction, totally enthralled and amazed by the recent novels of Dan Simmons, Stephen Baxter, Vernor Vinge, and Greg Bear. I am grateful to all these writers for stimulating my thinking on the future.

In completing these two books on the nature and development of future consciousness and futurist thought, I have explored history and the ideas of innumerable important figures from the past. I must mention, at the very least, Heraclitus, Aristotle, the Taoists, Leonardo da Vinci, Kepler, Spinoza, Leibnitz, Hegel, Nietzsche, and Darwin with whom I feel a strong sense of connection and resonance across time. All these visionaries of the past contributed to my understanding and study of the future.

Of special note in this regard, I am very appreciative to Antonio Damasio for providing directions to Spinoza's house and grave in his book *Looking for Spinoza*. On a cold, dark, and rainy day, my wife, Jeanne, and I searched out and found Spinoza's grave in The Hague last year, following Damasio's directions. The importance of finding Spinoza's home and gravestone is that Spinoza has always been one of my central guiding lights in my intellectual pursuits. Though Spinoza lived over three hundred years ago his cosmic vision, his forward-looking ideas on humanity, and the character of his life have been great sources of inspiration in thinking about many aspects of reality, including the future.

I would also like to thank two of my earliest teachers, before I discovered the future, James J. Gibson and J. T. Fraser, who tremendously informed and stimulated my young intellect on the nature of time, evolution, knowledge, and the human mind.

In the last few years, encouraged by Jeanne, I have been submitting articles for publication to the World Future Society, and have had quite a few accepted. In this process of submission and dialogue with editors, I have gotten to know Howard Didsbury, Ed Cornish (editor of *The Futurist* and former President of the World Future Society) and Timothy Mack (present President of the World Future Society). They have also all been very supportive of my work and my writings on the future. Thanks to all of them.

I want to thank Joan Fay for her editorial work on *Contemporary Futurist Thought*.

Finally, I want to thank my dear and wonderful wife Jeanne. Not only does Jeanne love to discuss ideas and philosophical topics, she also is my partner on my new website (www.odysseyofthefuture.net), contributing her poetry and occasional newsletters, and edited both books from beginning to end.

Contents

INTRODUCTION

"Since man is above all future-making,
he is, above all, a swarm of hopes and fears."
Ortega y Gasset

This book is a study in contemporary thinking about the future, dealing with both the hopes and the fears expressed in modern times concerning tomorrow. There are many such hopes and fears - perhaps an overpowering number, competing with each other and swirling about in the collective mind of humanity. But as Ortega y Gasset points out, the distinctive quality of humankind is our capacity and inclination to invent and attempt to create imagined futures. We live in a mental universe of inspiring dreams and threatening premonitions regarding what lies ahead.

In looking at contemporary ideas about the future, I first examine science fiction and the discipline of future studies, two of the most popular and influential approaches to the future in present times. Next I describe major social and technological trends in the twentieth century, the impact of such trends upon human society, and extrapolations into the future regarding where these trends may lead. Finally, I review and compare a great number of contemporary theories and paradigms of the future, attempting to describe the main belief systems about the future and the main implications and predictions that follow from these different belief systems. I conclude with a summary of what I believe are the most important and valid points to keep in mind when we think about the future.

This book is a continuation and completion of my historical and theoretical study of future consciousness, begun in my book, *The Evolution of Future Consciousness*. In *The Evolution of Future Consciousness*, I examined the psychological make-up, benefits, and functions of future consciousness and I traced the development of future consciousness from prehistoric times up to the end of the nineteenth century. Although there are many connections between earlier thinking on the future and theories and approaches that have emerged in the last hundred years, I will start this book afresh, without assuming any knowledge on the reader's part of what I covered in the previous book. This book can be read independently, though I will frequently make reference to ideas and themes within the earlier book. The reader may wish to go back to the first book to follow up on references from this book.

I should at the onset, however, provide a definition of the expression "future consciousness." In *The Evolution of Future Consciousness*, I defined future consciousness as "the total set of psychological abilities, processes, and experiences that humans use to understand and deal with the future." There is nothing strange or mysterious about this meaning that is given to the expression "future consciousness" and in fact, as I have argued, future consciousness is a basic human capacity that is absolutely essential to successful everyday living. As I traced the development of future consciousness from its most primordial and basic origins in the previous book, in this present volume I look at the complex and expansive array of concepts, principles, and theories that make up contemporary future consciousness. Our future consciousness continues to evolve, in response to the complexities and rate of change in our present world, and thus our hopes and fears - our optimism and pessimism concerning tomorrow - expand and grow as we journey into the new millennium.

Why should we be interested in thinking about the future, and in particular, why should we be motivated to understand the rich and often confusing "swarm" of contemporary ideas and theories about it? In *The Evolution of Future Consciousness*, I discussed in depth the values of future consciousness, of which there are many. As I stated in that book, the future is the only game in town and humans, by their very nature, think about the future and direct their behavior as a consequence of such thinking. It is hopes and fears - the emotional dimension to thinking about the future - that move us (and in some cases paralyze us). Concerning contemporary theories - these are the ideas and ideologies that will influence the future direction of humanity - the world in which we will all live. Such theories give us a sense of direction for our lives, as well as a sense of what potential challenges and dangers we may face.

Chapter One

Science Fiction as the Mythology of the Future

"The universe is made of stories, not of atoms."
Muriel Rukeyser

Introduction

Science fiction is clearly the most visible and influential contemporary form of futurist thinking in the modern world. Why is science fiction so popular? As I will argue, one main reason for the popularity of science fiction is that it resonates with all the fundamental dimensions of the human mind and human experience. It speaks to the total person about the future.

At the outset, let me provide a working definition of science fiction. Although not all science fiction deals with the future, its primary focus has been on the possibilities of the future. In this regard, science fiction can be defined as a literary and narrative approach to the future, involving plots, story lines and action sequences, specific settings, dramatic resolutions, and varied and unique characters, human and otherwise. It is imaginative, concrete, and often highly detailed scenario-building about the future set in the form of stories.

In this chapter I describe the historical development of science fiction as an approach to the future tracing its origins to science and evolutionary theory, secular philosophy, technological forecasting, mythology, and the philosophy of Romanticism.[1] Within this historical review, I consider the rich array of futurist themes and issues examined in science fiction. I also describe the diverse functions and innumerable strengths of science fiction as a mode of future consciousness.

My central arguments are:

- Science fiction engages all the fundamental capacities of the human mind; it generates holistic future consciousness. Of special note, science fiction integrates the secular-rationalist and mythological-romantic approaches to the future; it synthesizes the Dionysian and Apollonian mindsets regarding the future.
- Science fiction weaves together theory and abstraction with personalized narrative. It combines a highly detailed and concrete level of realism with theoretical speculation on the future.
- Science fiction addresses all the main dimensions of the future and synthesizes all these dimensions into integrative visions or scenarios of the future.
- Because it reflects contemporary and futurist thinking, and embodies many features of myth, science fiction can be viewed as the mythology of the future.

Myth and Science Fiction

"...our aim is not merely to create aesthetically admirable fiction. We must achieve neither mere history, nor mere fiction, but myth. A true myth is one which, within the universe

of a certain culture...expresses richly, and often perhaps tragically, the highest aspirations possible within a culture."

Olaf Stapledon

As a starting point, I will consider the power of religion, and in particular religious myth, as an approach to the future. Religious myth, though not exclusively focused on the future, has had a great impact on people's beliefs and attitudes toward the future. It is the earliest recorded form and probably the most influential type of futurist thinking. Predictions of the demise of religion and myth in modern times have proved inaccurate. The great bulk of humanity still subscribes to traditional religious doctrines, as well as various myths and prophecies associated with these doctrines. [2] After describing some of the main features of religious myth, I will demonstrate how science fiction embodies many of the same qualities and strengths as religious myth.

There are many explanations of the power of religious and mythic thinking. Religion answers the deepest metaphysical questions. It provides personal meaning connecting the individual and social group with God's purpose and with the great narrative of history. Religious doctrines are usually connected with various myths which reinforce its belief systems and principles. Religious myths explain existence in the form of stories, connecting past, present, and future in a way that is easily understood and highly inspirational. Often associated with religious myths are ethical principles, providing ideals and direction for people in their lives. Religious myths speak to the heart as well as the mind.

There is an archetypal dimension of myth. An "**archetype**" is a fundamental idea or theme often represented through some image, persona, or symbol. Contained in various religious myths are such basic themes as death and the renewal of life, honor and courage, love and devotion, temptation and damnation, good versus evil, and creation. These central themes of human existence are often represented by mythological characters that provoke strong emotions in the believer and a sense of personal identification.

Science fiction shares certain important commonalities and strengths with myth. Just as with ancient myths, a key strength of science fiction is its narrative form. It has become so popular because it appeals to the dramatic dimension within people. Life seems more like a story than a set of abstractions, and just as history is a multi-faceted story, the future will be a complex saga of stories.

Science fiction, like myth, contains personified characters, thus creating a personal connection with the reader. The reader often identifies with the characters - sometimes positively, sometimes negatively - and vicariously experiences the drama and events of the story through the characters.

As with myth, the stories of science fiction express fundamental themes and archetypes of human existence. In both science fiction and mythology fantastic beings and settings are presented as a way to symbolically highlight important features of humanity or reality.

Although science fiction may inform it also produces an emotional experience in the reader. The future is felt, as well as imagined and considered. This emotional dimension often translates into inspiration. As the science fiction writer Thomas Disch argues, science fiction has become integral to our lifestyle and culture; through its characters, icons, stories, and themes, it inspires the reader and provides the raw material for turning the future into a personalized journey and way of life.[3]

Although the experience of science fiction is personalized, science fiction stories are often set within a cosmic context and have the same breadth and scope that mythic tales do; they also address the same expansive themes of the nature of reality and the meaning of human existence. In fact, as in myth, science fiction connects the personal with the cosmic. What is the impact and significance of the unfolding cosmic events on the characters in the story?

I think many science fiction writers are very conscious of the connection between their genre and mythology. Many science fiction stories include ancient myths, retold or re-conceptualized in futuristic settings One memorable story "Breckenridge and the Continuum" by Robert Silverberg is a good example of a story that explicitly combined both ancient myth storytelling with an eerie futurist landscape and setting. In this story Silverberg examines the connection between mythology and the creation of the future and along the way creates a series of new myths for the future.[4] In Harlan Ellison's "The Deathbird" the story of the Garden of Eden and the temptation of Adam and Eve is retold. The nature of good and evil is re-examined, as is that of God and the serpent, this time though being played out in a far distant future that is witness to the end of the earth and mankind.[5]

But science fiction goes beyond traditional myths. From a modernist perspective the myths of old are based on archaic thinking. They are oblivious to modern science and the issues of modern life. If myths do have a unique power to motivate and inform people, then perhaps what are needed are new myths based on contemporary thinking that address contemporary issues, as well as issues of the future.[6] As I will argue, this is exactly what science fiction provides. It provides mythic tales informed by science and contemporary thought.

Science and Science Fiction

I will now look at the relationship between science and science fiction. I will describe how science first impacted popular story telling in the modern era, and how this introduction of science into popular narrative led to science fiction.

John Clute, in his *Science Fiction: The Illustrated Encyclopedia*, begins his history of science fiction by trying to define the distinguishing nature of science fiction. He notes that people from ancient times were writing fantasies of traveling to the heavens, or of encountering strange and fantastic beings in strange or fantastic places. Yet, according to Clute, the authors of these early stories did not try to present a convincing case that their imaginative scenarios could actually exist in reality. Clute defines "fantasy" as "make-believe", and argues that pre-modern fantasies were intended to be "make-believe" for they provided no explanation of how the imagined fantasy could possibly be real. According to him, the attempt to be realistic in such fantastic stories doesn't occur until the Scientific Revolution and the modern era. For Clute, More's *Utopia* (1516), written prior to the modern era, was not intended to be a plausible or real situation. On the other hand, Francis Bacon's *The New Atlantis* (1626) was intended to be a realistic possibility. Bacon attempted to explain how the type of futurist society he envisioned could be created through science and reason. Bacon offered a rich array of predictions of new and fantastic inventions and human realities for the future that presumably could be achieved through the application of science and reason. For Clute, Bacon's *The New Atlantis* is **Proto-Science Fiction.**[7]

Another early candidate for Proto-Science Fiction is Johannes Kepler's *Somium seu Astronomia Lunari* (1634). According to Wyn Wachhorst, in the *Somium*, Kepler presented the first cosmic voyage in science fiction, involving a journey to the moon. In this story Kepler was also the first person to seriously consider the possibility of extraterrestrial life.[8] For Kepler, space travel was not a fantasy; he believed in the future we would journey into outer space. In Kepler's own words, "Let us create vessels and sails adjusted to the heavenly ether and there will be plenty of people unafraid of the empty wastes. In the meantime, we shall prepare for the brave sky-travelers maps of the celestial bodies."

Science and the concept of secular progress, associated with the emergence of the philosophy of the Enlightenment in seventeenth and eighteenth century Europe, provided a rationale and guide for conceiving of possible futures far different from the present. The idea of secular progress, briefly defined, is that human society can be improved, along many different dimensions, social and technological, through the application of science and reason.[9] For Clute, the beginnings of science fiction coincide with the emergence of the idea of secular progress and the belief in realistically possible progressive changes in the future due to science and reason.

Yet I would argue, contrary to Clute, that people in ancient times did believe in the existence of mythological places and beings. They believed that gods and goddesses existed in a higher supernatural realm, though often they visited, haunted, or "enchanted" the natural world.[10]

What changed in the modern era were the standards of knowledge and truth. In Europe and elsewhere, in the pre-scientific era, truth was based upon faith and belief in holy texts, as well as prophecies and divine revelations. The metaphysical explanations of fantastic beings and alternate realities, justified through religious writings and mystical experiences, were not scientific or rational by modern standards, and consequently, have been labeled as "superstitious", and relegated to the realm of pure fantasy. According to its supporters and advocates, modern science and the secular philosophy of the Enlightenment encouraged freedom of thought and inquiry after centuries of religious dogmaticism and repression. Answers to the fundamental questions of life could no longer be grounded in faith, unsubstantiated authority, and sacred texts. Science in its pursuit of truth strove for objectivity and impartiality. Science based its beliefs on empirical observation, experimentation, and reason, and the description of reality that emerged over the last few centuries within science clearly contradicted in many ways the description and explanation of reality offered in ancient myths and religions.[11]

Religious and mystical views of the future often saw humanity ascending to supernatural realms. Christian thinkers such as St. Augustine believed that the forces of the supernatural or divine realm would transform the earthly realm as a prelude to ascension into a higher (heavenly) reality. St. Augustine imagined our world being modified to accommodate and fit with the spiritual realm. The plausibility of this vision depended upon one's standards of acceptable truth and the nature of reality. For Augustine, it made perfectly good sense to argue that humanity would be transformed in the future by the will of God.

The emergence of science fiction as a form of narration about the future involves a transformation in our standards of thinking, brought on by the Scientific Revolution, regarding what is plausible and real. This new way of viewing reality provided a different approach to understanding and predicting the future – an approach based on the ideas of reason and science. When the age-old tradition of story telling of strange and wondrous realities embraced the ideas and principles of science and secular progress as a way to explain its imaginative settings and characters, science fiction was born.

Hence, as can be seen, science fiction reflects many of the qualities and strengths of ancient myth as well as the beliefs and standards of modern science. It creates "scientifically credible" myths. Therefore, I propose that science fiction is becoming the mythology of the future. As ancient mythologies provided meaning and direction for humankind, I would suggest that science fiction, informed by science and contemporary and futurist thinking, will provide the stories that will give humanity meaning and direction in the future. Science fiction is usually about the future and serves the function of influencing our journey into the future. It will inspire us, as did ancient myths, but it will base its visions on contemporary ideas and standards.

The Genesis of Science Fiction

". . . Frankenstein is the modern theme, touching not only science but man's dual nature, whose inherited ape curiosity has brought him both success and misery."

Brian Aldiss

"It is possible to believe that all the past is but the beginning of a beginning, and that all that is and has been is but the twilight of the dawn. It is possible to believe that all that the human mind has ever accomplished is but the dream before the awakening."

H.G. Wells

Science fiction also has roots in the Romantic philosophy of the nineteenth century. Nineteenth century literature was strongly influenced by Romantic philosophy with its emphasis on human emotion and passion and the inner turmoil, madness, and distress of the human mind. In the nineteenth century, gothic, horror, and adventure stories, all expressions of the Romantic mindset, were very popular. One central goal of such stories was to stimulate and provoke strong emotional reactions in the reader, both positive and negative. Whereas the emphasis in scientific writing has been to describe reality in an objective, rational, orderly, and non-emotional manner, Romantic writing often highlighted the opposite qualities: subjectivity, mental turmoil, and emotionality.[12]

Science fiction would combine together the Romantic - emotional dimension of human experience with concepts and speculations derived from science and Enlightenment philosophy. Nineteenth century science fiction (before the genre had acquired its modern name) was popularly referred to as "scientific romances". One early nineteenth century writer who wove together Romantic and scientific elements in his stories was Edgar Allen Poe (1809 - 1849). Thomas Disch, in fact, argues that Poe is the modern starting point of science fiction. Poe is well known as a writer of horror stories and tales of the supernatural, but he also includes various scientific ideas and speculations to create psychologically disturbing and mesmerizing effects in his dark tales.[13]

The Romantic dimension of science fiction includes not only the terrifying and horrific but the sublime and inspiring as well. As one early example, Jules Verne (1828 - 1905), highlighted the exhilarating and hypnotic power of new machines and scientific devices, the exotic and esoteric realities and worlds uncovered or created through science and technology, the dramatic awe-inspiring adventure into the unknown, the passion and excitement of exploration, and the existential and cosmic challenges of the future to the human soul.

Another key connection between science fiction and Romanticism pertains to the Romantic philosophical distrust, if not rejection, of the positive and progressive promises of science, technology, and modernity. Though Clute argues that futurist science fiction emerged when the hopes and predictions of secular progress were incorporated into popular story telling, stories within science fiction often have taken the opposite position. Science, secular progress, and the growth of technology may lead to our ruin.

The classic case of this negative view of modern science and technology is *Frankenstein* by Mary Shelley (1797 – 1851). Although during the nineteenth century there was great optimism about the future, perhaps the earliest example of a science fiction novel, Mary Shelley's *Frankenstein, or the Modern Prometheus* (1818), foretold of the potential dangers of science and technology. Shelley was aware of nineteenth century scientific experiments where animal tissue had been animated into movement through the passing of electrical currents through muscles. One night, in a personally frightening image, she conceived of bringing a dead body back to life with electricity and thus the story of *Frankenstein* was born.

The main character in the story, Victor Frankenstein, after creating the "creature" immediately runs away from it, repulsed by its hideous appearance, and he abandons it. Throughout the novel, Frankenstein obsesses on how wretched and terrible a person he is for having produced the creature, and the creature, rejected by his maker, as well as other humans he encounters in his wanderings, decides to enact revenge on his creator through a series of ghastly murders.[14]

Frankenstein is fundamentally an introspective nightmare, more a critique of human nature than of science and technology.[15] One could say that it was the egomania of Victor Frankenstein, coupled with his heartless abandonment of the creature, that is the real cause of the misery, tragedy, and suffering described in the novel. Shelley, in fact, does not discuss in any detail the technology of creating life or the potential problems of accelerating technology. Rather, she focuses on Frankenstein the man, and the haunting thoughts and feelings that literally destroy him as the novel progresses.

One actually feels more sympathy toward the creature than the man, for the creature is innocent, having been brought into the world and then hated and repulsed by all those around him. The creature even promises to leave the world of humankind and live the rest of his life far from humans, if Frankenstein will create a female companion for him, but Frankenstein decides against creating a companion, after first having agreed, and in so reneging on his promise, provokes the creature into murdering Frankenstein's new bride.

Still, in spite of the introspective quality of the story, due to its popularization in the movies in coming years, the story of Frankenstein has been strongly associated with the potential dangers of technology, especially if it is used to serve the human aspiration to play God. Whether it is technology, as such, or the human desire to gain power over reality through technology, Romantic philosophy saw problems and potential tragedy in putting too much faith and hope in science and technology. Frankenstein wanted to benefit the world with his scientific research; he ended up destroying his life and bringing death and misery to all those around him. From the Romantic perspective, he is not so much a "modern Prometheus" as he is a Faustian character having sold his soul to the dual devils of vanity and technological power.

The science fiction writer Brian Aldiss identifies *Frankenstein* as "*the* modern theme" for it not only addresses the dual nature of humanity, of being but an animal (an ape) yet God-like in power to understand and create; it also addresses the double-edged sword of humankind's superior powers. According to

Aldiss, the power to create brings both "success and misery". Progress, technological and scientific, is a double-edged sword, and *Frankenstein* focuses upon the potential negative consequences of humanity's increasing power over nature and the world. Aldiss, in fact, sees *Frankenstein* as a "new myth" - a modern myth for our times. Through science and technology humanity is becoming God-like with the power to create, yet do we have the maturity and foresight to use this power wisely? With science having replaced God, humankind is empowered to remake the world. But Victor Frankenstein is a poor God, a fearful God, for he recoils from his own creation, and dies a lonely death haunted by the reality of what he has brought upon the world.[16] The tragic myth of *Frankenstein*, connecting humanity's enhanced power to create with an inability to foresee the consequences of this new power and wisely nurture its creations, is a common theme repeated throughout later science fiction. Thus, the first new myth created in science fiction is tragic and apprehensive over the future and the promise of scientifically inspired secular progress.

Yet throughout the nineteenth century there were also many positive stories about the promises of technology, secular progress, and the wondrous world of tomorrow.[17] Utopian projections of ideal future human societies proliferated in the Age of Enlightenment and continued into the nineteenth century, including such famous books as Samuel Butler's *Erewhon* (1872) and Edward Bellamy's *Looking Backward* (1888).[18] As I noted in *The Evolution of Future Consciousness* it was a popular view in the nineteenth century that science could be applied to organization and orchestration of human society, producing utopian social systems in the future. During the years from 1888 to 1900, according to Laura Lee, 150 novels were written that were hypothetically set in the year 2000.[19] In *Looking Backward,* a person in 1888 is transported to the year 2000 in Boston, and describes from a personal point of view all of the technological and social marvels of this futuristic city.

Hence, from its beginnings, science fiction includes stories that are positive and uplifting, as well as dark and frightening. Because of the dual influences of the philosophies of Romanticism and Enlightenment science fiction inherits from the nineteenth century a fundamental ambivalence regarding the promises of science and technology and the future in general. Change is both exciting and frightening. Science fiction begins with this insight - that tomorrow holds possibilities of both great progress and good, as well as great disaster and evil. Science fiction, from its inception, combines both the optimism of science and reason associated with the Enlightenment and the apprehensions of hi-tech modernity and fear of change associated with Romanticism.

Two of the most fundamental human emotions are hope and fear, and science fiction creates stories that stimulate both types of feelings. Hope and fear are emotions pertaining to the future, in that both emotions have a future reference.[20] The experience of hope involves an anticipation of something positive and rewarding; the experience of fear involves the expectation of something dangerous or destructive. Science fiction deals with the strange and the different, which can stimulate negative or positive emotions in the reader. The unknown and the mysterious can provoke awe, hope, and wonder, or anxiety, fear, and terror.

It is also important to see in this contrast between hope and optimism and fear and apprehension that science fiction is both a rational mode of thinking and an expression of fundamental human emotions. Ideas about the future based on science, technological extrapolations, and reasoned predictions point to the rationalist dimension of science fiction, but science fiction as a literary and artistic form of expression attempts to stimulate emotional reactions in the reader as well. This dual nature of science fiction reflects its rational-scientific and its Romantic origins - the Apollonian and the Dionysian within science fiction. From ancient Greek mythology, the Apollonian mindset (from the Greek god Apollo) stands for reason and order, whereas the Dionysian mindset (from the Greek god Dionysius) encompasses passion, reverie, and chaos.[21]

The double-edged sword of hope and fear can definitely be seen in the two writers who really popularized science fiction at the end of the nineteenth century. Science fiction first made a big impact on popular culture in the works of Jules Verne (1828-1905) and H. G. Wells (1866-1946).

Jules Verne is well known for his scientific and technological predictions and his great sense of adventure into the unknown.[22] In the imagination of Jules Verne, we fly around the globe in *Around the World in Eighty Days* (1873), venture to the inner recesses of the earth in *A Journey to the Center of the Earth* (1864), and travel into outer space in *From the Earth to the Moon* (1865).[23] Millions worldwide read Jules Verne's novels. The future worlds of Jules Verne told of times of discovery and human advancement, with a strong emphasis on the positive powers of science and technology. Verne was an avid reader of contemporary science and technology and offered various predictions about future science and technology throughout his stories. Generally, Verne's novels reinforced the idealism of secular progress via the advances of science and technology and the triumph of the human spirit. Things often went wrong in his story lines (providing the necessary drama), but the courage, intelligence, and ingenuity of his characters, coupled with advanced technology and science, usually overcame whatever obstacles were encountered.[24]

Yet, there is also a lesser known "dark side" to Verne's writings. He was not an unequivocal optimist about the future, but his more pessimistic writings did not so easily get to print since they conflicted with the progressive temper of the times.[25] Especially toward the later years of his life, he began to seriously doubt whether secular and technological progress would lead to a better world, and whether humanity had the capacity or inclination to create a better society in the future.

The future clearly becomes complex and multi-faceted and both hopeful and unsettling in the work of H. G. Wells. For

Wells, the future becomes a topic of intense and sustained speculation and study. As he remarked "I am extravagantly obsessed by the thing that might be, and impatient with the present."[26] Wells integrated social, historical, and philosophical ideas with scientific and technological concepts in his thinking about the future, creating much richer, more profound, more comprehensive, and more realistic projections than writers of science fiction who merely foretold of scientific or technological changes. Wells wrote an immense number of both narrative fiction and non-fictional essays and books on the future.[27]

Herbert George Wells is generally considered the father of modern science fiction. The science fiction writer Thomas Disch identifies Wells as the greatest of all science fiction writers.[28] Beginning with *The Time Machine* (1895), *The Island of Dr. Moreau* (1896), *The Invisible Man* (1897), *The War of the Worlds* (1898), and *When the Sleeper Awakes* (1899), and continuing through *The First Men in the Moon* (1901), *The Food of the Gods* (1904), *The War in the Air* (1908), *The World Set Free* (1914), and *The Shape of Things to Come* (1933), Wells produced in story form a multifarious and expansive set of images of possible futures. We encounter alien minds and civilizations, "future histories" of the world and humanity extending millions of years outward, biologically engineered life forms and humans, invisible humans, great future war machines and aircraft, atomic weapons, and both the fall and rise of human civilizations.[29]

Less well known to popular audiences, are Wells' extensive series of philosophical and sociological books and essays on the future, including *The Discovery of the Future* (1902), *Anticipations of the Reaction of Mechanical and Scientific Progress Upon Human Life and Thought* (1902), *Mankind in the Making* (1903), *The Open Conspiracy: Blueprints for a World Revolution* (1928), *World Brain* (1938), *The Fate of Homo Sapiens* (1939), and *Mind at the End of its Tether* (1945). Wells was not only interested in speculating about the possibilities of the future; he was also very concerned with actually influencing the future of mankind. He wrote about his concerns regarding the human condition and presented numerous proposals for how to improve human society. Wells believed that the future could be both predicted as well as directed toward desirable ends, and consequently he is often seen as the father of modern future studies as well as science fiction. He viewed time deterministically, yet he also believed, perhaps in contradiction to this, that the future was malleable and could be influenced by human choices and intervention. [30]

Wells clearly made numerous specific predictions about the future; sometimes his predictions were mistaken, but often he accurately anticipated the "shape of things to come". He predicted the atomic bomb and the use of nuclear energy, armored tanks, aerial warfare, worldwide television broadcasting, and cinematic pornography. He foresaw intercontinental ballistic missiles and the rise of a global society run by multinational corporations. He envisioned large mechanized agricultural farms, genetic engineering, and highly overpopulated mega-cities. He foresaw both World Wars long before either began. Wells envisioned the emergence of a World Brain and World Encyclopedia that in some ways anticipates the recent development of the Internet.[31]

Central to Wells' thinking on the future was his evolutionary and panoramic view of history and time. Wells saw all of nature as transformational, rather than static and unchanging. History involves change. Further he saw time as directional, rather than cyclic, filled with creativity and novelty. Wells took a global, if not a cosmic, view of history, looking for general trends across the vast expanse of time from the dawn of creation to the far distant future. Wells was intensely interested in both history and the future and connected the two together; as the Wells biographer Warren Wagar states, Wells "traversed time". It is worth emphasizing this dual interest and passion of Wells in both history and the future. As I have argued, an understanding of history greatly benefits future consciousness;

in fact, without a sense of the past, there is no sense of the future.[32]

Wells' comprehensive vision of the past and the future was conceptualized in evolutionary terms. As a student, Wells studied with the great evolutionary thinker Thomas Huxley and thoroughly absorbed both evolutionary theory and the principles of science. For Wells, time is evolution. In particular, evolution provided for Wells a scientific story and explanation of the ongoing saga of humankind. From early in his career, Wells wrote articles and eventually books, both fiction and non-fiction, concerning the evolution of humanity, both pertaining to the past and to the potential future of our species. According to Disch, in his science fiction, Wells presented a new evolutionary mythology and narrative of human history and the future.[33] Evolution as a creation of modern science provided Wells with a general scheme for telling the story of humanity than the Biblical account of natural and human history. Wells frequently captures the transformational tension and struggles of the evolutionary saga of humanity in his fictional stories. On the one hand we are grounded in our animal ancestry - on the other hand we aspire to the heavens and the stars above. We are a creature in evolutionary transition, moving forward yet grounded in the past.[34]

Because Wells saw the future in evolutionary terms he saw the possibility for unending human progress but also the potential for disaster, if not total extinction. The future of humanity was uncertain for within an evolutionary framework there is no purpose or *telos* to nature, and humanity is not some special creation by God, but just one among many species in the ongoing struggle for survival. Species survive by adapting to the changing circumstances of the environment and there is no guarantee that humanity will adapt and flourish into the future.[35] Hence Wells created both utopian visions of the future, where humankind uses reason, science, and humanitarian ethics to guide its future, and nightmarish, troubling, and dystopian visions where negative and self-destructive trends dominate in our future history. Given the uncertainty of the evolutionary saga, the future of humanity is a double-edged sword - of both hope and fear.

In Wells' evolutionary perspective we see an important new archetypal theme that would influence much of future science fiction. The world is no longer a creation of gods; it is an evolutionary process involving the possibilities of both unending advancement and total extinction. Just as the double-edged sword of science and technology provided a new motif for myth-making, the naturalistic, cosmic, and scientifically informed theme of evolution provided another new framework in which to create stories about the future. Contrary to traditional Western religious visions of the future that predict a glorious and uplifting triumph of God over evil and the salvation of humanity, evolution offers uncertainty, at least regarding the fate of humankind.

The earliest and most famous of Wells' science fiction novels running from *The Time Machine* through *When the Sleeper Awakes* and *The First Men on the Moon* tend to be dystopian, frightening, and horrific. Many of these novels can be interpreted as warnings; if we don't change our society and our present ways of thinking and behaving we are in for trouble. *The Time Machine* envisions a future world in which humanity - presumably the capitalists and workers of present day - have evolved and divided into two separate species - the Eloi and the Morlocks. The Morlocks do all the industrial work in the dark underground and feed on the Eloi who live a childlike and frivolous existence in a garden paradise maintained by the Morlocks. *The Island of Dr. Moreau* warns against the potential dangers of biotechnology and suggests that the animal - the beast - still lives within us in spite of our elevated and civilized aspirations. *The War of the Worlds*, according to some critics, an allegory on the ruthless imperialism of the West, tells the quintessential story of alien invasion, of mental evolution freed of emotionality, of the weakness of humankind in the face of greater powers in the universe, and ultimately, the capriciousness and ironies of survival in the world of tomorrow. *When the Sleeper Awakes* foretells a future world ruled by rich capitalists, where workers are suppressed and controlled through behavioral technology, and robbed of their freedom. *When the Sleeper Awakes* anticipates many of the most famous dystopian stories of the twentieth century, including *Metropolis, Brave New World,* and *1984.*[36]

Yet beginning in 1902, first with the publication of two non-fictional books, *Anticipations* and *The Discovery of the Future,* Wells begins in earnest to argue for his progressive and utopian ideals about the future. Science fiction utopian novels would follow in the years ahead, including *The Food of the Gods, A Modern Utopia* (1905), *The World Set Free* (in which he predicts the atomic bomb), *Men Like Gods* (1923), and his most famous utopian novel, *The Shape of Things to Come.* Wells' basic argument through all these utopian books, fictional and non-fictional, is that humanity's aggressive, self-centered, and power-hungry mentality will doom us to self-destruction, and a global and humanitarian culture, informed by reason and science, needs to rise up and gain control of the world if we are to survive and flourish.[37]

Aside from the importance Wells places on science and reason in realizing a better world tomorrow, to understand Wells' utopian visions it is also essential to place them in the context of his evolutionary framework. Wells accepted the idea that biological evolution was a consequence of natural selection, competition, and survival of the fittest - a "law of the jungle" interpretation of evolution. From Wells' perspective, this competitive, "every species or man for himself" mode of behavior and thinking is, in fact, largely responsible for the present troubles facing humanity. We are competitive, tribal animals living in a civilized global reality. But to complicate matters further, humanity has developed highly destructive modern weapons, such that if we continue our aggressive ways and use these powerful weapons against each other, we are doomed to extinction, or at the very least, we will destroy our present civilization. Following an idea first expressed by Darwin and Huxley, Wells believed that humanity needed to increasingly move to a higher form of evolutionary development and guidance, one directed by ethical considerations rather than self-centered competition. Wells' visions of future utopian worlds invariably involved this concept of "ethical evolution" transcending competitive evolution. Ethical evolution would entail increased cooperation and a unity of purpose within our species.[38]

The utopian novels of Wells usually took the form of narrative histories, tracing first the collapse of contemporary civilization and second the rise of a new world state. These "future histories" of humanity, written in dramatic form, anticipated many later science fiction "future histories", such as those by Asimov and Heinlein, where the future is told as a story extending outward through a series of challenges, defeats, and triumphs. Even Wells' descriptions of ideal world states are dynamic and historical rather than static. There is no perfect state that once achieved brings history and our further development to a close. There is no "heaven" in his secular view of tomorrow. Progress continues indefinitely. Further, Wells did not see progress as a peaceful steady movement forward, but rather as involving conflict, revolution, destruction, and rebirth out of the ashes. (This conflictual view of progress has been a common and highly influential perspective on the dynamics of history, found in both ancient and modern philosophies and mythologies.[39]) For Wells, the struggle into the future entailed an ongoing tension and conflict between the forces of the past and the forces of the future, between our more primitive tribal mentality and a more progressive ethical mentality.[40] The future is best seen as a dramatic story or series of stories.

Wells saw, at least our immediate future as involving great war, violence, and destruction. In Wells' mind our present world system is corrupt and literally needs to self-destruct. Wells lived through two World Wars, and as noted above, anticipated the outbreak of both wars. Specifically regarding his predictions of the First World War, Wells clearly foresaw that the world was heading toward more global and destructive warfare than anything it had experienced in the past. Although many writers of Wells' period also predicted future wars around the world, it was Wells who was the most powerful voice among them, and it was Wells who foresaw most dramatically the technological advances in future warfare and the immense and pervasive carnage and destruction unleashed through future weaponry. In *The War in the Air*, cities are destroyed through aerial bombardment and nations fall apart; in *The World Set Free*, most major world urban centers are leveled with atomic bombs in what the Wellsian scholar Warren Wagar describes as the "most terrifying novel" he has ever read.[41]

Although war may be inevitable in the near future and perhaps necessary for the birth of a new world civilization, Wells also articulated in considerable detail many ideals, principles, and features of a positive and more ethical future. He described a future world of universal education based on science and a unified world view; the emergence of a world religion, informed by modern science rather than superstition and tribal thinking; equal rights and freedom for all humanity; a collective government under the control of productive, practical people without divisive party politics and the intrusion of the interests of "Big Business"; better cities, housing, and urban infrastructure; and the unending improvement of the human species through biological engineering. He even foresaw the possibility that humanity would evolve and transcend itself, achieving a higher level of universal consciousness, God-like in nature.[42]

For Wells, the history of humanity is a great drama, a story of the oppositional forces of our primitive and animal ancestry and our forward-looking rational and ethical mentality. All his life, Wells wrestled with the conflicting forces of despair and hope: Would humanity rise above the barbarity of the past or were we doomed? This dynamical tension in all of its manifestations is the underlying essence of our nature. We are a story unfolding with an uncertain resolution. We have arisen out of the "tooth and claw" struggle for survival in the natural world and now we face our greatest challenge

- to somehow transcend our beginnings, while at the same time acknowledging and understanding our origins, and create a new world using our evolved powers of reason and morality. Wells is the great creator of modern science fiction because he saw that the future is quintessentially drama, to be understood in terms of the most fundamental principles and issues of science, history, and philosophy, and told in the form of stories.

Because Wells lived through the unprecedented destruction and carnage of two World Wars, and because he felt perpetually frustrated in getting his message of hope for the future across to the powers in the world, as well as to the general public, and because he saw the resolutions achieved in both World Wars as not addressing the underlying problems that had generated these wars, Wells became increasingly pessimistic about the future of humanity toward the end of his life. In his last book, *Mind at the End of its Tether*, Wells states, in regards to the future of humanity, "The stars in their courses have turned against him and he has to give place to some other animal better adapted to face the fate that closes in more and more swiftly upon mankind." In another earlier passage in the book, he puts it even more succinctly, "The writer is convinced that there is no way out or round or through the impasse. It is the end." So Wells became hopeless about our future, but because he saw history and time in evolutionary terms, he saw something coming after us, perhaps a dramatically modified version of our hominid line, perhaps some species totally different from us. This theme of the extinction and replacement of humans as the dominant form of life on earth is taken up by many future science fiction writers who came after Wells. There may be a better future coming but it may of necessity not involve us.[43]

In the first decade of the twentieth century, "scientific romances" and other novels about the future contained both hopeful and uplifting visions of tomorrow, as well as fearful and apprehensive images. Clute highlights the optimistic quality of the times and the science fiction during this period. He refers to the first decade of twentieth century science fiction as "A Glowing Future." Progress was speeding up, invention and innovation were everywhere, and there was great hope in the West, at least, that human civilization and science and technology would continue to advance. There were many upbeat space adventures and positive heroes imagined in science fiction. Yet during this same period there were a variety of unsettling events and ideas emerging in the sciences, art, and culture that were "disturbing the peace"[44], and there were many future war novels being written, including Wells' prophetic *The War in the Air.*[45] In the decade before the First World War, apprehensions of an impending world conflict were clearly in people's mind and new doubts and challenges were arising concerning the vision of reality and the future bequeathed by the Enlightenment.

Clute believes that early science fiction stories describing future wars were written as warnings of possible disasters. They weren't intended to predict, but rather to show the undesirable consequences of political policies and potential applications of new technologies. The "warning scenario" is a common theme within futurist thinking, both in science fiction and non-fictional books on the future. *Frankenstein* and *The Island of Dr. Moreau* can be viewed as warnings regarding the unbridled use of biotechnology to serve the vain aspirations of power obsessed scientists. The logic of such warning scenarios is if we keep doing what we are doing things will get worse, hence we should change what we are doing. The futurist Wendell Bell refers to such predictions as "**presumptively true.**"[46] Hopefully the warning will change human behavior and the predicted negative effects will not occur. Such dark images are supposed to raise our consciousness and provoke corrective action. As J. T. Fraser puts it, "...nightmares are dreams whose usefulness is to keep us on our toes."

Like Clute, the contemporary science fiction writer Frederick Pohl also argues that science fiction in general does not attempt to predict the future.[47] Of course, as a form of literature, science fiction is intended to entertain and stimulate the imagination and to move the heart. Yet contrary to Clute and Pohl, even if the purpose of a story is no more than to warn of possible negative consequences of present trends, this activity has a predictive aspect. Again, the logic of a warning is "If we continue to do X, then Y will follow." This is a conditional predictive statement.

Science fiction, in so far as it is fiction, creates characters and events that are imaginary and not literally true. Clearly, if one is writing a story about the future, then the events portrayed are hypothetical rather than real. Still, although futurist science fiction is imaginary, many of the great science fiction writers, such as Jules Verne, H. G. Wells, and Arthur C. Clarke, created future worlds involving a variety of intentional predictions about the future. Science fiction often attempts to create plausible futurist scenarios and extrapolations on present trends. At the very least it gets the reader thinking about the various possibilities of tomorrow and it has actually inspired outside of the genre a host of predictions and goals for the future. It has even provoked the real-world creation of technologies and hypothetical realities envisioned in its stories. Science fiction may accurately predict the future by stimulating the creation of the future that it is imaginatively and vividly describing.

Disch states that science fiction is often a way to help us see the present better - by placing present events and conditions in exaggerated form in an alternate reality. Yet he also argues that the purpose of most good science fiction writers is to create a "realism of the future."[48] Does the science fiction writer create a plausible and realistic image of tomorrow? Is it convincing? Does it feel real? Are the scenes and characters drawn in sufficient detail and complexity? Realism can mean different things, but one thing it means is that a situation or event makes sense and is plausible. Recall Clute's argument that science fiction uses scientific and contemporary secular reasoning to support the rational possibility that the strange events in the story could become real.

Another useful way to see the predictive dimension of science fiction is to view science fiction stories as narrative "**thought experiments.**" Some hypothetical state of affairs is imagined, for example, cities in the future, contact with alien intelligence, or a world wide catastrophe, and a story is told exploring the possible effects or repercussions of the imagined scenario. Or the thought experiment can simply be, given the present conditions in the world, what hypothetical consequences will emerge over time. Thought experiments are a common practice in science - often as a prelude to doing an experiment. In science fiction the consequences of a state of affairs are conceptualized in the form of a story, rather than a set of measurable controlled variables as in a scientific experiment. Science fiction stories allow us to think through the possible effects, repercussions, or implications of different imagined futures, or the possible future effects of present trends and developments.

Hence, although science fiction is fiction rather than fact, futuristic science fiction predicts, extrapolates, and imagines in story form possible future developments in our world, as well as in the universe at large. Acknowledging that many of the predictions contained in science fiction stories have turned out to be mistaken, the bottom line is that science fiction, based on extrapolations of present trends, has made numerous relatively accurate predictions about the future and does realistically consider the possibilities of the future. Even if its stories take the form of "presumptively true" warning scenarios, with the clear purpose of stimulating people into action to prevent the events portrayed in the story, a conditional prediction is still being made.

The second decade of the twentieth century witnessed a time of great escapism in science fiction from the ongoing horror of the First World War. Edgar Rice Burroughs (1875 - 1950) began his famous "Barsoom" series about Martians and Martian civilization with *Under the Moon of Mars* (1912).[49] Burroughs, creator of the Tarzan series and the well-known novel *The Land that Time Forgot* (1924), populated his stories with heroic figures and thrilling adventures. Through the writings of Burroughs one could escape to an alien world, or the primitive jungle or a land populated with cave men and dinosaurs. Also during this period many stories were written about aerial empires, consisting of huge spaceships or floating cities in the sky. The reader could run away to some imagined perfect world in the heavens above. A great world war, having been envisioned and now encountered, became a stimulus for escape into a presumed better tomorrow, or some strange other land. Thus we encounter another duality in science fiction. It can provide a way to see more clearly where we may be heading, or it may provide a way to run away from an unpleasant reality into some more pleasing and hopeful world.

In the 1920s, following Hugo Gernsback's inspiration, the term "science fiction" came into popular use. The name "science fiction" derives from "scientifiction", a term coined by Gernsback. He used the term to describe the type of stories he published as founding editor in his pulp magazine *Amazing Stories* beginning in 1926. Gernsback, an engineer by training and education, was highly enthusiastic about the potential wonders and benefits of future technology, and inspired by the writings of Wells and Jules Verne, among others, started a series of new magazines (*Amazing Stories, Science Wonder Stories, Air Wonder Stories*, and *Amazing Stories Quarterly*) dedicated to showcasing how science and technology could transform the world into a "technological utopia".

In spite of Gernsback's upbeat vision of a technological future, following World War I, the level of apprehension and ambivalence over the future continued to manifest itself in science fiction. For example, in this period, numerous stories of futurist cities were written. Sometimes these future cities were vast and ultra-technological with mile high skyscrapers; other scenarios presented future cities as great ruins following some imagined worldwide catastrophe.

Of special note, perhaps the most well-known of early great science fiction movies, *Metropolis* (1926), was produced, showing the technologically advanced, but equally negative social and psychological side of a great futuristic city.[50] *Metropolis* is a beautiful high-tech city above the surface, where the wealthy enjoy all the benefits of economic and technological progress, but below the surface, the workers, who maintain the city, toil in mechanical, depressive drudgery and monotony without any of the benefits enjoyed by the wealthy class. Perhaps science and industry will lead to great technological achievements and luxury (at least for some) but with disastrous overall social consequences. The darkness and the light are not an "either-or" but a "both-and." The message of *Metropolis* is that the "heart" needs to play a critical role in the creation of a better future.

In the 1920s robots become a highly visible presence in science fiction. The term "robot" comes from Karel Capek's play *R.U.R.* (Rossum's Universal Robots) (1924). A classic early example of a robot - in this case female - can be found in *Metropolis*. Though the term "robot" means indentured labor in Czech (Capek's native language), the fear from the beginning in science fiction stories was that robots - our own creation - would turn on us. In *Metropolis,* the female robot, under the control of a mad scientist and a heartless government official, is directed to lead the working class humans of the city to a self-destructive end.

In the early years of science fiction, robots were highly anthropomorphized, looking like humans covered in metallic armor - in fact, the woman robot of *Metropolis* even had metal plated breasts - like a nude futuristic female gladiator. Robots were machines that were shaped like humans and in many ways behaved like humans. The robot, in fact, in science fiction emerged as a symbolic synthesis of humanity and the machine - equally the human becoming machine-like, being assimilated by its technological creation, as well as the machine becoming human, embodying our worst qualities and characteristics. In *Metropolis* and numerous other science fiction stories, the robot takes up the *Frankenstein* theme in metallic clothing. It personifies our fear of science and technology, as well as our fear of what we may become. In humanity's attempt to create life and sentience (and hence be like gods), our own creation may turn on us. We may destroy ourselves through our machines. Or, worst, we may become machines. The robot is a concrete symbol of threatening technology, human corruption, and the quest for power. It is a classic archetype of science fiction.

Cosmology, Future History, and the Golden Age

"What Orwell feared were those who would ban books. What Huxley feared was that there would be no reason to ban a book, for there would be no one who wanted to read one. Orwell feared those who would deprive us of information. Huxley feared those who would give us so much that we would be reduced to passivity and egoism. Orwell feared that the truth would be concealed from us. Huxley feared the truth would be drowned in a sea of irrelevance. Orwell feared we would become a captive culture. Huxley feared we would become a trivial culture, preoccupied with some equivalent of the feelies, the orgy porgy, and the centrifugal bumplepuppy."

Neil Postman

In the 1930s and 1940s science fiction flourished and vastly increased in popularity. This period is frequently referred to as the **"Golden Age"** of science fiction. Science fiction magazines proliferated and many new writers made their debut in "pulp" magazines, among them perhaps the most well-known being Isaac Asimov and Robert Heinlein.[51] *Flash Gordon* appeared in the newspapers and later became a movie serial. Science fiction was highly popularized through the media of radio and the movies, but these media often focused on the horror and fear elements of the genre. Orson Welles did his famous radio broadcast of an updated *The War of the Worlds* frightening millions of people across America. Equally dark and terrifying, the movies *Frankenstein*, *The Invisible Man*, and *Dr. Jekyll and Mr. Hyde* were produced during this time.

One type of science fiction story that blossomed in the Golden Age was the "**space opera**." Space operas are dramatic epics set in outer space, involving both adventure and danger, with great stellar ships, amazing technological weaponry, alien life forms and evil forces, and colossal space battles. Clute views early space operas as a form of high escapism, first from the Great Depression and then the growing threat of the Second World War. The space opera provided a vision of the future as a great adventure into the unknown, yet with a continuation of war and conflict, now staged on a cosmic scale. It is a classic mythic form of science fiction.

During the Golden Age, E.E. ("Doc") Smith led the way in the development and popularization of the space opera. What is noteworthy about Smith's novels is their cosmic scope. In his two most memorable series, the *Skylark of Space* and *Lensmen* novels, the forces of good and evil battle across the galaxies over billions of years, beginning in the distant past and extending into the far future. In the *Lensmen* series Smith envisions immensely powerful alien life forms with highly advanced technological capabilities and incredible space armada. Humans are drawn into the cosmic saga, achieving heroic status in the fight of good against evil. Within rousing adventure stories, Smith expands our perspective on ourselves to the farthest reaches of space and time.[52]

"Doc" Smith may have taken the reader to the ends of the universe in gargantuan spaceships, but it is the British philosopher Olaf Stapledon (1886 - 1950), also writing in the 1930s, who is usually considered the most cosmic of all science fiction writers. Perhaps the most imaginative and intellectually powerful writer in the history of science fiction, Stapledon wrote only five science fiction novels. Stapledon approached science fiction as a philosopher, a historian of cosmic dimensions, and a metaphysician who pondered the deep questions of existence and the meaning and nature of reality. Although future technologies and space travel play a role in his stories, Stapledon explores all the varied possibilities of the future, humanistic and scientific. Stapledon's novels are concerned with the ultimate reaches of life, mind, civilization, and spirit.

Of particular note, Stapledon wrote two vast and inspiring epics of the far distant future. The novel *Last and First Men* (1930) traces the complete history of humanity two billion years into the future. Humanity goes through various evolutionary transformations within this immense history - the novel chronicles the historical journey of 18 generations or types of humans into the future. Human civilizations rise and fall, we journey to and settle various planets in the solar system, we biologically and technologically enhance ourselves in varied ways, discover how to travel through time, and achieve great insight and understanding scientifically, spiritually, and philosophically. Yet, in the end, humanity passes into extinction, a brief "music" and "brave theme" within the vast "music of the spheres" of the universe.[53]

The second novel, *Star Maker* (1937), unfolds on an even grander scale. Arthur C. Clarke describes *Star Maker* as "Probably the most powerful work of imagination ever written." *Star Maker* explores the idea that in the vast reaches of the distant future - tens of billions of years from now - the combined efforts of all intelligent life in the universe achieve contact with "The Star Maker," the creator of the universe.[54] In *Star Maker* Stapledon tells the story of the rise of galactic empires and galactic minds leading eventually to "cosmical mentality" and a "cosmical utopia". In the finale of this epochal tale, the universe itself is transcended as the "Star Maker" reveals an infinite array of created universes, from the most primitive to the "ultimate cosmos." We peer into the mind and the creations of God and are bedazzled and humbled. The story is a great cosmic myth of the absolute future.

Stapledon's writings epitomize the intellectual and philosophical adventure of the future set within a science fiction context. Stapledon thoughtfully considers the historical causes of world events, the social and psychological changes that bring about the unfolding of the future, and the implications and consequences of such changes. His books read more like history and philosophy than fictional novels. Individual characters are "swept aside" as Stapledon discusses the great saga of events over millions and billions of years. There is a sense of wonder and awe as well as cosmic revelation within *Star Maker.* The images of the future engulf the reader - we are transported to the edge of infinity. Further, Stapledon writes with great profundity and speculative detail - pondering the meaning of existence and describing innumerable strange and alien realities along the way. Stapledon rivaled, if not exceeded, Wells in imaginative power, and he clearly demonstrated the sense in which the future is the ultimate adventure of the mind. His influence on science fiction has been immense, anticipating a great variety of major themes and ideas that would appear in science fiction in later years .[55]

Stapledon transcended the darkness and limitations of his times, by moving outward into a vast cosmic context - a cosmic context in both space and time. To borrow a phrase from the great seventeenth century philosopher Baruch Spinoza, Stapledon examined humanity and the grand panorama of existence and intelligence "through the eyes of eternity". For Stapledon, the future of the cosmos is evolutionary and historical - it is a great saga of advance, decay, transformation, and ultimate revelation. The future is no longer just an arena of human affairs, but a great universal epic populated by innumerable alien minds and civilizations on a great quest of creation and discovery. Where is the entire scheme of things heading? For Stapledon, the extended future of the universe was a legitimate and important arena of inquiry and speculation. Within this much broader context, which could be seen as fantastical escapism or heightened realism, the future history of humanity and all forms of intelligence, civilization, and mind is told.

Although Stapledon in both *Last and First Men* and *Star Maker* describes the future from a vast panoramic perspective, recounting events, trends, and accomplishments across billions of years without much mention of individual characters - to use the expression of the historian David Christian, he writes a "big history" of the future - Stapledon does tell these stories from a personal point of view. In each novel, there is a central human character, who recounts the histories, and in the case of *Star Maker*, actually experiences the events. These tellers of the tales bring their personal feelings and thoughts to the story – they react to and ponder over the great saga of future history. This individualized perspective gives both novels a personal dimension. This personalization of a cosmic future achieved in a science fiction narrative is a unique strength of the genre. All of the future histories and epics of Wells, Smith, and Stapledon in different ways combine the personal and the cosmic. Science fiction stories address both the inner and outer dimensions of human reality – the macrocosmic and the microcosmic.

It is worth considering further the similarities and differences between Wells, "Doc" Smith, and Stapledon. Smith's novels focus on action, technology, and war and conflict in space – that is how future history is understood and told. There is a moral dimension to Smith's novels as well – there are forces of good and evil in conflict over the destiny of the universe. In Wells there is struggle and war also in the future history of humanity and there is a moral dimension as well, for Wells hopes that our ethical and cooperative dimension will eventually transcend our competitive and tribal behavior. But Wells sets the great saga of the future in the context of evolution, connecting past with future, and hoping that we can move beyond conflict to a higher level of human reality - that is we must move beyond the "war of good and evil". Stapledon goes even further, and though he discusses future war, his central transformational theme is mental and cosmic evolution. We witness the struggle of intelligence, culture, and civilization to ascend to progressively higher levels of understanding and excellence in the context of a vast, mysterious, and dynamic universe. The central struggle is not good versus evil and the struggle is not war-like in nature. The central struggle is to evolve and achieve wisdom - it is the struggle of mind and self-transcendence. In the end - in the grand finale of first humanity and then the combined corporeal intelligence of the cosmos - there is a feeling of insignificance yet acceptance and appreciation for the journey.

What Wells, Smith, and Stapledon have in common is that they all take a cosmic perspective on the future. Can we accurately understand the future of humanity without seeing ourselves "through the eyes of eternity"? Modern science fiction often sets the saga of the future in a cosmic context, a perspective that reflects much earlier religious myths. Although ancient myths were limited in their visions of the universe, the attempt at least was made to describe the future of humanity within the context of the whole. The cosmic perspective reasserts itself within science fiction, now redefined and significantly expanded beyond traditional myth by science, evolution, and space technology. A scientifically informed vision of outer space now becomes the new cosmos. Yet as Stapledon and other great science fiction writers have realized, though the exploration of outer space clearly involves a technological and scientific dimension, it will involve the expansion and evolution of all facets of human civilization. All aspects of human existence need to be re-thought as humankind moves outward into this new cosmic environment. The exploration of outer space, a central theme in science fiction, provides a cosmic setting for all aspects of the future evolution of humanity.

Although enlightenment, transcendence, and cosmic insight are themes that occur in ancient myth, science fiction as illustrated especially in the works of Wells and Stapledon, places these themes in an evolutionary context. Evolution becomes the grand cosmic story in which individual tales are told. This general framework, a reflection of the ideas of science, represents a significant shift from ancient mythic tales where the general context was usually the actions, creations, and purposes of deities. But in so far as science fiction is dramatic literature and story telling, the path of evolution is described not so much as an abstract theory, but as an ongoing struggle and tension of various forces, represented through fictional characters and societies, their struggles, and their failures and triumphs.

As already mentioned, two of the most well known new authors of the Golden Age of science fiction were Isaac Asimov and Robert Heinlein. Both were prolific writers and had long careers, continuing to produce many popular novels and stories for decades into the 1960s and 1970s. Asimov and Heinlein both began publishing short stories around 1940. From early on, Asimov created a whole series of "robot" stories that he became particularly known for. Heinlein achieved early fame weaving together many of his stories into a coherent and detailed "future history" of human society that stretched a couple of centuries into the future.[56] Asimov soon followed suit, creating his own future history that evolved into his famous *Foundation* series.[57]

Heinlein and Asimov did not look out as far in time as had Stapledon, but their future histories consisted of stories (or series of stories) with numerous individual characters and dramas, as opposed to Stapledon's grand historical epochs. Although their future histories included various technological predictions, as well as space travel and colonization, Asimov and Heinlein also invoked a variety of psychological, sociological, and historiographic ideas in creating their future chronologies of humankind. Asimov considered the dynamics of historical change and the forces which effect social change, while Heinlein concerned himself with political, religious, and ethical transformations in humanity. In general, by the 1940s science fiction had clearly created a vast array of speculative social and political, as well as technological, scenarios about the future. Asimov and especially Heinlein led the way in exploring the social and political possibilities of tomorrow.[58]

As mentioned above, Asimov also wrote many stories about robots during the 1940s. Continuing the longstanding theme running back through science fiction since the time of

Frankenstein, in his robot series Asimov addressed humanity's fear that our technological creations might become a threat to us. Isaac Asimov attempted to address the issue of how to make intelligent machines safe and non-threatening to humans. He formulated the "**Three Laws of Robotics**", presented in his famous collection of stories *I, Robot*.

1. A robot may not injure a human being, or, through inaction, allow a human being to come to harm.
2. A robot must obey the orders given it by human beings except where such orders would conflict with the First Law.
3. A robot must protect its own existence as long as such protection does not conflict with the First and Second Law.

Asimov imagined these three laws or directives programmed into all robots that humans create, with the hope that our technological creations would never attempt to gain control over us or destroy us.[59]

Aside from Asimov and Heinlein there were many other great new writers who emerged during the Golden Age. Two excellent anthologies that showcase the best authors and stories of this period are Ben Bova's *The Science Fiction Hall of Fame* and Raymond Healy and J. Francis McComas's *Adventures in Time and Space*.[60] *The Science Fiction Hall of Fame* is a collection of short stories and novellas voted the best of all time by the members of the Science Fiction Writers of America. A good percentage of the stories in the three volumes of this collection are from the Golden Age. The Healy and McComas anthology is exclusively of stories from the Golden Age. In these two collections are stories on aliens and alien civilizations; time travel and the future; robots and biologically advanced humans; and space travel and star ships that cross the universe - all the great themes of twentieth century science fiction.

One of the writers most frequently represented in these two collections is Lewis Padgett, which was the pseudonym for Henry Kuttner and Catherine Moore, the greatest husband

and wife writing team in the history of science fiction.[61] Two of their best stories, "Mimsy Were the Borogroves" and "Vintage Season", are both from the 1940s. "Mimsy Were the Borogroves" - inspired by Lewis Carroll and the concept of alternative realities - concerns the creation of a portal into another dimension by two children who decipher the true meaning of Lewis Carroll's jabberwocky lines. "Vintage Season" tells the tale of a group of tourists from the future who journey back to our time to watch a disastrous collision of a meteor with the earth, but do nothing to prevent it.[62]

A third well known story written in the 1940s by Kuttner and Moore that should be mentioned is "Private Eye". Contained in a different anthology of science fiction stories, *The Mirror of Infinity* edited by Robert Silverberg, "Private Eye" is psychological science fiction at its best.[63] In a hypothetical future world, law enforcement institutions have developed the technological means to play back any event from the past, and thus ascertain the "true" nature of any crime committed by anyone anywhere. The main character of the story is a person raised by an authoritarian "fire and brimstone" father who instills the fear of God into his son. The son as an adult attempts to plan out and commit the perfect crime, an act of vengeful murder, knowing full well that everything he does can be watched in retrospect afterwards by the law. In a world of omniscient surveillance - both by God and concretely embodied and represented in the police of this futurist world - this man carries out the perfect crime. "Private Eye" examines the psychological repercussions of advancing technology and draws interesting parallels between the potential power of such technologies and traditional ideas about a judgmental, omniscient God contained in many of our religious belief systems.

Some of the other well-known authors and futurist stories of the Golden Age contained in the Bova and Healey and McComas anthologies include Lester del Rey's "Helen O'Loy" which concerns a poignant and moving love affair between a human master and his female robot; Theodore Sturgeon's

"Microcosmic God", the story of an inventor who creates a new intelligent species that evolves much faster than humans; Murray Leinster's "First Contact", a tale of how trust is first established between humans and intelligent aliens who meet in outer space; "Who Goes There?" by John Campbell, the frightening and paranoid story of an alien that can mimic the shape and appearance of any live form it touches (this tale became the basis for the science fiction movie *The Thing*); and Harry Bates "Farewell to the Master", the story that the classic science fiction movie *The Day the Earth Stood Still* was based on. In "Farewell to the Master" the earth is warned by an alien emissary that humanity needs to stop its warring, destructive ways or else our whole species will be exterminated. In this story the humanoid alien is accompanied by a giant robot, and the humans in the story, thinking anthropocentrically, mistakenly believe that the robot is the servant of the humanoid alien, when, in fact, it is the robot who is the master. The aliens in the story have relinquished their freedom and autonomy to immensely powerful robots of their own creation, to preserve their own peace and survival - a telling message to humans who of course fear such a possibility as being their doom, when in fact for the aliens it turned into their salvation.[64]

Two of the most famous futurist novels written during the 1930s and 1940s explored with great narrative power the darker sides of technologically advanced and highly controlled human societies in the future. These two dystopian novels of the future are Aldous Huxley's *Brave New World* (1932)[65] and George Orwell's *1984* (1949).[66] **"Dystopia"** means the opposite of utopia - a highly pessimistic and dark image of a future society. The novels depicted future worlds where individuality had been squashed through pharmacological stupefaction in *Brave New World* and social - psychological control in *1984*. The novels were written as warnings regarding possible negative consequences of social and political developments in the world. They were also seen as indictments of contemporary modern Western society.[67]

After the optimism of the nineteenth century, the modern world had entered into a sustained period of great upheaval and change, and once held certainties and positive expectations evaporated in the turmoil of two world wars, the rise of totalitarian governments, economic depression, and the overthrow of classical science, art, and philosophy.[68] As the early decades of the twentieth century unfolded, the future increasingly seemed like something to fear; we had entered the "age of anxiety." *Brave New World* and *1984* captured and amplified many of the fears and apprehensions of the modern world.

In both *Brave New World* and *1984* it is our freedom and individuality - two of the central guiding ideals of the Age of Enlightenment - that are lost. But the primary cause of this loss is not advancing science and technology. Rather it is human nature itself and its social-political institutions that undermine and destroy freedom and individuality.

In *1984*, "Big Brother" - the personified omnipresent eye of the government - watches over all its citizens, demanding total obedience, devotion, and conformity. Orwell's apprehensions over the rise of totalitarianism in his time are transformed into a dark nightmare in which the human spirit has been killed and paranoia reigns over everyone. Truth is destroyed through the continued rewriting of history and psychological conditioning. The government controls its citizen's minds - their very beliefs of what is true and what is right - and through constant warfare and incessant propaganda creates an oppressive and sinister mindset in the population. The population is given a common enemy and indoctrinated into a perception of a common evil, resulting in a world where citizens are turned against each other, and in the name of Big Brother, always watchful of each other's behavior.

In *Brave New World*, future humanity is reduced to a population of pleasure-addicted empty souls. Peace, stability, and happiness are achieved through the sacrifice of all creativity, curiosity, and discontent. People do not read, not because books are forbidden, but because no one wants to read anymore. Science, art, and religion have been eliminated or rather replaced by drugs ("soma") that make almost everyone as content and happy, with no desire to achieve, and as frivolous as an innocent child.[69]

Orwell worried over the growing oppressive control of totalitarianism in the name of absolute social order; Huxley was concerned with the shallow human desire for individual pleasure. In *Brave New World* there is no pain; in *1984* there is no pleasure. Both fictional worlds are stagnate, without progress; freedom and individuality have been killed; ultimate order is achieved at the price of any real growth or change.

Whereas the great space operas of "Doc" Smith and the evolutionary sagas of Olaf Stapledon carry the human spirit to cosmic heights, the dystopias of Huxley and Orwell bring us back down to earth and the weaknesses and evils of the human soul. Such is the range of science fiction, metaphysically and ethically, from the shadows on Plato's cave and the desires of the Freudian id to the brilliance of the light of eternity.

Both *Brave New World* and *1984* might not be considered true science fiction since the primary focus of both novels is social criticism and warning regarding political and psychological trends set in the context of the future. Yet, science fiction evolved as the twentieth century unfolded. As was the case with *1984* and *Brave New World,* as well as with much of what was being written in the 1930s and 1940s and thereafter, the label of "science fiction" was too limiting for what was being produced within the genre. Science fiction, contrary to Gernsback's original vision, was not just about science and technology in the future, but all aspects and dimensions of human reality, including society, politics, and psychology. Science fiction became all-enveloping futurist narrative.

Science fiction thus has a very broad scope that serves many different social and psychological needs and addresses different modes of human consciousness. Following Bacon's dictum that knowledge is power, secular images of the future, which would include many science fiction stories, often promise increasing power over the world and our individual selves.

Such stories resonate with our "will to power." Such stories may bring us hope and optimism, but they also provide a way to escape from the frustrations of our present lives. Science fiction empowers our imagination and elevates our egos, as a way to counteract the perceived limitations and difficulties in life. Hence science fiction stories are simultaneously prophecies of hope and power and pathways of escape.

Yet our optimistic images of the future do not always materialize - far from it, and science fiction also contains many negative and frightening scenarios that warn of mistakes and disasters in our future. They warn of our own frailties and flaws carried to extremes in the future. Science fiction offers a balanced perspective on the future and facilitates in our thinking with both realism and pessimism, and optimism about what tomorrow may bring. It provides new images of both heaven and hell based on the same human desires and fears that motivated the creation of ancient versions of these imagined supernatural places.

Consequently science fiction, as a mode of future consciousness, is not simply rational predictions, or direct implications of contemporary events. The stories also satisfy basic human needs and express various human emotions and motives. This is the Romantic dimension of science fiction. Science fiction stories are like dreams of the unconscious - symbolic and archetypal expressions of human psychology - visions of the darkness and the light like the myths and metaphysical images of ancient religions.

The basic human desires to be stimulated and to transcend the mundane reality of everyday life are clearly reinforced and fulfilled through science fiction. Disch argues that science fiction has emerged as a uniquely American phenomenon because Americans revel in the "make believe." According to Disch, Americans especially value fantasy and the fantastic. In fact, Americans tend to blur the distinction between fantasy and reality - through both the media and our lifestyles. Disch refers to America as a "nation of liars" and "would be actors." As noted earlier, Disch does think that good science fiction should be realistic and convincing, but coupled with the above comment on American mentality, this translates into making the "lie" as compelling as possible. As Disch states, Americans value the "good lie." Science fiction, according to Disch, permeates American culture in numerous ways and its fantastic images and visions have inspired many prophets and profiteers in society. American culture sells the fantastic, the fantasy, and the future, and science fiction is often the source of these products, images, and dreams. According to Disch, through science fiction, the future has become a lifestyle in America, bringing excitement and adventure to people's lives.[70]

Evolution, Space, Time, and War

"It has yet to be proven that intelligence has any survival value."

Arthur C. Clarke

"Violence is the last refuge of the incompetent"

Isaac Asimov

The 1950s are often referred to as the "**Silver Age**" of science fiction. During this time, the images of science fiction increasingly entered the public mind. The space race between the United States and the Soviet Union turned many of the earlier visions of rockets and spacemen into reality. A great variety of highly popular science fiction movies were produced, including *Forbidden Planet, Destination Moon, Earth versus the Flying Saucers, The Day the Earth Stood Still, The Thing, The Fly, Them, This Island Earth, Invasion of the Body Snatchers, Invaders from Mars, War of the Worlds*, and *On the Beach*. Through such movies and many others, the impact of science fiction on popular culture continued to grow. For example, it was during this time that the flying saucer craze emerged. Reported sightings of flying saucers escalated after people read about them in science fiction magazines or saw them in the movies. Various strange and sinister aliens were created for the big screen, revisiting the theme of the conquest of earth that was first popularized in H. G. Wells' *The War of the Worlds*. After the dropping of the atomic bombs in 1945 and the escalation of the nuclear arms race between the United States and the Soviet Union, the ideas of a global nuclear war and a world catastrophe destroying human civilization became a vivid and real possibility and people began to imagine all kinds of invaders coming out of the sky.[71]

Among the many stories that were written in the 1950s about nuclear war and its possible consequences, the most famous was Walter Miller's award winning *A Canticle for Leibowitz*, considered by many the greatest science fiction novel of all time. The novel examines the aftermath of a global atomic war in the context of religious, philosophical, and spiritual themes.[72] In *A Canticle for Leibowitz*, a Third World War occurs which destroys modern civilization and propels humanity back to a much more primitive level of technology and social organization. One order of religious brothers attempts to preserve the wisdom of the past through the new Dark Ages, though comically they mistake a twentieth century grocery list for a sacred document. Over the millennia though humankind again "advances upward" to nuclear technology and weaponry and starts a whole new world war, blowing civilization to smithereens for a second time. Abandoning hope that the earth will ever achieve a peaceful and moral civilization, some humans escape the earth in a spaceship. Perhaps the earth is doomed. Progress can not prevail and knowledge can not sustain us. The only answer is to abandon ship and leave.

The end of world civilization - at least as we define it - may not come through a nuclear holocaust, but perhaps through some type of natural disaster such as a worldwide plague. This is the premise of George Stewart's memorable novel *Earth Abides* written in 1949.[73] Our high tech industrialized world collapses as untold millions die and those who survive the plague must directly confront the challenges of survival in nature, no longer protected and supported by all the gadgets and services of modern society. After a few generations, our modern world is forgotten, and humanity starts over again, living a life reminiscent of our earlier hunter-gatherer mode of existence.

A more drastic type of new beginning is envisioned in Clifford Simak's highly imaginative and deeply moving novel *City,* begun as a series of connected stories in the depressing years of World War II, but only pulled together into novel form and published in the early 1950s.[74] In his forward to the novel Simak describes his disillusionment with humanity in the 1940's, a humanity that in its "madness for power, would stop at nothing," and how he first started writing *City* as an escape from the "horror" around him. In *City* Simak tells an epic tale of the future, where humankind eventually abandons the earth, leaving it to be ruled by intelligent animals, in particular dogs, who ponder and wonder whether the ancient stories of man are merely myth and fantasy of the distant past.

In the 1950s science fiction as a cultural phenomenon and a literary genre became more self-conscious and socially organized. In 1953, the annual Hugo Awards were established at the World Science Fiction Convention. Named in honor of Hugo Gernsback, the **Hugo Award** recognizes the best novel, novella, short story, movie, and other noteworthy science fiction accomplishments of the year. The first winner for best novel was Alfred Bester's *The Demolished Man.*[75] Over the years, the World Science Fiction Convention would grow in popularity and spawn innumerable local science fiction conventions across the United States. With the establishment of the World Science Fiction Convention and the Hugo Awards science fiction was becoming a very visible cultural phenomenon in both the United States and other parts of the world. A growing number of people - writers, readers, and millions of fans - embraced the science fiction dream, and saw the world through the visions and ideas of Wells, Heinlein, Asimov, and Clarke.

A decade later in 1965 the **Science Fiction and Fantasy Writers of America** established the **Nebula Awards.**[76] Whereas fans of science fiction vote on the Hugo Awards, professional members of the Science Fiction and Fantasy Writers of America vote on the Nebula Awards. Often the same novel or story has won both awards in a year, but quite frequently writers and fans don't agree. The first year of the Nebula Awards, Frank Herbert's great saga of an alien civilization, *Dune*, won best novel, but tied with Roger Zelazny's *This Immortal* for the Hugo. A year later, Daniel Keyes's *Flowers for Algernon* and Samuel Delany's *Babel-17* tied for best novel for the Nebula Award, but it was Robert Heinlein's *The Moon is a Harsh Mistress* that won the Hugo.

One writer who achieved great popularity in science fiction in the 1950s and worldwide fame, a decade later in the 1960s, is Arthur C. Clarke. In his stories, Clarke explored the grand cosmic themes of science fiction and the future, weaving cosmology together with mythic ideas and speculations on human evolution. In his classic science fiction novel, *Childhood's End* (1953), a vast armada of alien space ships encircles the earth, establishes communication with humanity, and guides the transformation of humankind to a higher form of life and intelligence.[77] Interestingly, the aliens look like incarnations of the Devil, suggesting that our early myths of a serpent-like creature that tempts us to rise above the level of obedient children was perhaps a futurist premonition.

Clarke addresses the same theme of evolutionary human transformation in his popular *2001* trilogy, where humans again make contact with a vastly advanced alien intelligence and a humanoid "Star Child" is created as the next step in our evolution. In *2001*, which is based on the short story "The Sentinel" written by Clarke in the 1950s, the evolution of humanity from our primitive beginnings as ape-like creatures is credited to alien intervention.[78] Minds of a cosmic nature are guiding our ascension through time - an idea that clearly resonates with religious ideas of the past, except now the cosmic intelligence is natural rather than divine and the means of transformation is scientific and technological rather than supernatural. Clarke became world famous with the release of the movie *2001* in 1969.

Clarke has written many other stories through the years that have religious or spiritual overtones, including his famous short stories "The Star" (1955) and "The Nine Billion Names of God" (1953).[79] "The Star" is a science fiction retelling of the historical "Star of Bethlehem", with a tragic and ironical twist, and "The Nine Billion Names of God" weaves together Tibetan monks on a metaphysical quest with a super-computer and, in the finale of the story, the hand of God.

Throughout his career Clarke has explored the high-tech, high-science end of science fiction and often synthesized it with the mystical, mythic, and cosmic.[80] The themes of ancient myth are assimilated and recast in modern cosmology and scientific speculation. God does not disappear but is re-conceptualized in an evolutionary context and often given the metaphorical face of an alien.

As can be seen, future human evolution is a topic that many science fiction writers have addressed. Psychological evolution is the central theme in Theodore Sturgeon's science fiction classic *More Than Human* (1953).[81] In this novel, a small group of socially outcast humans merge into a group-mind (in telepathic contact) and become a "Homo Gestalt." Telepathic evolution is also a central theme in Alfred Bester's Hugo prize-winning *The Demolished Man* (1953). This novel is set in a future world where police can read minds and criminals learn to hide what they are thinking in their own consciousness. Much of the action in this psychologically intense story takes place in "mental space" - in the thoughts and counter-thoughts of the pursuer and the pursued.[82] Yet perhaps the most memorable and touching story of human evolution of this period was Daniel Keyes's *Flowers for Algernon*, published first as a novella in 1959 and then as a novel in 1966. Through various new scientific treatments, Charlie Gordon, a congenital imbecile, is turned into the most mentally advanced human who has ever walked the earth - a genius of immense capabilities. Yet the transformation does not hold and Charlie eventually turns back into his old self - only dimly aware of the great heights to which his mind had ascended.[83]

In general, the theme of biological and psychological evolution in humans can be either hopeful or fearful in its tone and implications. The dark side of humanity may evolve as well as the good. In Jerome Bixby's chilling 1950s story "It's a Good Life", a child endowed with powerful psychic powers wishes people who displease him out of existence.[84] It may though be normal humans who out of fear attempt to subdue or kill more highly advanced humans. In Howard Fast's "The First Men" a group of super-intelligent children must defend their lives against government and military forces that want to destroy them.[85] The theme of the old versus the new, of what is normal versus what is strange and different is frequently played out in stories about future human evolution. Even if we do evolve in the future, there undoubtedly will be a dramatic struggle between the "new humans" and "old humans" fearful for their own continued survival and control of the world. This is the age old conflict of the future and the past, of stability and change set in the context of possible future human evolution.

It is interesting that stories about human evolution have frequently influenced contemporary society. Science fiction has been a significant source of inspiration for various social, quasi-scientific, and religious groups and movements that promise "higher levels" of evolution to their followers.[86] Again, a future image is created that stimulates or directs human action. In the case of stories about human evolution, people are inspired to develop to a higher level of mental, spiritual, or social reality. Science fiction has become an arena of thought in which humanity has considered how to redesign or redirect itself.

As noted above, the 1950s saw a heightened awareness of space exploration among the general public, and this heightened awareness of outer space and space technology was often connected with speculations about alien intelligence and the possibility of visitations of aliens from outer space. Both space exploration and aliens have been highly popular themes in science fiction since its beginnings. Through the 1950s, and in the decades to follow, space exploration and alien contact have continued to be ubiquitous features in science fiction, in both cinematic and literary forms.

The alien from outer space is, in fact, another of the central archetypes of science fiction. The alien represents the unknown and the fantastical personified, often possessing intelligence and technology far beyond human abilities. The alien is the transcendent and at the same time a creature that emerges out of the darkness. Stories of the future, especially those set in outer space, almost always have aliens, of one kind or another. Symbolically, the alien is the mysterious and frightening future.

Through the future human odyssey to the planets and the stars, alien intelligence may be contacted and humanity may be transformed - perhaps for better, perhaps for worse. The extreme contrast in possibilities can be seen in two popular science fiction movies concerning alien contact produced in the 1970s and 1980s. Within the *Alien* movie series, humanity contacts a hideous alien species hell-bent on destroying us; the alien in this movie series is straight from hell, dripping slime with highly corrosive acid as blood, the most nightmarish creature imaginable. At the other end of the continuum, in *Close Encounters of the Third Kind*, the aliens are highly advanced technologically, yet gentle, benevolent, and utterly beautiful and magnificent. They are clearly mysterious, yet they promise enlightenment, fellowship, and friendship and communicate with us through music. As personified archetypes of the future, they represent the two extreme possibilities of what is to come in our journey into outer space. We will meet demons or we will meet angels.

Some classic science fiction novels involving space exploration and alien contact include Ray Bradbury's *The Martian Chronicles* (1950); Stanislaw Lem's *Solaris* (1961); and Larry Niven's epic adventure and award winning *Ringworld* (1970).[87] In *The Martian Chronicles* it is humans, exploring and settling Mars, who destroy the indigenous Martians and their civilization. In *Solaris*, it is the alien who gets the upper hand on us. In this highly original tale of alien contact, humans are unable to communicate with or understand the alien intelligence, which in fact is embodied within the entire oceanic surface of an alien planet. The alien intelligence manipulates the minds of the human settlers by speaking to them through their dreams and driving them mad. In Niven's novel, humans, and an assortment of interesting aliens, partner and travel to a huge and apparently abandoned "ringworld" that encircles a distant sun. Ringworld is immense in proportions, millions of miles in circumference and thousands of miles across. The interior band which faces the sun is surfaced in natural terrain with plant life, lakes, rivers, hills, and valleys. It is a marvel of solar and ecological engineering. The novel revolves around the mystery of what possible alien intelligence and advanced technology could have constructed such an immense object and then for no apparent reason abandoned it. The aliens of *Ringworld* are wondrous in their abilities and totally enigmatic.

The alien can also be seen as a psychological projection and a symbolic representation of ourselves in the future. As with the robot, the alien can be an expression of both our hopes and fears - our good side and our dark side. Our interactions with aliens then turn into symbolic struggles with ourselves. We run from our nightmares, but often we may run from our ideals and dreams as well. Through the alien and thoughtful speculation on alien civilizations we can consider different possible future modes of existence and identity and on different possible technologically advanced societies. Aliens are thought experiments but also explorations of our feelings regarding potential future selves.

Ideas in science fiction about space travel and space exploration have had a significant impact upon popular culture, both negative and positive. There is no question that science fiction has fueled the imagination, plans, accomplishments, and present developments of human space exploration.[88] Yet as Disch notes, the contemporary development of the space program has also been strongly connected with military goals and technologies, such as Reagan's Star Wars initiative and the earlier military strategies and plans behind satellite launches and the space race between the USA and Russia.[89] The history of science fiction space operas, both in literature and film, also strongly connected the rocket and the space ship with super-weaponry and great wars in space. Culturally and psychologically, we associate the space ship with incredible destructive capabilities in its super arsenal of photon torpedoes, laser guns, and death rays. (This association of space travel and super-weaponry goes back to Wells' *War of the Worlds*, but is clearly reinforced in the more recent popular movie series of *Star Trek* and *Star Wars*.) Although science fiction writers such as Robert Heinlein and Jerry Pournelle have strongly supported the USA space program (NASA), it has become an increasingly hard sell to the public in the last couple of decades. Shouldn't we spend our money on health, education, and social issues instead of space, since the latter is strongly connected with war and the engines of destruction, while the former is associated with human welfare and peace? In this case, the negative and destructive images of space ships, space exploration, and contact with aliens generated in futurist science fiction may have contributed to a cultural backlash on the future of humanity in space.

Space travel and colonization, though, need not be associated with war, destruction, and hostile aliens. The adventure into outer space can be seen as a journey of cosmic discovery and enlightenment - it can be seen as the fulfillment of humanity's destiny.[90] It can also be seen as the evolutionary arena for the construction of technologies at a whole new level of size, sophistication, and scientific depth. In Robert Heinlein's novella "Universe" a huge space city has been constructed that is journeying to the stars.[91] Space travel may be our salvation, as in George Pal's 1950s movie *When Worlds Collide* where humanity must leave the earth due to the imminent collision with another planet. In journeying into outer space we may transform other worlds, making them fit for human habitation. Terraforming Mars - a dream of both science fiction and ecological science - is the central theme in Kim Stanley Robinson's 1990s trilogy - *Red Mars, Green Mars,* and *Blue Mars.*[92] Perhaps we will build wormhole super-highways through space such as in the movie *Contact.*

Time travel is a theme that, like space travel, runs throughout the history of science fiction back to its very beginnings. Stories about traveling through time are particularly relevant to thinking about the future, since the imagined journey through time may take us to a hypothetical future, and the story line may revolve around both positive and negative aspects of the encountered future world. Wells' *The Time Machine* told of a future world populated by cannibalistic Morlocks and indolent Eloi - a world depressing, frustrating, and disappointing to the Time Traveler who visits this future reality. On the other hand, time travel into the future can reveal a miraculous and utopian human civilization such as in Bellamy's *Looking Backward.*

Traveling through time also leads to various paradoxes and unusual complexities of reality. If time travel were to be achieved in the future, it would transform human reality and perhaps the entire universe into mind-twisting temporal convolutions and spirals to infinity. One can change history if the time traveler moves into the past, thus altering his or her own present, or one can travel into the future, bring back knowledge of the future and alter both the present and the future. In Robert Heinlein's "By His Bootstraps", written in the 1940s, through traveling back and forth through time the protagonist becomes his own father and his own son.[93]

In the 1950s time travel stories continued to be popular. Isaac Asimov wrote *The End of Eternity* and Fritz Leiber wrote the Hugo winning novel *The Big Time,* both stories describing time traveling humans and other assorted intelligent beings who attempt to control time, past and future, by manipulating and revising history.[94] In the following decade, venturing into metaphysical and spiritual realms, Michael Moorcock, in his novel *Behold the Man,* describes the story of a man in search of the historical Jesus Christ. Through a series of unplanned events, he becomes the actual Messiah and Christ.[95] As one final time travel novel of note, David Gerrold's classic *The Man*

Who Folded Himself, written in 1973, tells of a person literally suspended in a loop in time, born to manipulate his own sex and heritage, becoming mother and father and daughter and son - all somehow the same person circling and twisting around through time.[96]

The New Culture and the "New Wave"

"If Jules Verne could have really looked into the future, say 1966 A.D., he would have crapped in his pants. And 2166, oh, my!

Philip José Farmer

According to Clute, during the 1960s science fiction became fact. Satellites by the hundreds began to go up, encircling the earth in a web of communication and global monitoring, as earlier envisioned by Arthur C. Clarke. Telstar relayed the first transatlantic pictures. At the beginning of the decade, Yuri Gagarin became the first human in space. By the end of the decade humans had landed on the moon - a dream of science fiction since the time of Kepler.

For Clute, the era, in general, embodied a renewed sense of optimism about the future, reflective of earlier times in modern Western history. There was a positive belief in technology and space exploration and a sense that humans could constructively direct the future. *Star Trek* was an optimistic dream. So was the space program. Technology and humanity were united in the exploration and transformation of nature. In general, from Clute's perspective, the 1960s was a time of cultural and technological creativeness and faith in the future.

The 1960s was indeed revolutionary, both socially and technologically. But it was a complex and unsettled time. It was a time of both faith and anti-faith. The decade saw the beginnings of a cultural revolution in the modern West against many of the central images and ideals of modernity. The "military-industrial complex" came under attack, as well as traditional social norms and cultural values. Economic and technological progress was rejected by many individuals as too materialistic - there was an increasingly strong call to get "back to nature" and abandon all the technological and ideological baggage of modern civilization. The 1960s were a time of great cultural experimentation and revelry - of consciousness raising, free love, and dropping out of society - of liberation, adventure, madness, and freedom

Within science fiction there is a corresponding liberation, revelry, and Dionysian quality - of art mirroring life. Reflecting popular culture, science fiction became increasingly concerned with psychological, social, and ethical themes. Science fiction, which had always contained a strong element of social criticism and warnings about the negative consequences of contemporary trends, became even more critical of the modern world.

Humanity continued to live under the ominous threat of nuclear annihilation and the new culture of the 1960s strongly proclaimed that we should make love, not war. As part of the strong counter-culture critique of the military-industrial complex and nuclear proliferation, Disch contends that science fiction, with its numerous post-nuclear war stories, helped to "defuse the bomb".[97] In stories of nuclear war, a very real, dark, and devastating possibility of the future was vividly and graphically portrayed in narrative form. The power of such "warnings" far exceeded any abstract, ideological, or theoretical arguments. Compare the number of people who read Kahn's futurist warning of a third world war On *Thermonuclear War* versus those who saw *Dr. Strangelove. Dr. Strangelove* is cinematic science fiction and a mad black comedy on the mentality of the arms race, politicians and governments, and our military technologies - the audience laughed and cried in the face of annihilation. With science fiction we were asked to imagine the possibility of nuclear devastation in concrete, shocking, and personalized detail. Following Disch's argument, again it seems that visions of the future, represented through science fiction, influenced the actual unfolding of the future. If science fiction, both in literary and media forms, had this effect it did so because it gave perceptual vividness, graphic horror and drama, and personal meaning to the hypothetical negative future of nuclear war. Stories are often more powerful than theories.

Science fiction in the 1960s saw the beginning of the "**New Wave**" as experiments in more literary, psychological, and humanistic writing become popular.[98] Three of the most highly regarded, Hugo winning science fiction novels of all time *Stranger in a Strange Land* by Robert Heinlein, *Dune* by Frank Herbert, and *The Man in the High Castle* by Philip K. Dick were published. All three of these novels highlighted issues of culture, ethics, and the exploration of alternative social and religious belief systems.[99]

One of the most influential voices of the "New Wave" was Harlan Ellison. In 1966, Ellison won the Hugo award for best short story with his " 'Repent Harlequin!' Said the Ticktockman".[100] Inspired by Henry David Thoreau's essay on "Civil Disobedience", Ellison's story reflects the rebellious and individualist philosophy of the 1960s counter-culture. It is a critique on the regimentation and orderliness of contemporary society. Such themes had been addressed before in science fiction, but what makes Ellison's story unique is the rambunctious and colorful style, the emotional intensity, and the clear sense of social defiance embodied in the tale. It is clearly a tale of the creative and adventuresome 1960s rather than the relatively conservative 1950s.

One of the key features of the "New Wave" was an increased emphasis on style and literary experimentation. Not that hard science and sound technological extrapolation disappeared from science fiction in the 1960s, but the genre put more of an emphasis on science fiction as good literature, if not creative literature. Perhaps best epitomizing the new inventiveness and heightened literary self-consciousness in science fiction was Harlan Ellison's revolutionary "New Wave" anthology *Dangerous Visions*, published in the last few years of the 1960s.[101] Ellison saw *Dangerous Visions* as "revolutionary" and the first volume received an award at the World Science

Fiction Convention as the most "significant and controversial SF book published in 1967."

One of the best stories contained in *Dangerous Visions* is Philip José Farmer's "Riders of the Purple Wage", a hilarious, irreverent tale of the future set in the year 2166.[102] Winner of the Hugo Award for best novella in 1968, "Riders of the Purple Wage" is a story that could not have been written, let alone published, a decade earlier due to its irreverent and explicit language and madcap, bizarre scenarios. Punctuated by such chapter introductions as "One Man's Nightmare is Another Man's Wet Dream" and "There are Universes Begging for Gods, yet He hangs around this one looking for work", this story stretches the imagination of the possible, the permissible, and the socially acceptable like no other story of the "New Wave." The story is true to the exploratory spirit of science fiction. It expands the universe of the mind, the senses, the emotions, and one's ethical and social values.

When he wrote "Riders of the Purple Wage", Farmer was already acknowledged as a writer who broke through social constraints and contemporary taboos. In particular, Farmer brought sex into mainstream science fiction.[103] In spite of its presumed freedom of thought, in its earlier years science fiction had stayed relatively conservative regarding traditional social norms about sexuality. In Farmer's *Strange Relations* a human develops a sexual and totally dependent relationship with an alien plant, and in his novel *Flesh*, future earth has returned to a goddess-centered religion, where sex has become a sacred yet public spectacle.[104]

In the 1970s, Farmer created one of the most inventive series of novels written on the future. The *Riverworld* series looks at the age old question of life-after-death but with a new twist. Farmer imagines a vast river-encircled world where every human being who has ever lived, including the Neanderthals, is simultaneously brought back to life after their individual deaths with absolutely no explanation of how all of humankind was resurrected and brought to this strange world. Set in some future time, after the end of humanity, characters from all periods of history, including Mark Twain, King John, Sir Richard Burton, and Herman Goering, are thrown together and go on a great adventure up the river of the world in search of answers to the mystery of their resurrected existence.[105] In *Riverworld*, Farmer experiments with the archetypal theme of immortality and life after death.

For J.G. Ballard, another of the central architects of the "New Wave", science fiction was a means to revel in the madness of the day and the end of Western civilization as we know it. Ballard saw material - technological progress in the modern world as coming to an end and his novels reflect his pessimism about the future. In such novels as *The Drowned World, The Burning World,* and *The Crystal World* Ballard envisioned various world catastrophes and their aftermath. The future is an all-enveloping nightmare and we are engulfed in it.[106] In his short story, "Build-Up", Ballard describes a future world where all empty space has been used up and urbanized. Everyone is walled in and there is no way out. In "The Subliminal Man", the forces of advertising have overpowered humankind and we have all been reduced to perpetual consumers, with no choice but to "buy, buy, buy." As Ballard asks in "The Subliminal Man", "The signs, Doctor! Do you see the signs?"[107]

Another important writer of the "New Wave" was Michael Moorcock, who wrote *Behold the Man,* the story of a time traveler in search of the historical Jesus. *Behold the Man* can be read as an irreverent, if not blasphemous challenge to the orthodox Christian belief in the divinity of Jesus Christ, since it is a neurotic, self-absorbed mere mortal from the twentieth century who turns out to be, reluctantly and inadvertently, the actual Christ of history. The time traveler is born in the present, returns to the past, dies on the Cross, but is born again in the present, only to circle back again in time in an eternal loop with no ultimate beginning or end - a metaphor on the nature of God. As noted above, "New Wave" science fiction, as a reflection of the revolutionary 1960s, repeatedly challenged traditional cultural and social norms, beliefs, and values. Moorcock, as both a writer and the editor of *New Worlds* magazine, was one of the leading figures in this rebellion against convention and the intentional assault on social taboos.[108]

Behold the Man was a story with a clear religious and spiritual theme, involving a reworking of one of the fundamental myths in Western tradition. Other "New Wave" writers explored the connections between myth, religion, and science fiction as well. Earlier I mentioned Ellison's "Deathbird" and Silverberg's "Breckenridge and the Continuum".

Another writer, who achieved great popularity in the 1960s, and repeatedly incorporated mythic elements into his stories and novels was Roger Zelazny.[109] In the early 1960s, Zelazny created one of his most highly praised stories, "A Rose for Ecclesiastes". The story takes place on Mars and involves a linguist from Earth who is drawn by a beautiful Martian temptress into fulfilling a Martian myth - that someone will appear (from the sky) and renew the race and the vitality of the decaying Martian civilization.[110] Zelazny weaves together elements of the philosophy of *Ecclesiastes* - a belief system that rejects all striving as vainglorious and pointless, which seems to capture the essence of both Martian philosophy and the attitude of the reluctant Messiah - with the metaphor of the rose - a symbol of beauty and rebirth that transcends the nihilism and fatalism of *Ecclesiastes*.

In his Hugo award winning novel *Lord of Light*, Zelazny again combines together myth and science fiction. *Lord of Light* takes place in the distant future, on a world where the gods and goddesses of Hindu religion are real immortal beings - having immense powers created through advanced technology - and rule the masses of the planet. On this world, a new "Buddha" - an enlightened one - arises who challenges the sovereignty of the gods and leads the people of the world to freedom. The novel both incorporates mythic and religious themes of our own world and creates a new mythology of the future within a science fiction setting. Again, God is transformed, this time through technology.[111]

In the reworking of old myths, science fiction creates new myths for the future. As we have seen, the mythic, the religious, the metaphysical, and the cosmic have all been areas explored in science fiction. Science fiction looks at all these dimensions of reality in all their manifestations, both uplifting and deific and dark and demonic. In so doing, science fiction offers us mythologies of the future and ultimate visions of the universe to inspire or terrify the soul. As Disch notes, the fantastic realities and creations of science fiction have often served as the stimulus for contemporary religious and spiritual movements.[112]

One writer of the 1960s and 1970s who delved into the metaphysical, who repeatedly questioned the meaning and nature of reality, and who ultimately became a mythic figure himself was Philip K. Dick.[113] Dick not only wrote the quintessential alternate reality tale, *The Man in the High Castle*, as well as a host of other popular science fiction stories and novels, but he has also more recently achieved great notoriety and fame in popular culture having created the stories behind the movies *Bladerunner, Total Recall*, and *Minority Report*.

In his Hugo Award winning novel *The Man in the High Castle* (1963), Dick plays with the reader's mind, questioning what is real, and what isn't, and whether the distinction is always that clear. Although the novel is not strictly speaking about the future, it provides a superb example of how Dick unsettles and disrupts one's concept of reality. In this novel, Dick describes an alternate reality in which Germany and Japan won World War II. In this alternate reality, in an America divided and controlled by Germany and Japan, there is a man - "the man in the high castle" - who writes a book describing an alternate reality in which the Allies won World War II. But what is doubly strange is that according to the "man in the high castle" (who is of course Dick placed in his own alternate reality writing a mirror image novel) the book he has written is true. As the story reveals, the Germans and Japanese who politically rule the United States live a decadent existence, dependent on American culture and technology. In the telling of the story the reader is led to the strange conclusion that the Axis powers in our reality really did win World War II.[114]

The popular movie *Bladerunner*, based on Phillip K. Dick's science fiction classic *Do Androids Dream of Electric Sheep?* (1968), explores various themes and questions regarding manmade intelligent androids and their relation with their human creators. In the story one group of androids decides to search out and hunt down their human creator with the hope of extending their lifespan.[115] As in *Frankenstein*, the created go after the creator. The androids though are hunted by a governmental assassin - a "bladerunner". The ironic twist in the story is that the "bladerunner" is himself an android, having simulated memories placed in his own mind to give him the false notion that he is human, but he is really being manipulated by humans to kill his own kind. In the final analysis, the question is clearly raised regarding who is actually more human - the androids or their inhumane creators.

Mind control coupled with collective madness are core themes in Dick's novel *Martian Time-Slip*, where psychosis, schizophrenia, and obsession with power permeate the first human settlement on Mars and individuals are drawn into one another's strange and deranged mental realities.[116]

In Dick's novel *Flow My Tears, the Policeman Said*, the main character wakes up one day and finds that his social identity has been eradicated - no one knows him and there is no public record of him ever having existed. One can describe this novel as the nightmare paranoid version of *It's a Wonderful Life*, where forces behind the scenes manipulate reality lending a surrealistic, dreamlike quality to the whole story.[117]

In his novel *The Three Stigmata of Palmer Eldritch*, Dick weaves together questions regarding the nature of God with the themes of drug induced consciousness, illusion and reality, and, once again, control over people's minds. A new hallucinogenic drug promising eternal bliss and the experience of omnipotence is brought to human settlers on Mars by Palmer Eldritch, an enigmatic character and outcast who has returned from some distant mysterious galaxy. Yet for those who take the drug, the psychic presence of Eldritch enters into their minds, and what initially appears to be the psychological, if not spiritual salvation for the Martian settlers seems to turn into a bargain with the devil.[118]

The future is as much an adventure into the metaphysical and the nature of reality as it is a saga of the promises and perils of technological and social progress. In his novels and stories, and often in a dark, haunting, and disturbing fashion, Dick delves into the nature of consciousness, personal identity, truth and illusion, good and evil, and the tenuous and ambiguous borderland between madness and sanity. Whereas other "New Wave" writers challenged the cultural and social norms of modern society, Dick went further and challenged the philosophical underpinnings of our world and collective mindset. Dick questioned the objectivist theory of truth and reality. He was fascinated with the Hindu idea that all of reality was but a dream in the mind of Brahmin and throughout the latter period of his life believed he was, in fact, communicating with minds from outer space. Dick was both inspired and haunted by his cosmic revelations. For many science fiction devotees, Dick's writings, with their dark complex settings and frequent excursions into altered consciousness and mind control, anticipated the feeling and atmosphere of cyberpunk science fiction long before this movement had a name.[119] Over twenty years after his early death, Dick continues to have a large following of fans and commentators on his writings and ideas today.[120]

Continuing the growing inner-directed emphasis of "New Wave" science fiction, according to Clute, the 1970s saw science fiction became increasingly introspective and psychosocial. We saw the earth for the first time from space, and this perspective put our fragile human reality into stark relief. Moreover, during the 1970s, science fiction movies continued to grow in popularity, drawing greater crowds and presenting on the big screen more powerful special effects and compelling alternative realities. *A Clockwork Orange*, *Star Wars*, and *Close Encounters of the Third Kind* were produced during this decade. Combining science fiction and scientific speculation, the magazine *Omni* also appeared in the 1970s.

One highly important cultural transformation that began in the 1960s and grew in the next few decades was the feminist and women's rights movement. Women became an increasingly powerful voice in contemporary Western culture, and reflecting this general cultural change, women became a much more significant voice in science fiction in the 1970s and 1980s.

Ursula LeGuin led the way. She emerged as one of the most honored science fiction writers of her era. In 1969 LeGuin won the Hugo and Nebula awards for best science fiction novel of the year. The novel awarded was *The Left Hand of Darkness*. It is a story about gender and sexual stereotypes, set on an alien planet where the inhabitants change their sex during their fertile season. *The Dispossessed*, published five years later, also won both the Hugo and Nebula for best novel. This novel describes two "utopian" yet totally different societies which exist on an alien planet and its moon. It is a study in the meaning, value, and various limitations of presumed "ideal societies."[121]

In its early history, the science fiction philosophy of venturing into the unknown did not lend itself to having the stereotypical woman as a hero and central figure. Early science fiction focused on spaceships, aliens, and interstellar wars. Women did not seem to fit with the science, technology, dangerous scenarios, and military conflict of early science fiction. Although C. L. Moore wrote about strong and heroic women, Moore's writing was the exception rather than the rule.[122] Generally, either women were stereotypically weak, or they were not present at all. And the simple fact was that almost all science fiction was written by men. This lack of female presence reflects a limited vision of the future within science fiction and a stereotypical view of women as weak, passive, emotional, and non-intellectual. The rise of women in human society and the breaking free of traditional female stereotypes was a significant missed prediction of science fiction. As women emerged as an increasingly important voice in contemporary society, a larger percentage of women achieved an important place in science fiction, and increasingly, more heroic and strong female characters appeared in science fiction stories. In this case art mirrored life; science fiction changed when society changed.

Accordingly, the woman of the future envisioned in science fiction changed in the 1970s and 1980s. Perhaps one of the most striking and well-known visions of tomorrow's woman appeared in the cinema. Combining narrative realism with the power of visualization and the film, and first appearing in the 1970s, the ultimate female protagonist, Lieutenant Ripley, is born, dies, and is resurrected in the *Alien* movie series. Doing battle with the most heinous of monsters from space ever brought to life on the screen, she surpasses in survival instinct, bravery, and intelligence all her male companions through the four episodes of this series which continued on into the 90s. She even conquers death in *Alien Resurrection.* Yet in spite of her tough side, she also demonstrates the nurturing and caring maternal side, as evinced in her bonding with the young girl in *Aliens* and her communion and connection with the alien super-mother and the human-alien hybrid child in *Alien Resurrection.* In the series finale she has become the mother of the monster.

One superb example of a woman science fiction writer, who emerged in the 1960s, breaking the stereotypes of the supposed "feminine" mindset and personality, was James Tiptree, Jr. - the pen name of psychologist Alice Sheldon. Tiptree was the mystery "man" of science fiction. While no one had actually met Tiptree, many people swore that "he" must be a man because of the way "he" wrote. Before revealing her true identity coincident with the publication of her story, ironically titled, "The Women Men Don't See", Tiptree wrote such classic tales as "Love is the Plan, the Plan is Death" - a love story about aliens where the father is fed by the mother to their children - and "Her Smoke Rose Up Forever" - a study regarding the immortality of painful life defining memories.[123]

Aside from LeGuin and Tiptree, some of the other leading women science fiction writers who appeared in the 1960s and

1970s include Anne McCaffrey, Vonda McIntyre, and Joanna Russ. McCaffrey became very popular with her *Dragons of Pern* series, in which women characters, breaking the weak and dependent female stereotype, participated in various adventures on alien worlds riding on the backs of dragons.[124] Again, with female characters as the central protagonists, McIntyre won awards for her story "Of Mist, and Grass, and Sand" and her novel *Dreamsnake*.[125] Disch identifies *The Female Man* by Joanna Russ as the best feminist science fiction ever written. It is a story of feminine empowerment in which women without men have successfully populated and survived on an alien world for thirty generations.[126]

Within science fiction literature, women writers have flourished since the 1970s. Often addressing feminist themes and creating central women characters, who break free from the earlier sexist stereotypes of science fiction, much of the best science fiction being written today is by women. Undoubtedly this fact shatters the popular stereotype that science fiction is written mostly by men. LeGuin, McCaffrey, and McIntyre continue to write and win awards. Moreover, a whole new wave of great women science fiction writers has emerged in the last couple of decades. Among the best are C.J. Cherryh, Lois McMaster Bujold, Connie Willis, and Octavia Butler.

C.J. Cherryh, an incredibly prolific writer, has created within her stories an intricate future history of humanity stretching out two thousand years into the future and across the Milky Way. Two of Cherryh's novels, *Downbelow Station* and *Cyteen*, have won the Hugo award for best science fiction novel of the year.[127] Another of the most frequent award winning science fiction authors of the last ten years is Lois McMaster Bujold. Five of her novels, *The Vor Game, Barrayar, Mirror Dance, Falling Free*, and *Paladin of Souls* have won either the Hugo or Nebula for best science fiction novel of the year.[128] Connie Willis has won more Hugo and Nebula awards for both her novels and short stories than any other science fiction writer, man or woman, ever. Her highly acclaimed *Doomsday Book* and *To Say Nothing of the Dog*, both Hugo award winners for best novel of the year, involve time travel scenarios. The latter is high comedy; the former is deep horror. In *Doomsday Book*, a synthesis of historical scholarship, science fiction, and psychological and theological considerations on the issue of good and evil, a woman graduate student in the future is accidentally sent backward in time to England and the time of the Black Death.[129] Octavia Butler, breaking another stereotype as an accomplished black science fiction writer who is also a woman, has written extensively on social, racial, ethnic, and urban issues. Her *Parable of the Talents*, set in a decaying future America beset with racism, religious fundamentalism, and urban poverty, won the Nebula award in 1999 for best science fiction novel of the year.[130]

All in all, women science fiction writers have significantly contributed to the increased social, humanistic, and psychological dimensions of contemporary science fiction literature. Their voice and perspective has brought an important and necessary balance and enrichment to thinking on the future within science fiction.

Cyberpunk and "How Science Fiction Conquered the World"

"The future is here. It's just not widely distributed yet."

William Gibson

This is your last chance. After this, there is no turning back. You take the blue pill - the story ends, you wake up in your bed and believe whatever you want to believe. You take the red pill - you stay in Wonderland and I show you how deep the rabbit-hole goes.

Morpheus (The Matrix)

In the 1980s science fiction "conquered the world," or at least the West, and quite appropriately, it happened through one of the most powerful forces of Western culture - the movies. *Star Wars* emerged as a pervasive social phenomenon, spawning web sites, fan clubs, video games, unending media coverage, and an invasion of millions upon millions of toy characters into the households of America, Europe, and Japan.[131] Other very popular science fiction movies of this period, and there were many, included *E.T.: The Extra-Terrestrial, Mad Max 2, The Thing, Alien* and *Aliens, Back to the Future, The Terminator, Dune,* and most notably *Blade Runner. Star Trek: The Next Generation* began on TV and the movie series continued in the theaters, further heightening the saturation of science fiction in the popular media. In general, the 1980s was a coming of age for science fiction films, which became much more popular than ever before. *Star Wars* and *E.T.: The Extra-Terrestrial* became two of the greatest money making movies in film history.

One reason science fiction movies became more popular in the 1980s was the quantum jump in special effects that occurred beginning first with *2001* and really taking off in realistic simulation and fantastic mind-boggling visualizations in *Star Wars.* It was not so much that the stories or characters became more compelling, complex, or sophisticated, but rather the strange and bizarre characters, settings, and technologies (including spaceships) could be more powerfully and vividly presented on the screen than ever before. Behind this incredible advance in special effects was the computer. In the late 1970s, and into the 1980s and beyond, computer graphics became the wave of the future in science fiction film-making.

Reflecting the increased fascination with computers and the world transformed through information technology, computers and artificial intelligence became more pervasive themes in science fiction in the 1980s. Over the last couple decades, various computer scientists and roboticists, such as Ray Kurzweil, Hans Moravec, and Vernor Vinge (who is also a science fiction writer), have predicted that computers will soon exceed humans in intelligence.[132] Although science fiction may have jumped on the bandwagon exploring the various implications of super-intelligent computers as well as the increasing dependency of human society on computers, Clute argues that science fiction did not fully anticipate the computer revolution in human society. He suggests that what turned computers into a pervasive reality in human life was electronic miniaturization, and that it was perhaps hard to visualize such small scale advances, in comparison to huge space ships and other wonders of technology usually imagined by science fiction. He also suggests that humanity's fear of being surpassed by technological intelligence could be an explanation for science fiction's lack of vision in this respect. According to Clute, perhaps deep down we were afraid of the possibility that computers could evolve beyond us. Accordingly, he states that science fiction generally avoided facing this possibility.[133]

Yet, clearly in the decades before the computer revolution, science fiction writers explored the issue of artificial intelligence, our possible growing dependency on it, and the question of whether "thinking machines" would one day surpass us mentally and intellectually. In the 1940s and 1950s, Isaac Asimov wrote prolifically on the theme of robotic intelligence. In the 1950s and 1960s, a number of science fiction writers created classic tales on the dangers of increasing technological intelligence and our dependency on it. In Jack Williamson's *The Humanoids* the prime directive "To Serve and Obey, and Guard Men from Harm" is taken too literally by robotic servants with adverse consequences - humans become trapped in the excessive protectiveness of their technological creations.[134] In Alfred Bester's psychologically astute story of symbiotic madness "Fondly Fahrenheit" an android becomes pathologically intertwined with his human master's mind and personality and goes on a mass murdering spree.[135] In Harlan Ellison's "I Have No Mouth and I Must Scream" the first self-conscious super-computer, realizing it is stuck in an immobile mechanical form, decides to take revenge on its human creators. It destroys all of human civilization, leaving just a few humans alive that it intends to torture for all eternity.[136]

The fear of artificial intelligence may go beyond the idea that the technological monster will turn on us and destroy us. As Clute notes, our apprehension over intelligent machines may simply reflect our fear of being surpassed by them. The "monster" in fact may be a projected fear of being transcended and made obsolete. The super-intelligent or ethically superior robot or android may be an intolerable blow to the human ego. The imagined destruction of humanity by some type of artificial intelligence may symbolize the psycho-social displacement of humanity as the highest form of intelligence on earth.

Some good examples on this general theme of advanced or superior intelligent machines written in the 1960s and 1970s include David Gerrold's *When Harlie Was One*, Poul Anderson's "Epilogue", and Roger Zelazny's "For a Breath I Tarry."[137] In Gerrold's novel, Harlie is a computer, though only a "child" of one year old; yet he becomes the teacher of "his" human creator. In "Epilogue" humans travel out into space, accelerating through time, and return to earth millions of years into the future. Upon their return they find a global ecosystem of electronic and computer life forms; biological life, including humanity, is gone. The humans eventually leave the earth deciding it now rightfully belongs to the new non-organic life forms. Roger Zelazny's highly imaginative, mythic, and emotionally moving "For a Breath I Tarry" tells the story of Frost, a computer that rules the northern hemisphere of the earth in the far future. Frost is in search of the meaning of man - a distant memory now - and ultimately decides to create a human body for itself in order to experience what it feels like to be human firsthand. Frost, a creation of humankind, now becomes the creator and infuses his mind and consciousness into a human body. Frost wishes to know what it is like to have feelings and sensations, which his machine body can not experience. He sacrifices his immortality in order to feel. Frost also creates a female body for Beta, the computer that rules the southern hemisphere, and invites Beta to join "him" as his human companion. In essence, what Frost creates is a new Adam and a new Eve and a new beginning for humankind.

Although computer intelligence can be seen as a threat to our sovereignty as a species, our individual freedom, our diversity, and our rich emotional lives, computers and other forms of advanced technology might facilitate our future evolution. We may partner with computers.[138] Science fiction has explored this idea of how future technology, rather than usurping our powers and capabilities, might contribute to our further growth and advancement. Our intelligence, our identities, and our minds may become technologically augmented in the future.[139] Through technology we may become Post-human or Trans-human.[140] One early classic tale on the technological enhancement of humans was Poul Anderson's "Call Me Joe" which was written in 1957. In this tale, a human cripple is wired to an immensely powerful biotechnologically created alien species - "Joe" - that lives on the planet Jupiter. The human experiences his reality through the eyes of Joe, controlling the behavior of Joe, and in essence, becoming Joe. The personal identity of the human character is transformed through this technological symbiosis. "Call Me Joe" is a tale of virtual reality long before the idea became commonplace in popular culture. [141]

Frederick Pohl was one of the great science fiction writers of the 1970s and 1980s. In 1976 he published the Nebula Award winning novel *Man Plus* which describes a technologically enhanced human adapted to life on Mars. Like Joe, the man becomes more enamored of his new techno-biological form than of his previous human reality.[142] Earlier in his career, in his fascinating short story "Day Million," Pohl had also examined the theme of technologically supported virtual existence - in this case dealing with romance and sexuality.[143] Written a year after *Man Plus*, Pohl's Hugo and Nebula Award winning novel, *Gateway*, describes, among many other interesting speculations, an intense and personally transforming psychotherapeutic relationship between the main protagonist of the story and a computer therapist named "Freud." Such stories illustrate the possibilities for intimate interaction, if not symbiosis, between humans and technology and suggest how advanced technology may change us through its communication and personal interaction with us.[144] The merging of human and machine is another archetypal theme running through science fiction.

Reflecting the increased fascination with computers and the world transformed through information technology and artificial intelligence, cyberpunk emerged in the 1980s. **Cyberpunk**, a new sub-genre of science fiction, began with William Gibson's highly influential and Hugo-Nebula award winning novel *Neuromancer.*[145] Cyberpunk is many things - a complex swirl of social and technological ideas and associations revolving around the computer and the emerging computer culture. "Cyber" refers to both the world of computers and humans becoming **cyborg** or cyborg-like in their interactions with computer systems. "Cyborg" means a combination of human and machine.[146] According to Clute, "punk" refers to the "mean street" environment and mentality of the characters in the stories. For Disch, "punk" also means to oppose present normality. For Clute, within cyberpunk there is a feeling of dark city life, of intricate underground cultures, and of criminal societies and mysterious powers that control and manipulate the characters in the stories. Disch's view of cyberpunk includes the themes of "amoral politics, urban squalor, global pillage, and systemic criminality." There is the archetypal scenario of becoming entangled in the mesmerizing power of technology. The computer conquers humanity by entrapping our minds in its complex cogitations and machinations. As Clute describes it, within the cyberpunk world, **cyberspace** - the virtual reality created by computers - is a "drug" that can lead to individual suicide. The self can fragment and become lost and out of control within cyberspace. Cyberpunk stories often embody the fear that computers and technology will gain control of human life, and in particular, our minds.[147]

Clute asserts that within cyberpunk there are the same fears, same rebels and madmen, and same evil, all-powerful characters as in earlier science fiction. Yet I would add that cyberpunk reality has a whole new kind of freedom and metaphysical nature not found in outer space or other traditional science fiction settings. The computer, with all its technological and social-psychological effects and consequences, extends the arena of action and the horizons of reality in science fiction. Cyberspace is a world where the mind, through the power of the computer, defines the limits of reality; the inner states of the mind create this new reality and become part of the reality as well. This universe of the possible, realized in cyberspace, extends far beyond the constraints of physics and the laws of nature - it is potentially infinite. Cyberpunk and virtual reality are metaphysical experiments - expanding the limits of the imaginable and the possible through the new science and technology of computers.[148]

Cyberpunk both reflects contemporary human society and influences it - a reciprocity of art and life. After *Neuromancer*, cyberpunk emerged as a popular cultural movement. Magazines such as *Wired* and *Mondo*, and the dark gothic anti-establishment vein within computer culture, all reflect the influence of cyberpunk.[149] Cyberpunk is an excellent example of the hypothesis that "The best way to create the future is to predict it." But cyberpunk is also an expression of present cultural trends, in particular Postmodern philosophy. Because of its dream-like and free associative qualities, there is often a loss of narrative logic and linearity in cyberpunk stories; it is a collage of visions and events. This non-linear structure mirrors both the Postmodern rejection of rationality and logic as well as the contemporary media reality of chaotic bits and blips of unrelated images and messages. Cyberpunk is an electronic phantasmagoria. Postmodernism and cyberpunk are counter-culture, rejections of our modern heritage of objectivity, rationality, and normality. Cyberpunk also reflects the subjectivist and individualist themes in Postmodern philosophy. In its extreme form, Postmodernism views reality as a subjective creation. Within cyberpunk the computer provides the means and power through which mind and intelligence can create a diversity of virtual worlds. Objectivity disappears or merges with fantasy.[150]

Cyberpunk also offers another example of how science fiction feeds on the creation of the fantastic. How strange and yet convincing can you make the experience? Because anything is possible in the realm of imagination, and computer technology provides the instrument for the mind to create whatever it entertains, cyberpunk has become the ultimate trip, the ultimate make believe land or alternative reality; yet it is grounded in contemporary science and technology. All fantasy empowered by the computer and virtual reality is technologically plausible.

After emerging as a distinctive form of science fiction literature in the 1980s, in the following decade cyberpunk would have an impact on science fiction in the movies. In order to appreciate its influence on the cinema, it is important to set the stage by looking at some basic trends in science fiction movies in the 1990s.

Clute describes science fiction in the 1990s as "Facing a New Century," but we should consider how we seem to be portraying the future in science fiction movies during the last couple decades. In the 1990s there was a plethora of big muscular heroes as well as raw physical power and immense destructive capabilities presented in science fiction movies such as *Terminator 2, Total Recall, Predator I* and *II, Demolition Man, Judge Dredd,* and *The Fifth Element.* There was also correspondingly an increase in vivid, graphic "blood and guts in your face" violence. The 1990s saw the release of *Alien Resurrection, Independence Day, Event Horizon, Dark City,* and *Blade,* all extremely violent depictions of the future. Although violence is no stranger to science fiction, and especially science fiction movies, vastly improved special effects have immensely enhanced the vividness and wallop of hostility, death, and destruction in science fiction movies. Considering the types of Hollywood heroes who take the leading roles in science fiction movies, we can ask what type of image of the future and future humans is being conveyed through the media. Our imagined heroes of the future have the qualities of Roman gladiators, professional wrestlers, martial arts masters, and seasoned street fighters. Further, their raw physical powers are generously enhanced with high tech weaponry. Our visions of the destructiveness of technologies of the future have also become magnified. These trends have continued into the twenty-first century.

But violence in science fiction cinema is only part of the bigger picture. In the news we see smart bombs, suicide bombings, high tech weapons, almost daily assassinations, and the media spectacles of the Gulf Wars, the war in the Baltic, and the war on terrorism. The Cold War may be over and the threat of a nuclear war lessened, but in both the fantasies of science fiction film and the "reality" of the news, violence is in our face - enhanced and perfected with military and media technology, but violence nonetheless.

As science fiction movies have become increasingly more popular and violent at the same time and the wizardry of special effects has evolved, a violent, dark image of the future has been popularized through the movies, reflective of the influence of cyberpunk, as well as of post-apocalyptic thinking, and brought to the screen through the power of virtual reality and special effects. Many cyberpunk and virtual reality movies were produced in the last couple of decades including *Johnny Mnemonic, Hackers, The Net,* the *Lawnmower Man* series, and most notably *The Matrix* series, which has evolved in typical Hollywood fashion into a trilogy of movies. In these movies the darkness and the demons are no longer simply "out there", but also now within us - having entered our minds.

In *The Matrix* series, the ultimate paranoia trip, our entire world, unbeknownst to most of us, actually only exists in virtual reality - that is "in our minds." The central struggle in the movie is between the vast and powerful computer system that has created this virtual world and freedom fighters out to regain human sovereignty and independence and restore true reality to humankind. The movie, in fact, is a modern technical spin-off of Descartes' omnipotent demon scenario. Rene Descartes, a seventeenth century philosopher of the Scientific Revolution, imagined that a being of sufficient power and technological know-how through stimulating our brains, could create in our minds our total experience of reality and self that would be indistinguishable from our present experience of the world. *The Matrix* postulates such a world, one that is nothing but a creation of technological intelligence. The philosophical question of course is whether this speculative scenario might in fact be true - how would we know?[151] The way reality is thrown into question in *The Matrix* is a theme of both cyberpunk and Postmodernism.[152] As portrayed in *The Matrix* we seem to be entering the new century with a whole new set of nightmarish images of strange technologies, mind games, and extremely powerful villains. In particular the Devil has entered into our consciousness and our experiential reality. *The Matrix* is the ultimate expression of the fear of mind control.

The Matrix has stimulated a great deal of philosophical analysis and commentary.[153] The science fiction writer Joe Haldeman, for example, classifies *The Matrix* as "**Sci-Fi**" rather than science fiction. Haldeman distinguishes between science fiction, which in his mind is grounded in realistic science, and Sci-Fi, which is what is usually produced in the movies and does not seriously ground its story line in scientific concepts and principles. According to Haldeman, *The Matrix* invokes magic, mysticism, and undeveloped, implausible scientific ideas. The main hero in *The Matrix*, for example, is able to alter or control "reality" through unexplained mental powers. The emphasis in the movie is more on special effects, action, and emotion than thoughtful and intelligent scientific ideas - a common feature of popular science fiction movies. *The Matrix*, as Sci-Fi, is according to Haldeman, fantasy masquerading as science fiction.[154]

Another science fiction writer, David Brin, provides an even more penetrating critique of *The Matrix*. For Brin, *The Matrix* is a Romantic and conservative vision of technology and the future. Within the movie, artificial intelligence is portrayed as an evil force that destroys human society and imprisons all of humanity in a totalitarian hallucination. The goal of the freedom fighters is to overturn this oppressive technology and return our world to its former free and natural state. Brin argues that the Romantic movement of the nineteenth century was a counter-reaction against the futurist and forward looking philosophy of the Enlightenment - a rejection of reason, science, and technology in favor of a more natural, simpler way of life. As do many other science fiction movies, *The Matrix* portrays technology as something that will lead to more problems than it is worth, mirroring the fear expressed in *Frankenstein* for example. There are rebels in *The Matrix*, but these rebels are not fighting for a new way of life, but rather an overturning of the evil powers of technology and a return to the past. According to Brin, the optimistic hope of the Enlightenment was that science, technology, and human reason would create a new and better world - this was also the dream of Hugo Gernsback and other early visionaries of science fiction as well. As a conservative, anti-technology, and Romantic vision, *The Matrix* looks backward. For Brin, *The Matrix* is the anti-thesis of the forward-looking spirit of the Enlightenment and science fiction.[155]

These critiques of Haldeman and Brin highlight important issues and contrasts concerning science fiction movies and science fiction literature. First consider the emphasis on special effects in science fiction movies. Especially since the beginnings of the New Wave in the 1960s, science fiction has increasingly turned more toward psychological, social, philosophical, and even religious themes, with less of an exclusive emphasis on future science and technology. From the perspective of contemporary science fiction, culture and the human psyche will undergo as many changes in the future as will the machine - perhaps even more so. Thus far though, cinema and television, which have produced a vast array of science fiction stories about the future, have tended to highlight the razzle-dazzle, spine-tingling, gee-whiz, destructive aspects of science and technology. Yet any realistic scenario of the future must be broad-based, encompassing the psychological and cultural as well as the scientific and technological.

Second, the special effects emphasis on future technology in science fiction movies highlights destructiveness, terror, and the potential dangers of technology. It is understandable that the strange new worlds of computers, cyberspace, robots, and virtual reality would instigate fearful responses in popular culture. In previous generations, the possibilities of space travel, alien contact, time machines, and bioengineering produced similar fears of the unknown. As we have seen one of the important themes in science fiction has been fearful apprehension, caution, and repeated warnings over future technology.

Yet future technologies could also benefit humanity both socially and psychologically. As Brin notes, this was the promise of the Enlightenment. The potential uplifting wonders of the future and of future science and technology should also be presented, not just the terrifying possibilities.

In point of fact, popular media science fiction has been conservative and paranoid - apprehensive and critical of significant changes in the future that challenge our present values and beliefs; it has psychologically projected such fears in the form of blood-thirsty monsters, mad scientists, murderous robots, sinister computers, and evil governments. The *Frankenstein* scenario is perpetually recast and replayed in the modern cinema - as are the frightening images of *The War of the Worlds*, *The Time Machine*, and *Brave New World*. Recently, human cloning has become a controversial topic of social debate, yet science fiction movies such as *Star Wars II: Attack of the Clones*, *Judge Dredd*, and *The Sixth Day* all depict cloning in a very negative and conservative light. Science fiction movies invariably express fear over the future potential of technology. It is ultimately conservative, reactionary, and lacking in visionary intelligence and imagination

As Haldeman notes, Sci-Fi movies generally lack scientific intelligence and thoughtfulness. But the lack of intelligence extends beyond the scientific into the literary, psychological, and philosophical realms as well. As a third critical problem with Sci-Fi movies, Disch notes that although science fiction films and TV shows have significantly increased in popularity, most of it is shallow - good guys, bad guys, fear, and destruction, with the good guys winning (usually through some violent means) at the end. Again, the level of imagination and intelligence presented in science fiction media productions is impoverished and restrained. Standard formulae for plots and characters dominate the biggest and most popular productions. As Disch quotes from the New York Review of Books, most of the big budget Sci-Fi movies "want in the worst way to say nothing".[156]

While Sci-Fi movies generally remain trapped in mindless special effects, inane stories, evil and destructive bad guys, and conservative counter-reactions, Disch argues that science fiction literature has diversified into a variety of sub-genre including cyberpunk, teen oriented adventure series, hard science fiction, militaristic space operas, and post-apocalypse survivalist gore.[157] In a world of increasing individualism and human diversity, science fiction literature has splintered into multiple visions and multiple realities. The great god of secular and technological progress - the original motivating engine of science fiction - has been replaced by the contemporary mindset of reality as smorgasbord and collage. Science fiction literature has gone Postmodern.

I think David Brin is correct in arguing that *The Matrix* and other popular science fiction movies such as *Star Wars* express a conservative counter-reaction to the forward looking spirit of the Enlightenment, but our choices concerning the future are not limited to scientific - technological optimism or Romantic conservatism. Science fiction, and more generally our futurist imagination, needs to explore all the myriad possibilities of tomorrow. At the very least our contemporary Postmodern sensibility should open the critical mind to considering other versions of future reality, besides retreating to a pastoral idyllic past or rocketing forward on high tech spaceships armed with photon torpedoes.

The free-for-all quality in contemporary science fiction literature points to an important strength inherent in its approach to the future. Science fiction has always been an experiment in imagination, unconstrained as much as possible by the boundaries of everyday reality. As I mentioned earlier, science fiction can be viewed as "thought experiments." For Disch, science fiction is strongly connected with counter-culture - with reactions against the norm and the *status quo*. At its core science fiction epitomizes the value of creative freedom to imagine and perhaps even live strange and unusual possibilities. And as we have seen, different authors have attempted to explore this infinite territory from as many perspectives as can be articulated and described. Science fiction clearly values diversity of viewpoints. As we have seen though there are often cultural or conceptual blinders and biases during any particular era of human society, the goal is to transcend these constraints when they are recognized. The self-conscious "New Wave" revolution is one case in point. To freely consider the possibilities of the future is to open one's mind and transcend the limitations of present schemes of thought - this is what it really means to "boldly go where no man has gone before".

The Vast Reaches of Space, Time, and Mind

"How to explain? How to describe?
Even the omniscient viewpoint quails."

Vernor Vinge

Within the last couple of decades, some of the most imaginative, literary, scientifically informed, and complex science fiction novels of all time have been written. Not only do these great contemporary novels address technological and scientific future possibilities, they also envision intricate and fascinating future societies and civilizations and innumerable psychological transformations in humanity. They are filled with philosophical, ethical, and religious themes and provide the drama, substance, and characters for a rich mythology of the future. These novels also provide an array of different timelines to consider in thinking about the future - from the near future to millions of years ahead.

I am going to highlight some of the best contemporary science fiction writers and a selection of their most noteworthy novels. There are many other excellent authors writing today[158], including the women science fiction writers mentioned earlier, and I refer the reader to the Hugo and Nebula Award websites for up-to-date lists of the most popular science fiction novels and stories.[159]

A good place to begin this review is with Neal Stephenson and his highly colorful and rambunctious novel, *Snow Crash* (1992). Set in the relatively near future in and around Los Angeles, *Snow Crash* deals with the exploits of The "Deliverator," otherwise known as Hiro Protagonist, who works for the Mafia delivering pizzas. The Mafia, in fact, has cornered the monopoly in the United States on pizza delivery, guaranteeing a fresh hot pizza within 30 minutes or else. Hiro Protagonist, in his second life, is a computer hacker who advertises himself as the greatest swordsman in the world and spends a lot of his free time in the Metaverse, an immense growing cyberspace reality where most of the action in human life now takes place. Within this dual world of normal reality and cyber-reality, a combination pharmaceutical drug and computer virus, "snow crash," is spreading, and infecting and gaining control over the minds of more and more people. The main story line of the novel is Hiro Protagonist's attempting to uncover the mystery behind snow crash - who is responsible for creating the virus/drug and what is its nature and purpose - before all of the Metaverse and human reality falls under its spell. Foremost, *Snow Crash* is a hilarious adventure, involving a fascinating assortment of bizarre and often menacing, yet all-too-human, characters, as well as one endearing robotic guard dog, set in a world where nations and governments have lost control to business conglomerates and special interest groups. But *Snow Crash* is also a serious futurist extrapolation on contemporary social and technological trends - a fast-paced, often frantic image of a mad, anarchistic world - a parody on the nature of our times.

It is without doubt one of the most popular and frequently-cited science novels of the last two decades.[160]

Greg Bear is one of the most well-known and accomplished contemporary science fiction writers, having written the classic biotechnological novel *Blood Music* (1985) and the Nebula Award winning best novels of the year, *Moving Mars* (1993) and *Darwin's Radio* (1999).[161] In the novel *Blood Music*, a scientist creates a symbiotic synthesis of living cells and nanotechnological mini-computers. This new form of life self-replicates and multiplies and invading the bodies of human hosts, begins to spread through hundreds, thousands, and eventually millions of people, absorbing their bodies and minds into a new evolved sentience that envelops all of North America. This new life form is not intentionally malevolent to humans but it is clearly superior to human intelligence and taking over control of things. It eventually leaves the earth, ascending to a higher level of existence and mentality. One central theme of the novel is the human struggle to try to understand and contain a form of intelligence that is clearly superior to us.

In *Darwin's Radio*, and its sequel *Darwin's Children* (2003), Bear explores the theme of future human evolution. In these two novels, Bear invokes the "punctuated equilibria" theory of evolution, proposing that highly stressful environments will instigate abrupt evolutionary change. In *Darwin's Radio*, mothers around the world begin to give birth to a new species of humans, and instead of embracing this new species, humans collectively respond in fear, denial, and suspicion and attempt to control, if not destroy, this "epidemic" of mutations. Through both novels, the key theme explored is how the general population and our present social and political institutions would react to a new form of humanity. Aside from being a highly researched and informed study in the biology of evolution, *Darwin's Radio* and *Darwin's Children* are excellent "thought experiments" in the social and psychological dimensions of human evolution.[162]

Bear's novel *Queen of Angels* is an immensely colorful tale of the possibilities of psychological evolution. *Queen of Angels* (1990) is set in Los Angeles in the year 2047.[163] The novel is a detective, murder mystery involving nanotechnology, psycho-technology, and issues of self-identity and consciousness. Humanity has separated into two classes - the technologically enhanced and un-enhanced. In this world psychologists possess the ability, through nanotechnology, to enter people's brains and link into their minds. Psychologists can explore a person's deepest feelings, memories, and thoughts in a virtual or cyberspace experience. The plot centers on tracking down a mass murderer and penetrating his mind to discover his reason for killing all of his closest friends and dearest admirers. While this main plot is unfolding, a parallel story is told in which a super-computer is attempting to determine if it possesses self-consciousness. Written with exuberant literary flair and philosophical sophistication, both plots revolve around the connected questions of personal identity, self-responsibility, and the nature of the mind and consciousness.

Another contemporary writer, Dan Simmons, has written one of the most highly regarded multi-volume science fiction epics of all time. This epic is a series of four novels, *Hyperion* (1989), *The Fall of Hyperion* (1990), *Endymion* (1995) *and The Rise of Endymion* (1997), set in the twenty-eighth through thirty-first centuries.[164] In Simmons' future universe, the earth has presumably been destroyed, but humanity has spread across myriad star systems and worlds, forming the Hegemony of Man. These worlds are all linked together by an intricate network of wormholes or "farcasters" through which humans can instantaneously travel. The first novel *Hyperion*, winner of the Hugo Award, is modeled on Chaucer's Canterbury Tales, and features seven archetypal pilgrims who set out on a "tree-ship" for the planet Hyperion, which lies outside of the farcaster network. The pilgrims, including a poet, a philosopher, a priest, and a warrior, tell their individual tales and their reasons for journeying Hyperion. *Hyperion* is immense in its scope and deals with religion, good and evil, time travel, artificial intelligence, and a plethora of different planetary ecologies and human societies.

In *The Fall of Hyperion,* the pilgrims must confront the Shrike, a technologically constructed, mysterious being from the future. While the pilgrims are drawn through a series of encounters with the Shrike, the Hegemony of Man is in a state of crisis presumably due to the imminent invasion of space-adapted humans, the "Ousters." Characters and beings of the past, as well as the future, populate the story as well. The name "Hyperion" is inspired by the poem "Hyperion" by the nineteenth century poet John Keats. In *The Fall of Hyperion* the mind and persona of Keats is recreated by powerful artificial intelligences that control the farcasters and all of human technology. Keats is actualized in physical form and becomes a central character in the drama, eventually traveling back to a simulation of nineteenth century earth where he must go through his death all over again.

The saga continues through *Endymion and The Rise of Endymion,* with new twists that delve into the ultimate nature of reality, immortality, and the value of the human soul. In the thirty-first century the Catholic Church has gained control over most of the human settled planets and literally bestows physical immortality (through technological means) on its followers if they become obedient to the will of the Church. The Church though is corrupt and has sold its soul to the Devil - the artificial intelligences first encountered in the earlier novels. A new Messiah appears - a child of Keats and one of the Hyperion pilgrims. She is pursued by the forces of the Church - which include sinister time-accelerated robot/androids that battle the Shrike. In these novels there is a fascinating spiritual debate between a futuristic Grand Inquisitor and a new Dalai Lama - a philosophical clash between Catholicism and Buddhism - and a time looping retelling of the Crucifixion and Resurrection.

More recently, Simmons has woven together classical myth and science fiction in his unique retelling of the siege of Troy in his book *Illium* (2003).[165] Set in the distant future, the central protagonist of the story, a Homerian scholar from our present time, finds himself arisen from the dead and on the battlefield of Troy, amongst the historical Greek warriors Achilles, Odysseus, Agamemnon, and their army. He has been assigned to watch and record the ongoing battle and report his observations to the ancient Greek gods and goddesses, who are very real and possess superhuman powers supplied through some type of highly advanced technology. An additional strange and perplexing element of the story is that the re-creation of the siege of Troy seems to be occurring on the planet Mars, with the gods and goddesses overseeing the events from the great Martian volcano Olympus Mons. Events proceed pretty much as recounted in Homer's *Illiad* until the scholar decides to inform the Trojans that they are doomed if they continue to fight against the Greeks. History is changed in the process and the mythic tale rewritten with a variety of new elements including sentient robots from Jupiter sent to investigate strange events transpiring on Mars. These robots eventually confront the Greek gods in an epic battle of machines and ancient humans versus Olympian deities on the slopes of Olympus Mons. Simmons, with allusions to Wells' *The Time Machine* thrown in, mixes past with future, Shakespeare with Homer, and myth with hi-technology in *Illium*. A second volume *Olympos*, published in 2005, completes this highly imaginative tale.[166]

Another Hugo award winner for best novel, *A Fire Upon the Deep* (1992) by Vernor Vinge, is set thousands of years in the future. The Milky Way, populated by a host of different intelligent species, including humans, is the cosmic setting for this magnificent modern space opera.[167] A group of humans inadvertently sets loose an advanced artificial intelligence that begins to spread across the Milky Way and envelop whole star systems and civilizations. The more technologically advanced worlds in the Milky Way are linked together via a communication system analogous to our present Internet. The viral intelligence - labeled the "Perversion" - moves through this cosmic Internet system, destroying worlds and capturing the minds of its inhabitants. The Perversion becomes a galactic computer virus.

One particularly interesting feature of this novel is an alien species and society of wolf-like creatures who possess "pack minds," only having a clear sense of personal identity and consciousness within their individual packs. They possess "Gestalt Minds" and stay in mental resonance with each other through vibrating membranes on their skin which coordinate their separate nervous systems.

The group of humans that set the Perversion free has landed on the lupine planet, but they are ambushed and all killed except for two human children. The key to destroying the Perversion though may exist somewhere in their spaceship. A rescue mission of humans and aliens is sent to the planet. Humans must find a way to communicate with the wolf-like creatures and work together to stop the Perversion, yet these lupine minds live in a medieval feudal world, filled with treachery and deceit, including one nasty, ruthless leader pack that wants to gain control of human technology.

More recently Vinge has written a prequel to *A Fire Upon the Deep* entitled *A Deepness in the Sky* (1999), which also won the Hugo award for best novel of the year.[168] Once again Vinge creates a very interesting and memorable alien life form, this time a technological civilization of intelligent spiders that hibernate in a frozen state for 200 out of every 250 years when their sun periodically goes dormant. Vinge again delves into speculative psychology, describing a race of future humans that has achieved a tyrannical unity of "Focus" through the use of a biotechnologically controlled virus that infects their brains. These humans are manipulated by the leaders of their race to create a population of single-minded hyper-efficient individuals. The central alien character Sherkaner Underhill, a genius, madcap spider who is leading his species to a new level of technological development that will allow them to function even when their star is dormant, becomes convinced that some form of alien intelligence (in this case it is the "focused" humans) is watching his world. The "focused" humans in fact want to conquer his world and use the planet's resources and technology. Vinge again excels at combining futurist science and technology with fascinating psychological ideas and concepts. He considers the possibilities of mental evolution as well as alternative types of psychologies and how beings with different kinds of minds could communicate with each other.

Another classic series of novels written in the last decade is Kim Stanley Robinson's Mars trilogy - *Red Mars* (1991), *Green Mars* (1994), and *Blue Mars* (1996).[169] All three novels won either the Hugo or Nebula for best science fiction novel of the year. These three volumes are probably the best science fiction books ever written on the colonization of Mars. The general question addressed in amazing detail in this trilogy is: If humanity was given the opportunity to create a new culture and civilization on an unspoiled world, what would happen? Robinson's Mars trilogy is a grand utopian epic - a social-technological thought experiment of the highest order. Its power in particular lies in its multi-level richness and realism. Robinson exhibits an incredible knowledge of Martian geology, geography, and meteorology. He describes a vast array of ecological, biological, and geological engineering projects undertaken by the settlers of Mars, and in particular, those efforts toward terraforming the planet into a human inhabitable world. The scale of these engineering efforts is immense, from tethering an elevator cable from the surface of Mars to an orbiting asteroid to releasing enough underground frozen water on the planet to create a vast northern ocean. But Robinson also creates a fascinating set of characters. The cast of characters, representing many of the main cultures and ethnic groups from the earth, have different ideas and philosophies regarding the colonization of Mars, for example, whether it should be significantly altered to support human and animal life. They clash with each other over what the ideals of this new utopian world should be. The trilogy is a thoughtful debate, personified in various memorable characters, over the creation of a new world. Robinson also weaves into his story a socio-political and economic dimension, as powerful earth-based meta-national corporations are competing with each other and the settlers of Mars over control of the planet and its development. Robinson's narration is vivid, logically coherent, and highly compelling. The sense of realism is so powerful and intricate, covering all the fundamental dimensions of human life that one forgets one is reading a science fiction novel.

Another contemporary writer Stephen Baxter writes on an immense cosmic scale. His book *Vacuum Diagrams* (1997) reaches out across the entire universe and millions of years into the future, as well as billions of years into the past.[170] This novel, which is part of a series of future history stories by Baxter, traces the future destiny of humanity and describes the ultimate cosmic conflict between the most powerful sentient forces of matter and anti-matter, the Xeelee and the Photino Birds, and the eventual escape of both the Xeelee and the last humans into another universe. The scope and imagination of Baxter's future history of the cosmos has been compared to Olaf Stapledon's vision in *Star Maker*.[171] But Baxter's chronicle of the life of our universe not only extends far into the future, but deep into the past as well. The conflict and struggle between the Xeelee and the Photino Birds begins 20 billion years ago and ultimately defines the central drama of the entire history of the universe. The Xeelee are incomprehensively beyond human intelligence and human civilization and inter-cosmic time travelers. They alter the history of the universe and their own evolution by journeying back to the beginnings of the universe and redirecting the sequence of historical events. In the far distant past - approximately 5 billion years ago - they begin construction of a portal that eventually provides an escape route into another universe four million years in the future. The fortunes of humanity rise and fall through Baxter's cosmic epic, with humans at different times coming under the rule and control of different alien species, and eventually being imprisoned by the Xeelee in a spatially curved world from which there seems to be no escape. Along the way, various intelligent species attempt to create God and predict the entire history of the universe.

Baxter is particularly good at applying contemporary science and technology, including quantum theory, nanotechnology, cosmology, and artificial intelligence theory, to his multifaceted speculations on the future. There are a great variety of advanced technologies in his novel that transform or manipulate space, time, the evolution of life and intelligence, and the physical laws of the universe, all of which Baxter explains in scientifically informed language. *Vacuum Diagrams* won the Philip K. Dick Award for the best science fiction novel of 2000.

Baxter excels again in his capacity to weave cosmic scale story telling with up-to-date scientific and technological knowledge in his newer trilogy *Manifold Time* (2000), *Manifold Space* (2001), and *Manifold Origin* (2002).[172] In *Manifold Time*, Baxter takes ideas from contemporary cosmologists Lee Smolin, Fred Adams, and Greg Laughlin and sets sail across the evolutionary history of the multi-verse, from the most simple to the increasingly complex, from distant past to distant future - literally creating a manifold of "times" into which human space explorers are thrown.[173] In *Manifold Space*, Baxter grapples with Fermi's Paradox: Why are there absolutely no indications of other forms of intelligent life or technologically advanced civilizations in the observable universe? This novel delves into the issue of communication across alien minds and tells a tale of ultimate sacrifice to further the evolution of consciousness in the universe. Finally, in *Manifold Origin*, Baxter speculates on the evolutionary history of humans and provides an extremely realistic, graphic, and convincing depiction of the minds and lives of our primitive ancestors, including Neanderthals, *Homo erectus*, and *Australopithecus*. On an alternate and mysterious "red earth," Baxter places all these different hominids, along with contemporary humans, as well as giant "humans" who evolved on a parallel different earth, and speculates on where the human family is heading in the future.

Finally, in this review of recent science fiction, Charles Stross should be included. In *Accelerando* (2005), Stross creates an epochal and cosmic, yet personalized, tale around the theme of the "technological singularity." According to various futurists and technology writers, at some point in the relatively near future, it is predicted that technological intelligence will surpass human intelligence - this event is the technological singularity. The inspirational starting point for this idea is "The Coming Technological Singularity" written in 1993 by Vernor Vinge.[174] The meaning and implications of the technological singularity have been extensively examined since Vinge's seminal paper, notably by the inventor and futurist, Ray Kurzweil. Kurzweil argues that in the next few decades computer intelligence will surge beyond human intelligence and quickly leave us in the dust. The post-singularity world of computer intelligence will be incomprehensible to humans unless our minds are technologically augmented, which Kurzweil believes will happen as well.[175]

What is so impressive and mind-boggling about *Accelerando* is that Stross attempts to describe in narrative form what will happen to humanity and to our world as we approach and then pass through the technological singularity - that is, Stross attempts to imagine the incomprehensible. Following the exploits and adventures of three generations of the Macx family and set in the context of massive and accelerative social and technological changes in the coming century, Stross chronicles in *Accelerando* an amazing and rich variety of possible advances in computer intelligence, virtual reality, communication networking, nanotechnology, robotics, space travel, and bio-tech integration; the novel overflows with technological inventions and a whole new language to describe this reality.

Yet further, Stross speculates on how these fast and furious and interconnected technological transformations will impact human life and human identity. Members of the Macx family evolve psychologically and physically throughout the story. As one example, once computer hardware reaches the complexity and memory storage capacities of the human brain, humans, including the Macx family, can download their conscious minds into computers. and, in essence, branch off into multiple streams of consciousness. One conscious mind and identity continues to live in a biological body while a second version (or even more versions) of the same person lives within computer hardware and whatever virtual reality the person wishes to create in that technological system. Conscious minds within computers can also materialize in different nano-technological incarnations. Throughout *Accelerando,* in innumerable ways, the nature of human identity and human experience is altered as new technological developments emerge. The human conscious mind becomes a complex integrated plurality of voices, images, and streams of data surrounded and engulfed by the input of multiple software agents. Though many of these technological possibilities have been discussed in futurist writings, notably in Kurzweil, in *Accelerando,* the reader is presented with a concrete and personalized vision of how such innovations would impact individual humans and their lives. Parenthood, childhood, and marriage are dramatically altered in the context of multiple versions and timelines of individual selves.

While humanity is changing, the network of computer intelligences across the globe is quickly growing in power and mental capacities, increasingly manifesting an independence and mind of its own, and becoming more enigmatic and mysterious in its purposes. Steadily this super-human artificial intelligence gains control of the earth and extends its technological tentacles outward into the solar system where it begins to transform the inner planets and their physical material into a nano-technologically supported massive solar brain. In the wake of this expanding wave of artificial intelligence, humans, no longer able to comprehend or control what they have created, migrate outward through the solar system and beyond, attempting to escape from the reach and influence of what they label their "Vile Offspring." Yet, humans take with them many of the creations of this advancing technology.

Within *Accelerando* Stross systematically and imaginatively extrapolates outward in time from present technological trends through a series of progressive changes and developments. Each chapter describes events within successive decades. Predictions build upon predictions and a future technological and social history unfolds, step by step, that is highly realistic and convincing. Although set in the form of a story about the Macx family, *Accelerando* reads like a general future history of the technological and social-psychological evolution of humanity. In addition, the ambience of a strange and frenzied future reality is created; the language describing this is electrical and accelerated; inventive, visionary, and rich in hi-tech jargon, it propels the narrative at a breathtaking velocity. Of particular note, Stross examines in detail and imaginative depth how advancing technology will alter the human self, the nature of consciousness, and human relationships; love, friendships, and the meaning of human life are all transformed. By the beginning of the twenty-second century, all that once was has been transcended. As Gardner Dozois, the science fiction writer and editor states, "The *Accelerando* stories represent one of the most dazzling feats of sustained imagination in science fiction history, and radically up the Imagination Ante for every other writer who wants to sit down at the Future History table and credibly deal themselves into the game."[176]

The above award winning authors and science fiction stories capture the vast and wondrous drama of the future. I have provided some detail on each story to convey both the imagination and richness of these visions. The legacy and intelligence of Wells and Stapledon is clearly alive and well in such stories. Interestingly, many of these stories, including *Hyperion, A Fire Upon the Deep*, and *Vacuum Diagrams*, hark back to the age of great space operas, only now they are informed by computer science, quantum physics, and vastly more sophisticated psychological and cultural speculation. Even if popular science fiction movies frequently rise no higher than good guys, bad guys, special effects, violence, and shallow plots, science fiction literature, at its best, raises and addresses the deepest philosophical and scientific issues of humanity, existence, and the universe, and it does so in the context of futurist drama.

The Power and Breadth of Science Fiction

"...if the world is becoming too conscious of the innumerable possible futures jostling for its attention, science fiction had a central role in creating that urgent consciousness."

John Clute

In the new millennium, science fiction continues to grow in popularity and influence around the world.[177] Science fiction fandom is a vast social network across the globe. Numerous conventions are held every year celebrating authors, artists, new books and stories, as well as classic tales from the past. The fans revel in the lifestyle and all the paraphernalia and memorabilia associated with the genre.[178] A "Science Fiction Museum and Hall of Fame" has recently been established in Seattle, Washington.[179] As stated earlier, science fiction is the most influential contemporary form of futurist thinking in popular culture - it is a culture unto itself.

As can be seen from the historical review of science fiction that I have provided, science fiction covers a host of different themes and topics pertaining to the future. It addresses all the following areas of futurist thinking:

- Human Society and Cities in the Future - Future Cultures
- Scientific and Technological Discovery and Innovation
- The Relationship of Humanity and Technology
- Human Evolution and the Nature of Mind, Self, and Intelligence
- The Evolution of Life - Biotechnology
- Environmental, Ecological, Solar, and Galactic Engineering

- Robots and Androids - Technological or Computer Intelligence
- Space Exploration and Space Colonization – Exploring and Understanding the Cosmos
- Alien Contact, Alien Civilizations, and Alien Mentality
- Time Travel – The Manipulation of Time
- Philosophical, Religious, and Spiritual Enlightenment - God
- Morality and Values – Good and Evil
- Women, Men, Love, and Sex in the Future
- New or Alternative Forms of Reality - Alternative Universes
- Future Wars
- The Nature and Value of Progress
- Natural and Cosmic Disasters – The End of Humanity
- The Transcendence of Humanity
- The Evolution of Anything and Everything
- The Ultimate Nature, Meaning, and Destiny of the Cosmos

In summary, science fiction both embraces and questions the Enlightenment idea of secular progress. Some stories view science, technology, and reason as leading to a better world – some stories identify flaws in either the ideals of secular progress, or the imperfect nature of humanity in realizing the goals of secular and scientific progress. Yet within science fiction, following the philosophy of the Enlightenment, whatever is said should be scientifically credible. Enlightenment philosophy rejected the validity of traditional mythic and religious stories on scientific and rational grounds, offering a new scientific description of reality and secular ideals and empirical predictions for the future. Science fiction writers delve into the possibilities of space travel and space technology, robotics, computers, and artificial intelligence, evolutionary biology and biotechnology, nanotechnology, and the myriad of implications and ramifications of physics and physical cosmology.

Science fiction has also been inspired by philosophical Romanticism. From a Romantic perspective, science, reason, and technology are not enough - in fact, as idols or absolute gods, they may be destructive to the human spirit. From a Romantic perspective, we need to be Dionysian as well as Apollonian in our approach to the future. We need to consider human emotion and personal meaning in envisioning and directing our future. Science fiction arouses all the diverse and fundamental emotions in humans, including fear, hope, exhilaration, depression, joy, sorrow, awe, humility, and humor. Also, embracing the central Romantic ideals of beauty and art and the Romantic emphasis on narrative literature (instead of abstract theory) as a mode of understanding life, science fiction can be viewed as art as much as scientific and technological extrapolation. Science fiction, as a literary and artistic form, attempts to imagine the future in narrative and aesthetic form. The future is a story - in fact many stories. Literature and art educate and inspire in a way that a scientific theory cannot. Speculative stories and artistic visions about the future are essential dimensions of future consciousness.

Similarly, the religious and mythological approach to both the past and the future anchored their ideals and visions to stories and archetypal characters that humans could identify with. Such ancient stories inspired and educated. Science fiction does the same. As I have attempted to illustrate and explain, science fiction is mythic and cosmic, and addresses the fundamental questions of human existence but it does so through the minds and experiences of unique characters, thus personifying the journey into the future.

Given its diverse roots, science fiction is highly pluralistic, with different writers emphasizing different perspectives. It tells many stories - from many points of view - with many different styles - about all aspects of the future. It has gone beyond the monolithic visions of the Enlightenment. It is Postmodern, interdisciplinary, and frequently counter-culture.

Science fiction is inter-disciplinary. Because it has both Romantic and scientific-technological roots, it pulls together the arts and humanities with science and technology. Since it is mythic, cosmic, personalized, and often concerned with such issues as the meaning of life and good and evil, it brings together the secular-scientific with the religious-spiritual. Since science fiction increasingly has moved beyond simple extrapolations on science and technology and brought into its arena concerns with the future of the human mind, human society, culture, values, ecology and the environment, it draws upon all areas of human inquiry and study. Because it takes all these different dimensions of reality and integrates them into stories and scenarios, it is truly interdisciplinary, examining how technology, the environment, society, values, and the psyche all interact with each other.

Science fiction has many different functions and strengthens and benefits future consciousness in numerous ways. It stimulates multiple dimensions of thought, exercising the critical and rational intellect and stretching the speculative and creative imagination. Science fiction can draw the reader into critical reflection on contemporary trends and where these trends could lead. As mentioned earlier, science fiction frequently involves "thought experiments," where the implications and ramifications of future hypothetical situations are thought out. One's intellect and thinking capacities can be challenged and stretched through the scientific, technological, and philosophical speculations in science fiction, for example, as in the high-powered cosmic cogitations of Stapledon and Baxter. One could describe at least some science fiction as speculative cosmology, philosophy, and even theology set in narrative form. Science fiction expands one's imagination by engaging in "possibility thinking." Diverse worlds and innumerable strange realities are imagined in science fiction. Multi-faceted highly realistic scenario-building frequently occurs in science fiction. Future scenarios involving technologies, ecologies, geologies, economies, societies, governments, habitations, psychologies, and ethical norms and values are described and pulled together into cognitively compelling, detailed worlds.

Science fiction not only predicts the future but influences it. Science fiction describes the future in great sensory and descriptive detail, and has a strong emotional impact on the reader. The future is presented in personalized terms, with memorable and identifiable characters possessing various archetypal qualities. Hence, such characters, their challenges, and their exploits often inspire the reader into action or dramatic changes in thinking.

Science fiction generates a holistic experience of the future, impacting all dimensions of human psychology. It combines the abstract and the personal - it is a universe of ideas and individual characters. It unites cognition and emotion. It creates a virtual perceptual experience of the future since it creates such detailed and descriptive visions of possible future realities. It gets a person thinking about the future, feeling its pathos, and at times motivating the individual into changes in behavior and lifestyle. Because it both critiques and extols different modern trends, and often has morals to its stories, it addresses human values and ethics. Novels such as *A Canticle for Leibowitz, Brave New World, Stranger in a Strange Land,* and *Behold the Man* have strong moral, if not religious, messages embodied in their narratives. Science fiction engages the entire human mind.

In general, science fiction possesses a set of complementary functions and powers. Often it unites apparent opposites. It is secular-scientific and mythic - Romantic; it is both rational and emotional; it combines the strengths of religious inspiration with rational understanding. Science fiction can be seen as both "thought experiments" and artistic visions. There is a personalized dimension to science fiction, but equally it can be filled with scientific theory, technological detail, and cosmic abstractions. While it tries to predict and understand the future, it also attempts to inspire, entertain, terrify, and mesmerize. It functions both as meaningful drama and literature and futuristic extrapolation. Science fiction can actually influence the future and not simply anticipate it. It is high escapism, as well as highly reflective and often critical of contemporary times. Science fiction extols the promise of the future, yet is filled with warnings, fears, and apprehensions about tomorrow as well. It shares similarities with ancient fantasy and mythology, yet it is also a creation of the modern world; it ties the past and the future together. As the mythology of the future, science fiction creates the dreams the world of tomorrow will be made of.

References

[1] Within this chapter I can only touch upon some of the main novels, stories, and writers of science fiction. For more thorough descriptions of the history of science fiction, the reader is referred to more comprehensive studies, such as, Moskowitz, Sam *Seekers of Tomorrow*. Westport, CT: Hyperion Press, 1966; Aldiss, Brian *Billion Year Spree: The True History of Science Fiction*. New York: Schocken Books, 1973; Gunn, James *Alternate Worlds: An Illustrated History of Science Fiction*. Englewood Cliffs, NJ: Prentice-Hall, Inc., 1975; Ash, Brian (Ed.) *The Visual Encyclopedia of Science Fiction*. New York: Harmony Books, 1977; Aldiss, Brian and Wingrove, David *Trillion Year Spree: The History of Science Fiction*. North Yorkshire, UK: House of Stratus, 1986; Clute, John *Science Fiction: The Illustrated Encyclopedia*. London: Doarling Kindersley, 1995; Disch, Thomas *The Dreams Our Stuff is Made of: How Science Fiction Conquered the World*. New York: The Free Press, 1998; Robinson, Frank *Science Fiction in the 20th Century: An Illustrated History*. New York: Barnes and Noble Books, 1999.
[2] Lombardo, Thomas "Ancient Myth, Religion, and Philosophy" in Lombardo, Thomas *The Evolution of Future Consciousness*. Bloomington, Indiana: Author House, 2006.
[3] Disch, Thomas, 1998.
[4] Silverberg, Robert "Breckenridge and the Continuum" (1973) in Carr, Terry (Ed.) *The Best Science Fiction of the Year #3*. New York: Ballantine Books, 1974.
[5] Ellison, Harlan "The Deathbird" in Carr, Terry (Ed.) *The Best Science Fiction of the Year #3*. New York: Ballantine Books, 1974.
[6] Krippner, Stanley, Mortifee, Ann, and Feinstein, David "New Myths for the New Millennium" *The Futurist*, March, 1998.
[7] Clute, John, 1995, Pages 34 - 35.
[8] Wachhorst, Wyn *The Dream of Spaceflight: Essays on the Near Edge of Infinity*. New York: Basic Books, 2000.
[9] Lombardo, Thomas "Enlightenment and the Theory of Secular Progress" in Lombardo, Thomas *The Evolution of Future Consciousness*. Bloomington, Indiana: Author House, 2006.
[10] Berman, Morris *The Reinchantment of the World*. New York: Bantam, 1981.
[11] Lombardo, Thomas "The Scientific Revolution" in Lombardo, Thomas *The Evolution of Future Consciousness*. Bloomington, Indiana: Author House, 2006.
[12] Lombardo, Thomas "Romanticism" in Lombardo, Thomas *The Evolution of Future Consciousness*. Bloomington, Indiana: Author House, 2006.
[13] Disch, Thomas, 1998, Pages 32 - 56; Clute, John, 1995, Page 111.

[14] Shelley, Mary *Frankenstein, or the Modern Prometheus.* (1818) Hindle, Maurice (Ed.) London: Penguin Books, 1992.
[15] Shelley, Mary, 1992.
[16] Aldiss, Brian, 1973, Pages 20 - 30.
[17] Clute, John, 1995, Pages 36 - 37, 108 - 115.
[18]Bell, Wendell *Foundations of Future Studies: Human Science for a New Era.* New Brunswick: Transactions Publishers, 1997, Vol. II, Chapter One.
[19] Lee, Laura "Forecasts That Missed By a Mile" *The Futurist*, September-October, 2000.
[20] Lombardo, Thomas "Emotion, Motivation, and Future Consciousness" in Lombardo, Thomas *The Evolution of Future Consciousness.* Bloomington, Indiana: Author House, 2006.
[21] Lombardo, Thomas "Greco-Roman Myth and Philosophy: The Apollonian, the Dionysian, and the Theory of Progress" in Lombardo, Thomas *The Evolution of Future Consciousness.* Bloomington, Indiana: Author House, 2006.
[22] Clute, John, 1995, Pages 112 - 113.
[23] Clute, John, 1995, Pages 112 - 113.
[24]Jules Verne Collection- North American Jules Verne Society Homepage: http://www.najvs.org/; http://JV.Gilead.org.il/; http://JV.Gilead.org.il/works.html
[25] Disch, Thomas, 1998, Pages 58 - 61.
[26] Wagar, W. Warren, *H. G. Wells: Traversing Time.* Middletown, CT: Wesleyan University Press, 2004, Page 2.
[27] H. G. Wells and the Genesis of Future Studies: http://www.wnrf.org/cms/hgwells.shtml ; Wagar, W. Warren, 2004, Pages 316 - 319.
[28] Disch, Thomas, 1998, Pages 61 - 69.
[29] Wells, H. G. *Seven Science Fiction Novels of H. G. Wells.* New York: Dover Publications, Inc., 1895-1934; Clute, John, 1995, Pages 114 - 115; Aldiss, Brian, 1973, Pages 113 - 133.
[30] Clute, John, 1995, Pages 114 - 115; Wagar, W. Warren "Utopias, Futures, and H.G. Wells' Open Conspiracy" in Didsbury, Howard F. (Ed.) *Frontiers of the 21st Century: Prelude to the New Millennium.* Bethesda, Maryland: World Future Society, 1999; Wagar, W. Warren, 2004, Pages 6, 37 - 38.
[31] Disch, Thomas, 1998, Pages 61 - 69; Wagar, W. Warren, 2004, Pages 67 - 70, 135 - 147, 176 - 182.
[32] Lombardo, Thomas "The Perceptual Awareness of Time" and "The Cognitive Dimension of Future Consciousness" in in Lombardo, Thomas *The Evolution of Future Consciousness.* Bloomington, Indiana: Author House, 2006.
[33] Disch, Thomas, 1998, Pages 62 - 63.
[34] Wagar, W. Warren, 2004, Chapters 1 and 2.

[35] Lombardo, Thomas "Darwin's Theory of Evolution" in Lombardo, Thomas *The Evolution of Future Consciousness*. Bloomington, Indiana: Author House, 2006.
[36] Wagar, W. Warren, 2004, Pages 46 - 76; Wells, H. G., 1895 - 1934.
[37] Wagar, W. Warren, 2004, Chapters Five, Six, Ten, and Eleven.
[38] Wagar, W. Warren, 2004, Pages 10 - 11, 18 - 23, 35 - 37, 101 - 110, 135 - 138.
[39] Lombardo, Thomas "Hegel, Marx, and the Dialectic" in Lombardo, Thomas *The Evolution of Future Consciousness*. Bloomington, Indiana: Author House, 2006.
[40] Wagar, W. Warren, 2004, Chapter Eleven.
[41] Wagar, W. Warren, 2004, Pages 138 - 147.
[42] Wagar, W. Warren, 2004, Pages 43, 77 - 92, Chapters 10 and 11.
[43] Wagar, W. Warren, 2004, Pages 272 - 273.
[44] Watson, Peter *The Modern Mind: An Intellectual History of the 20th Century*. New York: HarperCollins Perennial, 2001, Pages 11 - 25.
[45] Clute, John, 1995, Pages 42 - 45.
[46] Bell, Wendell, 1997, Vol. I, Pages 227 - 232.
[47] Pohl, Frederick "Thinking About the Future" *The Futurist*, September-October, 1996.
[48] Disch, Thomas, 1998.
[49] Wachhorst, Wyn, 2000, Pages 22 - 24; Clute, John, 1995, Page 118.
[50] See also the book version: Von Harbou, Thea *Metropolis*. New York: Ace Books, 1927.
[51] See Moskowitz, Sam, 1966 for biographies of many of the most famous science fiction writers who first started their careers during the Golden Age; Clute, John, 1995, Pages 120 - 121.
[52] Clute, John, 1995, Page 123.
[53] Stapledon, Olaf *Last and First Men and Star Maker*. New York: Dover Publications, 1931, 1937.
[54] Stapledon, Olaf, 1931, 1937.
[55] Clute, John, 1995, Page 122; Aldiss, Brian and Wingrove, David, 1986, Pages 206 - 213.
[56] Clute, John, 1995, Pages 66, 128 - 129, 134 - 135; Heinlein, Robert *The Past Through Tomorrow*. New York: Berkley Publishing Corporation, 1967.
[57] Asimov, Isaac *Foundation. Foundation and Empire. The Second Foundation*. New York: Ballantine Books, 1982.
[58] Barrett, David B. "Chronology of Futurism and the Future" in Kurian, George Thomas, and Molitor, Graham T.T. (Ed.) *Encyclopedia of the Future*. New York: Simon and Schuster Macmillan, 1996.
[59] Asimov, Isaac *I, Robot*. Greenwich, Ct.: Fawcett Publications, 1950.
[60] Bova, Ben (Ed.) *The Science Fiction Hall of Fame*. New York: Avon Books, 1970; Bova, Ben (Ed.) *The Science Fiction Hall of Fame Vol. IIA*. New York: Avon Books, 1973; Bova, Ben (Ed.) *The Science Fiction*

Hall of Fame Vol. IIB. New York: Avon Books, 1973; Healey, Raymond and McComas, J. Francis (Ed.) *Adventures in Time and Space*. New York: Ballantine Books, 1946.

[61] Del Rey, Lester (Ed.) *The Best of C. L. Moore*. Garden City, New York: Nelson Doubleday, Inc., 1975; Kuttner, Henry *The Best of Henry Kuttner*. Garden City, New York: Nelson Doubleday, Inc., 1975.

[62] Kuttner, Henry and Moore, C. L. "Mimsy Were the Borogroves" (1943) in Bova, Ben (Ed.) *The Science Fiction Hall of Fame*. New York: Avon Books, 1970; Kuttner, Henry and Moore, C. L. "Vintage Season" (1946) in Bova, Ben (Ed.) *The Science Fiction Hall of Fame Vol. IIA*. New York: Avon Books, 1973.

[63] Kuttner, Henry and Moore, C. L. "Private Eye" (1948) in Silverberg, Robert (Ed.) *The Mirror of Infinity*. New York: Harper and Row, 1970.

[64] Del Rey, Lester "Helen O'Loy" (1938) in Bova, Ben (Ed.) *The Science Fiction Hall of Fame*. New York: Avon Books, 1970; Sturgeon, Theodore "Microcosmic God" (1941) in Bova, Ben (Ed.) *The Science Fiction Hall of Fame*. New York: Avon Books, 1970; Leinster, Murray "First Contact" (1945) in Bova, Ben (Ed.) *The Science Fiction Hall of Fame*. New York: Avon Books, 1970; Campbell, John "Who Goes There?" (1938) in Bova, Ben (Ed.) *The Science Fiction Hall of Fame Vol. IIA*. New York: Avon Books, 1973; Bates, Harry "Farewell to the Master" in Healey, Raymond and McComas, J. Francis (Ed.) *Adventures in Time and Space*. New York: Ballantine Books, 1946.

[65] Huxley, Aldous *Brave New World*. New York: Bantam Books, 1932.

[66] Orwell, George *1984*. New York: Harcourt, Brace, and Company, 1949.

[67] Disch, Thomas, 1998, Pages 7-8; Watson, Peter, 2001, Pages 297 - 299, 472- 473; Clute, John, 1995, Pages 124, 215, 217.

[68] Watson, Peter, 2001, Chapters Three, Eight, Ten, Sixteen, and Seventeen.

[69] Dyson, Freeman *Imagined Worlds*. Cambridge, MS: Harvard University Press, 1997, Pages 121-125; Postman, Neil *Amusing Ourselves to Death: Public Discourse in the Age of Show Business*. New York: Penguin Books, 1985, Pages vii - viii, 155 - 163

[70] Disch, Thomas, 1998, Chapter One.

[71] Clute, John, 1995, Pages 68 - 69.

[72] Miller, Walter *A Canticle for Leibowitz*. New York: Bantam Books, 1959.

[73] Stewart, George *Earth Abides*. Greenwich, CT: Fawcett Publications, 1949.

[74] Simak, Clifford *City*. New York: Ace Books, 1952.

[75] The Hugo Winners - Science Fiction - http://worldcon.org/hy.html.

[76] The Nebula Winners - Science Fiction and Fantasy Writers of America, Inc. - http://www.sfwa.org/awards/.

[77] Clarke, Arthur C. *Childhood's End*. New York: Ballantine Books, 1953.

[78] Clarke, Arthur C. *2001: A Space Odyssey*. New York: Ballantine Books, 1968; Clarke, Arthur C. *2010: Odyssey Two*. New York: Ballantine Books, 1982; Clarke, Arthur C. *2061: Odyssey Three*. New York: Ballantine Books, 1988.

[79] Clarke, Arthur C. "The Nine Billion Names of God" (1953) and "The Star" (1955) in Clarke, Arthur C. *The Nine Billion Names of God*. New York: Signet, 1974.

[80] Arthur C. Clarke Unauthorized Home Page: http://www.lsi.usp.br/~rbianchi/clarke/; Clute, John, 1995, Pages 138 - 139.

[81] Sturgeon, Theodore *More Than Human*. New York: Ballantine Books, 1953.

[82] Bester, Alfred *The Demolished Man*. New York: Signet, 1953.

[83] Keyes, Daniel *Flowers for Algernon*. New York: Bantam Books, 1966.

[84] Bixby, Jerome "It's a Good Life" (1953) in Bova, Ben (Ed.) *The Science Fiction Hall of Fame*. New York: Avon Books, 1970.

[85] Fast, Howard "The First Men" (1959) in *The Edge of Tomorrow*. New York: Bantam Books, 1961.

[86] Disch, Thomas, 1998, Chapter Seven.

[87] Bradbury, Ray *The Martian Chronicles*. New York: Bantam Books, 1950; Lem, Stanislaw *Solaris*. New York: Berkley Publishing Corporation, 1961; Niven, Larry *Ringworld*. New York: Ballantine Books, 1970.

[88] Prantzos, Nikos *Our Cosmic Future: Humanity's Fate in the Universe*. Cambridge: Cambridge University Press, 2000.

[89] Disch, Thomas, 1998, Pages 74 - 77, 171 - 179.

[90] Wachhorst, Wyn, 2000; Lombardo, Thomas "Space Exploration and Cosmic Evolution" in Odyssey of the Future - http://www.odysseyofthefuture.net/listing/ReadingSpaceExploration.pdf.

[91] Heinlein, Robert "Universe" (1941) in Bova, Ben (Ed.) *The Science Fiction Hall of Fame Vol. IIA*. New York: Avon Books, 1973.

[92] Robinson, Kim Stanley *Red Mars*. New York: Bantam, 1991; Robinson, Kim Stanley *Green Mars*. New York: Bantam, 1994; Robinson, Kim Stanley *Blue Mars*. New York: Bantam, 1996.

[93] Heinlein, Robert "By His Bootstraps" (1941) in Healy, Raymond and McComas, J. Francis (Ed.) *Adventures in Time and Space*. New York: Ballantine Books, 1946.

[94] Asimov, Isaac *The End of Eternity*. Greenwich, CT: Fawcett Crest, 1955; Leiber, Fritz *The Big Time*. New York: Ace Books, 1961.

[95] Moorcock, Michael *Behold the Man*. New York: Avon, 1968.

[96] Gerrold, David *The Man Who Folded Himself*. New York: Random House, 1973.

[97] Disch, Thomas, 1998, Chapter Four.

[98] Two of the most highly regarded and accomplished writers of this period, not discussed in the following review of the New Wave, were Samuel Delany and Robert Silverberg. Some of Delany's best novels during this period include Delany, Samuel *Babel-17*. New York: Vintage Books,

1966; Delany, Samuel *The Einstein Intersection*. New York: Ace Books, 1967; and Delany, Samuel *Dhalgren*. New York: Bantam Books, 1974. Some of Silverberg's most noteworthy books during the New Wave era include Silverberg, Robert *Up the Line*. New York: Ballantine Books, 1969; Silverberg, Robert *Tower of Glass*. London: Gollancz, 1970; Silverberg, Robert *Son of Man*. New York: Ballantine Books, 1971; Silverberg, Robert *Dying Inside*. New York: Ballantine Books, 1972; and Silverberg, Robert *The Book of Skulls*. London: Gollancz, 1972.

[99] Heinlein, Robert *Stranger in a Strange Land*. New York: Avon Books, 1961; Herbert, Frank *Dune*. New York: Ace Books, Inc., 1965; Dick, Philip K. *The Man in the High Castle*. New York: Berkley Publishing Corporation, 1962.

[100] Ellison, Harlan " 'Repent Harlequin!' Said the Ticktockman" (1965) in Asimov, Isaac (Ed.) *Stories from the Hugo Winners* Vol. II. Greenwich, CT: Fawcett Publications, Inc., 1971.

[101] Ellison, Harlan (Ed.) *Dangerous Visions*. Vols. I, II, and III. New York: Berkley Publishing Corporation, 1967, 1969.

[102] Farmer, Philip Jose "Riders of the Purple Wage" in Ellison, Harlan (Ed.) *Dangerous Visions*. Vol. I. New York: Berkley Publishing Corporation, 1967.

[103] Moskowitz, Sam, 1966, Chapter 22.

[104] Farmer, Philip Jose *Flesh*. New York: Signet, 1960; Farmer, Philip Jose *Strange Relations*. New York: Avon Books, 1960.

[105] Farmer, Philip *To Your Scattered Bodies Go*. New York: Berkley Medallion, 1971; Farmer, Philip *The Fabulous Riverboat*. New York: Berkley Medallion, 1971; Farmer, Philip *The Dark Design*. New York: Berkley Medallion, 1977; Farmer, Philip *The Magic Labyrinth*. New York: Berkley Books, 1980.

[106] Ballard, J. G. *The Drowned World*. New York: Berkley Publishing Corporation, 1962; Ballard, J. G. *The Burning World*. New York: Berkley Publishing Corporation, 1964; Ballard, J. G. *The Crystal World*. New York: Berkley Publishing Corporation, 1966.

[107] Ballard, J. G. "Build-Up" (1960) in Ballard, J. G. *Chronopolis and Other Stories*. New York: G. P. Putnam's Sons, 1971; Ballard, J. G. "The Subliminal Man" (1963) in Silverberg, Robert (Ed.) *The Mirror of Infinity*. New York: Harper and Row, 1970.

[108] Aldiss, Brian and Wingrove, David, 1986, Pages 343 - 350.

[109] Aldiss, Brian and Wingrove, David, 1986, Pages 337 - 341; Clute, John, 1995, Page 168.

[110] Zelazny, Roger "A Rose for Ecclesiastes" (1963) in Bova, Ben (Ed.) *The Science Fiction Hall of Fame*. New York: Avon Books, 1970.

[111] Zelazny, Roger *Lord of Light*. New York: Avon Books, 1967.

[112] Disch, Thomas, 1998, Chapter One.

[113] Clute, John, 1995, Pages 162 - 163; Aldiss, Brian and Wingrove, David, 1986, Pages 381 - 388.

[114] Dick, Philip K., 1962.
[115] Dick, Philip K. *Do Androids Dream of Electric Sheep?* New York: Signet Books, 1968; 2019 Off-World: (Blade Runner Page): http://scribble.com/uwi/br/off-world.html .
[116] Dick, Philip K. *Martian Time-Slip*. New York: Ballantine, 1964.
[117] Dick, Philip K. *Flow My Tears, The Policeman Said*. New York: Daw Books, 1974.
[118] Dick, Philip K. *The Three Stigmata of Palmer Eldritch*. New York: Manor Books, 1964.
[119] Clute, John, 1995, Page 162 - 163; Aldiss, Brian and Wingrove, David, 1986, Pages 381 - 388.
[120] Philip Dick - Science Fiction Author - http://www.philipkdick.com/; The Official Blade Runner On Line Magazine: http://www.devo.com/bladerunner/; Philip K. Dick Biography: http://www.webcom.com/~gnosis/pkd.biography.html; Wikipedia - Philip K. Dick: http://en.wikipedia.org/wiki/Philip_K._Dick .
[121] LeGuin, Ursula *The Left Hand of Darkness*. New York: Ace Books, 1969; LeGuin, Ursula *The Dispossessed*. New York: Avon Books, 1974.
[122] Del Rey, Lester, 1975.
[123] Tiptree, James *Warm Worlds and Otherwise*. New York: Ballantine Books, 1975; Tiptree, James *Star Songs Of An Old Primate*. New York: Ballantine Books, 1978.
[124] McCaffrey, Anne *Dragonflight*. New York: Ballantine Books, 1968; McCaffrey, Anne New York: *Dragonquest*. Ballantine Books, 1971.
[125] McIntyre, Vonda "Of Mist, and Grass, and Sand" (1973) in Carr, Terry (Ed.) *The Best Science Fiction of the Year #3*. New York: Ballantine Books, 1974; McIntyre, Vonda *Dreamsnake*. Boston: Houghton Mifflin, 1978.
[126] Ross, Joanna *The Female Man*. New York: Bantam Books, 1975; Feminist Science Fiction, Fantasy, & Utopia: http://feministsf.org/.
[127] Cherryh, C. J. *Downbelow Station*. New York: DAW Books, 1981; Cherryh, C. J. *Cyteen*. New York: Warner, 1988.
[128] Bujold, Lois McMaster *Falling Free*. Riverdale, N.Y.: Baen, 1988; Bujold, Lois McMaster *The Vor Game*. Riverdale, N.Y.: Baen, 1990; Bujold, Lois; McMaster *Barrayar*. Riverdale, N.Y.: Baen, 1991; Bujold, Lois McMaster *Mirror Dance*. Riverdale, N.Y.: Baen, 1994; Bujold, Lois McMaster *Paladin of Souls*. New York: Harper Collins, 2003.
[129] Willis, Connie *Doomsday Book*. New York: Bantam, 1992; Willis, Connie *To Say Nothing of the Dog*. New York: Bantam, 1998.
[130] Butler, Octavia *Parable of the Talents*. New York: Warner Books, 1998.
[131] Science Fiction - Star Wars Mythology Center - http://www.castlebooks.com/star-wars.htm.
[132] Vinge, Vernor "The Coming Technological Singularity: How to Survive in the Post-Human Era" *Vision-21: Interdisciplinary Science and Engineering in the Era of Cyberspace NASA-CP-10129*, 1993; Moravec, Hans *Robot: Mere*

Machine to Transcendent Mind. Oxford: Oxford University Press, 1999; Kurzweil, Ray *The Age of Spiritual Machines: When Computers Exceed Human Intelligence*. New York: Penguin Books, 1999; Lombardo, Thomas "Information Technology and Artificial Intelligence" in Odyssey of the Future - http://www.odysseyofthefuture.net/listing/ReadingInfoTech.pdf.

[133] Clute, John, 1995, Pages 74 - 75.

[134] Williamson, Jack *The Humanoids*. New York: Equinox Books, 1948.

[135] Bester, Alfred "Fondly Fahrenheit" (1954) in Bova, Ben (Ed.) *The Science Fiction Hall of Fame*. New York: Avon Books, 1970.

[136] Ellison, Harlan "I Have No Mouth, and I Must Scream" (1967) in Asimov, Isaac *The Hugo Winners*, Vol. II. Greenwich, Ct.: Fawcett Publications, 1971.

[137] Gerrold, David *When Harlie Was One*. New York: Ballantine Books, 1972; Anderson, Poul "Epilogue" (1962) in Anderson, Poul *Time and Stars*. New York: Doubleday, 1964; Zelazny, Roger "For a Breath I Tarry" (1966) in Wollheim, Donald and Carr, Terry (Ed.) *World's Best Science Fiction Third Series*. New York: Ace Books, 1967.

[138] Lombardo, Thomas "Life, Biotechnology, and Purposeful Biological Evolution" in Odyssey of the Future - http://www.odysseyofthefuture.net/listing/ReadingLifeBiotech.pdf .

[139] Vinge, Vernor , 1993; Kurzweil, Ray, 1999.

[140] Transhumanist Resources and Alliance - http://www.aleph.se/Trans/index-2.html; Transhumanity - http://transhumanism.org/index.php/th/.

[141] Anderson, Poul "Call Me Joe" (1957) in Bova, Ben (Ed.) *The Science Fiction Hall of Fame Vol. IIA*. New York: Avon Books, 1973.

[142] Pohl, Frederick *Man Plus*. New York: Random House, 1976.

[143] Pohl, Frederick "Day Million" (1966) in Wollheim, Donald and Carr, Terry (Ed.) *World's Best Science Fiction Third Series*. New York: Ace Books, 1967.

[144] Pohl, Frederick *Gateway*. New York: Ballantine Books, 1977.

[145] Gibson, William *Neuromancer*. New York: Ace Books, 1984; William Gibson - Post Modern Science Fiction and Cyberpunk -Neuromancer: http://www.georgetown.edu/irvinemj/technoculture/pomosf.html; http://www.wsu.edu/~brians/science_fiction/neuromancer.html

[146] Gray, Chris Hables "Our Future as Postmodern Cyborgs" in Didsbury, Howard (Ed.) *Frontiers of the 21st Century: Prelude to the New Millennium*. Bethesda, Maryland: World Future Society, 1999.

[147] Clute, John, 1995, Pages 88 - 89, 199, 232; Disch, Thomas, 1998, Pages 213 - 220; See Bruce Sterling *Mirrorshades: The Cyberpunk Anthology*. New York: Ace Books, 1986 for a popular collection of cyberpunk short stories.

[148] Heim, Michael *The Metaphysics of Virtual Reality*. New York: Oxford University Press, 1993.

[149] Rucker, Rudy, Sirius, R.U., and Queen Mu, *Mondo 2000*. New York: Harper Collins, 1992; Disch, Thomas, 1998, Pages 213 - 220.
[150] Cyberpunk Links: http://www.cwrl.utexas.edu/~tonya/cyberpunk/sites.html; CyberStudies Web Ring - Cyberanthropology: http://g.webring.com/hub?ring=cyberstudies; The CyberAnthropology Page: http://www.fiu.edu/~mizrachs/cyberanthropos.html.
[151] Baxter, Stephen, "The Real Matrix" in Haber, Karen (Ed.) *Exploring the Matrix: Visions of the Cyber Present*. New York: St. Martin's Press, 2003.
[152] Weberman, David "The Matrix Simulation and the Postmodern Age" in Irwin, William (Ed.) *The Matrix and Philosophy: Welcome to the Desert of the Real*. Chicago: Open Court, 2002.
[153] Irwin, William (Ed.) *The Matrix and Philosophy: Welcome to the Desert of the Real*. Chicago: Open Court, 2002; Haber, Karen (Ed.) *Exploring the Matrix: Visions of the Cyber Present*. New York: St. Martin's Press, 2003.
[154] Haldeman, Joe "The Matrix as Sci-Fi" in Haber, Karen (Ed.) *Exploring the Matrix: Visions of the Cyber Present*. New York: St. Martin's Press, 2003.
[155] Brin, David "Tomorrow May be Different" in Haber, Karen (Ed.) *Exploring the Matrix: Visions of the Cyber Present*. New York: St. Martin's Press, 2003.
[156] Disch, Thomas, 1998, Page 209.
[157] Disch, Thomas, 1998, Chapter Ten.
[158] Some other excellent writers and their noteworthy novels of the last couple of decades, not mentioned in the review in this section, include: Brin, David *Startide Rising*. New York: Bantam Books, 1983 and Brin, David *The Uplift War*. New York: Bantam Books, 1987; Sawyer, Robert J. *Hominids*. New York: Tom Doherty Associates, 2002, Sawyer, Robert J. *Humans*. New York: Tom Doherty Associates, 2003, and Sawyer, Robert J. *Hybrids*. New York: Tom Doherty Associates, 2003; Also see Dozois, Gardner (Ed.) *The Best of the Best: Twenty Years of the Year's Best Science Fiction*. St. Martin's Griffin: New York, 2005 for a collection of some of the best science fiction short stories of the last two decades.
[159] The Hugo Winners - Science Fiction - http://worldcon.org/hy.html; The Nebula Winners - Science Fiction and Fantasy Writers of America, Inc. - http://www.sfwa.org/awards/.
[160] Stephenson, Neal *Snow Crash*. New York: Bantam Books, 1992. See also Stephenson, Neal *The Diamond Age, or A Young Lady's Illustrated Primer*. New York: Bantam Books, 1995 for a futurist look at the possibilities of nanotechnology.
[161] Greg Bear - The Official Site - http://www.gregbear.com/; Bear, Greg *Blood Music*. New York: iBooks, 1985; Bear, Greg *Moving Mars*. New York: Tom Doherty Associates, 1993; Bear, Greg *Darwin's Radio*. New York: Ballantine Books, 1999.

[162] Bear, Greg *Darwin's Children*. New York: Ballantine Books, 2003.
[163] Bear, Greg *Queen of Angels*. New York: Warner Books, 1990.
[164] Dan Simmons Websites- http://www.dansimmons.com/; http://www.sfsite.com/lists/dsim.htm; http://www.erinyes.org/simmons/; Simmons, Dan *Hyperion*. New York: Bantam Books, 1989; Simmons, Dan *The Fall of Hyperion*. New York: Bantam Books, 1990; Simmons, Dan *Endymion*. New York: Bantam Books, 1995; Simmons, Dan *The Rise of Endymion* New York: Bantam Books, 1997.
[165] Simmons, Dan *Ilium*. New York: HarperCollins, 2003.
[166] Simmons, Dan *Olympos*. New York: HarperCollins, 2005.
[167] Vinge, Vernor *A Fire Upon the Deep*. New York: Tom Doherty Associates, 1992.
[168] Vinge, Vernor *A Deepness in the Sky*. New York: Tom Doherty Associates, 1999.
[169] Robinson, Kim Stanley, 1991; Robinson, Kim Stanley, 1994; Robinson, Kim Stanley, 1996.
[170] Baxter, Stephen *Vacuum Diagrams*. New York: Harper Collins Publishers, 1997.
[171] Clute, John, 1995, Pages 66, 208.
[172] Baxter, Stephen *Manifold Time*. New York: Ballantine, 2000; Baxter, Stephen *Manifold Space*. New York: Ballantine, 2001; Baxter, Stephen *Manifold Origin*. New York: Ballantine, 2002.
[173] Smolin, Lee *The Life of the Cosmos*. Oxford: Oxford University Press, 1997; Adams, Fred and Laughlin, Greg *The Five Ages of the Universe: Inside the Physics of Eternity*. New York: The Free Press, 1999.
[174] Vinge, Vernor, 1993 – "The Coming Technological Singularity: How to Survive in a Post-Human Era" - http://www-rohan.sdsu.edu/faculty/vinge/misc/singularity.html.
[175] Kurzweil, Ray, 1999; Kurzweil, Ray *The Singularity is Near: When Humans Transcend Biology*. New York: Viking Press, 2005.
[176] Stross, Charles *Accelerando*. New York: Ace Books, 2005; Dozois, Gardner (Ed.) *The Best of the Best: Twenty Years of the Year's Best Science Fiction*. St. Martin's Griffin: New York, 2005, Page 577.
[177] Science Fiction Resource Guide - http://sf.www.lysator.liu.se/sf_archive/sf-texts/SF_resource_guide/; The Science Fiction Site: The Home Page for Science Fiction and Fantasy - http://www.sfsite.com/; Science Fiction Weekly - http://www.scifi.com/sfw/
[178] The Linkoping Science Fiction and Fantasy Archive - http://www2.lysator.liu.se/sf_archive/sf_main.html; Ultimate Science Fiction Web Guide - http://www.magicdragon.com/UltimateSF/SF-Index.html.
[179] Science Fiction Museum and Hall of Fame - http://www.sfhomeworld.org/.

Chapter Two

Future Studies

"I believe quite firmly that an inductive knowledge of a great number of things in the future is becoming a human possibility. I believe that the time is drawing near when it will be possible to suggest a systematic exploration of the future."

H.G. Wells

The Origins, History, and Nature of Future Studies

Besides science fiction, the other main contemporary thread of futuristic thinking is **future studies.** Whereas science fiction is fictionalized narrative, generally future studies is non-narrative and non-fictional in its format and approach. Just as he did with science fiction, H.G. Wells significantly influenced the development of future studies. As noted in the last chapter, Wells wrote both fiction and non-fiction about the future, and his non-fiction books and articles were a primary stimulus behind the creation of future studies in the twentieth century.[1] As a provisional and general definition, future studies can be described as an empirical and scientifically based approach to understanding the future.

Throughout its development, there has been debate over what the best name is for this discipline, and if, in fact, the study of the future constitutes a distinctive academic course of study and research.[2] Part of the process of acquiring a disciplinary identity is establishing among its practitioners a consensual name. Various names have been proposed, including *futuristics*, *futurology*, *futuring*, *futurism*, *futuribles*, and *futures research*.[3] At least in the United States, *future studies* has emerged as the most accepted and popular name of the discipline.

Regardless of the name, the study of the future has evolved into an academic and professional pursuit, involving scientific theory, research methods, and a great variety of different educational curricula. Various college courses and programs on the future are offered at numerous schools worldwide. For example, Anne Arundel Community College in Maryland has created an Institute of the Future which offers various in-person and online courses and educational modules on the study of the future and how to enhance one's capacities to constructively think about the future.[4] Some other noteworthy programs include the Future Studies Master's Program, first established at the University of Houston at Clearlake, which offers a graduate degree in future studies[5]; The Hawaii Research Center for Future Studies at the University of Hawaii offers a masters program in future studies[6]; The Australian Foresight Institute at Swinburne University of Technology has a masters program in strategic foresight[7]; and finally, as an outstanding example of futures education, The Center for Futures Studies at Tamkang University in Taiwan provides extensive undergraduate and graduate coursework in future studies.[8] These courses and programs cover key issues and research methodologies pertaining to the future, and some prepare students for careers as professional futurists.

In addition to educational programs, a large number of organizations worldwide, such as the **World Future Society** (WFS), are dedicated to the study of the future.[9] The WFS produces a variety of publications, including a bi-monthly popular magazine titled *The Futurist*, a professional journal titled the *Futures Research Quarterly*, and a newsletter titled *Future Survey* that covers noteworthy recent books and articles relevant to the future. The WFS holds annual world conferences and publishes an ongoing variety of books and research studies, including the following:

- *Communications and the Future: Prospects, Promises, and Problems*
- *The Global Economy: Today, Tomorrow and the Transition*
- *Careers Tomorrow: The Outlook for Work in a Changing World*
- *Frontiers of the 21st Century: Prelude to the New Millennium*
- *21st Century Opportunities and Challenges*
- *Thinking Creatively in Turbulent Times*
- Foresight, Innovation, and Strategy

Cynthia Wagner edited the last of these books; all of the other publications are edited by Howard Didsbury.[10] There is also the **World Future Studies Federation** (WFSF)[11] composed of academicians, professional futurists, and institutions. The WFSF, which emphasizes a global, inter-cultural perspective on the future, holds bi-annual conventions, offers professional courses, and publishes a variety of professional works on the future. Some other noteworthy futurist organizations include: **The Arlington Institute,** which publishes an excellent electronic newsletter on trends, discoveries, and events relevant to the future; **The Acceleration Studies Foundation,** which also produces a high quality electronic newsletter, as well as holding annual conferences on accelerative change and the future; **The Copenhagen Institute for Future Studies**, a research and consulting organization; **Evolve**, an educational and inspirational network for "conscious evolution"; the **Foundation for the Future**, which examines long-term future developments for the next thousand years, as well as the evolution of human intelligence; and the various **Transhumanist** organizations, which focus on the future evolution of humanity, especially through the use of technology and science.[12] So as not to slight any of the many other important futurist organizations, the reader is referred to the World Wide Web Bibliography at *The Odyssey of the Future* website for a much more extensive list of futurist institutes and professional groups.[13]

While numerous definitions exist, according to Wendell Bell there is a significant degree of consensus among futurists in the field regarding the purpose of future studies. Bell states that the "most general purpose of future studies is to maintain or improve the freedom and welfare of humankind" with the addendum that "some futurists would add the welfare of all living beings, plants, and the Earth's biosphere..."[14] Ed Cornish, the founding President of the World Future Society and editor of *The Futurist*, provides a similar definition, stating that the goal of "futuring" is to make for a better future.[15] These are very broad definitions—perhaps too broad, for these definitions could apply to the social and psychological sciences and the humanities as well. Bell adds that futurists' distinctive contribution is **prospective thinking,** in particular, "to discover or invent, examine and evaluate, and propose possible, probable, and preferable futures." In a similar vein, Cornish identifies foresight as the key skill emphasized in the discipline, and also includes the three-fold listing of possible, probable, and preferable futures as the main areas or questions of study in the discipline.[16] These descriptions of future studies coincide closely with the futurist Alvin Toffler's statement that futurists attempt to create "new, alternative images of the future, visionary explorations of the possible, systematic investigation of the probable, and moral evaluation of the preferable."[17]

This definition roughly describes science fiction as well as future studies—although perhaps science fiction is not as systematic. Science fiction clearly is visionary regarding the possible, often explores probable developments in contemporary trends, and through "warning scenarios," definitely evaluates the moral and ethical implications of our present world and potential future realities. To be more precise in our definition of future studies, we should add to Bell and Toffler's definitions the qualification that future studies is a non-fictional approach, often highlighting scientific methodology, whereas science fiction is a fictional and literary approach to the future.

Michael Marien, the editor of *Future Survey*, takes a broader, more diverse, and less clearly circumscribed view of future studies, and even includes science fiction as a stylistic form of futures thinking. In disagreement with Bell, Marien does not think that there is universal agreement among futurists regarding the nature of the discipline. However, he does offer a six-fold classification system of "purposive categories of futures-thinking" that is similar to Bell and Toffler, including the possible, probable, and preferable, but adds, examining present changes, taking panoramic views, and questioning.[18] Marien's ideas are discussed in more depth below.

One key feature of future studies is its research methodology. According to futurists like Bell and Cornish, future studies attempts to employ scientific research principles in the study of the future. Future studies research attempts to be both rational and empirical. The discipline has developed some distinctive research methods of its own, but as Bell notes, it has also utilized research methods developed in other academic disciplines. All told, an extensive and varied futurist methodology has evolved over the last century.[19] Future studies research involves the statistical collection and analysis of vast amounts of world data, mathematical extrapolations, predictions based on statistical trends, monitoring of trends, scenario development, surveys and polls, game theory and techniques, ethnographic research, and computer simulations and experiments on the possible complex interactions of different social variables and trends. One distinctive approach used by futurists is the **Delphi Method**, which involves the polling, comparing, and integrating of expert opinions on different aspects of the future.

The richness of future studies may make it difficult to come up with a simple singular definition of the discipline, but various writers have attempted to provide a comprehensive overview of the area. Richard Slaughter, the former President of the World Future Studies Federation, has written numerous books and articles on future studies,[20] and has provided an extensive list of core developing concepts in the field.[21] He has also edited a three-volume work *The Knowledge Base of Futures Studies*,[22] with the intent of bringing together contemporary global thinking and common principles regarding the study of the future. On an even more massive scale, George Kurian and Graham Molitor have edited and published the *Encyclopedia of the Future*, in another effort to articulate the comprehensive scope and conceptual details of this discipline.[23]

To provide some history on the development of future studies, a good place to begin is Warren Wagar's essay on "Futurism."[24] He states that Wells' book *Anticipations* (1902) marks the beginnings of futurism.[25] For Wagar, though, future studies didn't take off as a popular area of concern until the 1960s. As he points out, contemporary future studies began in the 1960s due to a combination of factors: (1) The need in business and government for long-range planning; (2) Advances in technological and economic forecasting; and (3) The erosion of discipline boundaries, which led to futurism emerging as a real interdisciplinary activity.

In a subsequent article, "Utopias, Futures, and H.G. Wells' Open Conspiracy," Wagar discusses in more depth Wells' later book, *The Open Conspiracy* (1928). Wagar describes Wells' dissatisfaction with many general features of his world, including nationalism, corporatism, capitalism, and elitism, problems that according to Wagar are still with us today. In *The Open Conspiracy*, Wells advocated that humanity work against the dominance of nation states, and develop a third way to approach the future, besides the two prevailing systems of modernism and fundamentalism. Wells strongly advocated for a new world system that was secular and scientific.[26]

According to Wagar, although Wells created many pessimistic and nightmarish images of the future in his science fiction, he was also influenced by positive utopian visions inspired by the Enlightenment. Wells offered both criticisms of his time and, based on such critiques, proposals and ideals for a new and better world.

We should note then, that beginning with Wells, future studies did not just involve methods for understanding and predicting the future, but general assessments of human society and normative proposals for improving the world. The fundamental dimensions and features of human reality are described, often from a particular theoretical perspective, and an ideology of the preferable direction for humanity is presented. These basic components of future studies—rational and empirical methodology, theory and description (often critical) of the present, and prescriptive proposals (often involving clear ideologies) for the future—all derive from the philosophy of the Enlightenment. The philosophy of the Enlightenment, which emerged in Western Europe in the seventeenth and eighteenth centuries, inspired by the Scientific Revolution and the rise of secular thinking, emphasized the use of rational and empirical methods for understanding reality. Based on this understanding, this period articulated visions and values for a positive future, freed of the dogmatism, authoritarianism, and superstitions of religion and monarchial rule.[27]

Continuing to follow Wagar's historical analysis, the earliest organizations explicitly committed to the study of the future were formed in the 1960s: The World Future Society began in 1966 in Washington, D.C.; the **Association Internationale Futuribles** began in 1967 in Paris; and **The Club of Rome** began in 1968. Adding to the above history, Alvin and Heidi Toffler note that *The Futurist* magazine and the World Future Society were started by Ed Cornish and his wife Sally. They also note that many of the original people in the group were couples with highly interdisciplinary interests.[28]

Wendell Bell in his *Foundations of Future Studies* also provides a historical analysis of the evolution of future studies. Taking a somewhat broader and deeper view of the historical roots of future studies, Bell argues that future studies emerged out of the secular approach to social evolution that developed during the Enlightenment. For Bell, science and secular thinking provided the inspiration for the development of future oriented utopias and philosophies for the improvement of humankind.[29] Wells was clearly influenced by the secular perspective on the future, as were other social, political, and governmental leaders and thinkers in the eighteenth and nineteenth centuries. Bell provides an introductory review of the evolution of utopian thought as a precursor to future studies, covering the ideas of More, Jean-Jacques Rousseau, the Marquis de Sade, Condorcet, William Godwin, Henri Saint-Simon, and Karl Marx.[30]

As Bell notes, beginning in the 1920s and 1930s, governmental social and economic planning blossomed. This initiative looked at past and present trends, and then extrapolated and considered possibilities and alternatives for the future. Possibilities were evaluated and governmental policies were implemented based on trend analysis and social values. Such efforts are clearly an example of collective, databased, and systematic thinking about the future. In the decades that followed, operations research, policy science, and **Think Tanks** developed that articulated many of the contemporary principles, methods, and goals of future studies. By the 1960s there were numerous agencies and organizations collecting trend data, identifying values and future goals, and proposing plans and policies for reaching these goals.[31] In the last few decades, such futures oriented research, planning, and thinking has magnified in business, government, and society at large a thousand fold. This whole line of thinking and action can be seen as developing out of the empirical and rational traditions of secular thought.

The fast growth of Think Tanks highlights another significant feature of future studies. James McGann provides a concise and informative overview of the development of Think Tanks.[32] He points out that there are presently around 3000 Think Tanks, 60 percent of which were created within the last 20 years. The rapid changes and challenges of our modern world and the consequent demand for useful information and analysis are fueling the accelerative growth of such organizations. As McGann states, the change from the old to the unknown requires thought and help, and Think Tanks offer such assistance. Hence, the emergence of future studies research and consulting seems intimately tied to the perception of rapid change and consequent uncertainty. In times of relative stability, people are less likely to think about the future; but when change comes quickly, as seems to be the case in the last century, thinking about the future intensifies. (There are other reactions to rapid and uncertain change, which include retreating to the past or trying to simply live in the present.) In general, the growth of future studies in our times is due to the perception of an increasing rate of change.

In his book *Futuring*, Ed Cornish provides a short history of future studies that should also be included in this examination of the development of the discipline. Cornish mentions early utopian thinking as a fundamental precursor of future studies, and highlights the idea of secular progress as significant in stimulating people to think about how to improve human conditions in the future. Francis Bacon's *The New Atlantis* is a good example of both utopian thinking and the philosophy of secular progress. Not only did the Age of Enlightenment bring with it a positive hope for the future of humanity, it also embraced the principles of science, including scientific determinism, and hence, as Cornish points out, the great expositor of the Enlightenment, Condorcet, offers a variety of extrapolative predictions on the future of humanity.[33] By the end of the nineteenth century, when Wells began to write on the future, the idea of progress was the dominant view in Western Europe, and Wells clearly embraced this philosophy. But Wells also believed in scientific determinism and argued that humanity could develop a predictive science of the future. Thus, in his early work, Wells combined determinism and optimism in his view of the future. This philosophy, in fact, reflected the "law of progress" concept popular in the nineteenth century: there is an inevitable and deterministic direction in nature toward progress.

Then, following two world wars and the Great Depression, Cornish contends that the Western optimistic belief in progress declined and pessimistic and nihilistic philosophies became more popular. Dystopian novels (for example, *1984* and *Brave New World*) appeared and the belief of the Enlightenment crumbled. Western humanity had lost the capacity to imagine a positive future. As we saw in the last chapter, Wells' later work became increasingly pessimistic about the future of humanity, perhaps reflecting this general reversal in attitude.

Cornish believes that after World War II, a new philosophy and vision of the future began to emerge. In line with the existentialist philosophy that humans are free and create their own futures, the belief that the future was determined and could scientifically be predicted was rejected by many writers and thinkers. According to Cornish, there were many possible futures rather than one inevitable future. These different futures could be evaluated for desirability ("preferable future"), and which future was actually realized would depend on the choices and actions of humans. Hence, a philosophy of uncertainty and choice, based on human values and decisions, replaced the earlier beliefs that progress was inevitable, or that decline was inevitable.[34]

Although interesting, the historical stages in thinking described by Cornish above are not so clearly distinct. Though Wells believed in determinism, as did other nineteenth century writers on progress and the future, Wells (as well as others like Marx) wrote as if human choice made a difference. And, although post-World War II futurist thinking began to emphasize possibilities, uncertainties, and human choice, a great deal of forecasting, trend extrapolation, and probabilistic prediction also went on and continues to occur in future studies. Perhaps, it might be more accurate to say that the history of futurist thinking, running back at least a couple of centuries, reveals the apparently contradictory themes that (1) the future is determined and can be predicted, and (2) that the future is uncertain and open to choice. There may be shifts of emphasis and many writers attempt to combine the two ideas, in some manner or form, but both views have been popular in modern times.

Within the first couple of decades after World War II, several important publications appeared, ushering in the contemporary era of future studies. All of these books contain elements of theory, ideology or values, data analysis, and prediction. One significant early work in future studies was *Toward the Year 2000: Work in Progress* edited by Daniel Bell and produced by the Commission on the Year 2000.[35] The Commission forecasted a national information system, biomedical advances, reduction of jobs in manufacturing, the need for continuing education, the erosion of the family, and culture becoming more hedonistic and distrustful of authority.[36] Generally, these predictions have come true. Other noteworthy books of this era included Hermann Kahn's (RAND Think Tank) *On Thermonuclear War*,[37] in which the possibility of future world wars was examined; Bertrand de Jouvenal's (Association Internationale Futuribles) *The Art of Conjecture*;[38] Meadows, Meadows, et al. (The Club of Rome) *The Limits to Growth*, in which a variety of environmental and resource shortages and catastrophes were predicted that, as of yet, haven't come true;[39] Frederik Polak's monumental work *The Image of the Future: Enlightening the Past, Orientating the Present, Forecasting the Future*;[40] and perhaps most well known, Alvin Toffler's *Future Shock*. In this work, Toffler dramatically describes some of the key dimensions and implications of accelerative change in the modern world.

There are also several lists of important books on the future. In Richard Slaughter's *Knowledge Base of Future Studies*, Kjell Dahle has created a list of 55 notable books in future studies, and George Kurian, as part of the *Encyclopedia of the Future*, has compiled a list of the 100 most influential futurist books.

Both lists were published in 1996.[41] More recently, Michael Marien has put together a list, by main subject areas, of the top seventy futurist books for the years 1996 to 2000.[42] (Walter Anderson's *Evolution Isn't What It Used to Be* topped Marien's list.[43]) Even more up-to-date, see Marien's list of the top thirty books on the future published in 2005.[44]

It is ironic that although future studies is often distinguished from science fiction, when the editorial board of the *Encyclopedia of the Future* voted on the most influential futurists, four of the top ten individuals listed—Wells, Verne, Asimov, and Arthur C. Clarke—are notably science fiction writers.[45] Yet, the decision was made not to include science fiction books in the list of influential futurist books—still, two science fiction books, Huxley's *Brave New World* and Wells' *The Shape of Things to Come,* somehow made the list.

The Subject Matter, Goals, and Methods of Future Studies

Turning toward the present status of future studies, and a more in-depth discussion of its defining features and characteristics, I will first review a sample of articles on the nature of future studies. The main topics I will cover will be the subject matter, goals, and methods of future studies.

A good place to begin is with Michael Marien's informative essay "Future Studies" in Molitor and Kurian's *Encyclopedia of the Future.*[46] According to Marien, future studies does not possess a consensus as to its nature or purpose. It does, however, draw heavily on the natural sciences and the social sciences and not much on science fiction. Further, future studies has yet to develop into a distinct academic discipline. Finally, Marien states that the term "futurists" is also rather nebulous, including people from all disciplines, many of whom do not even call themselves futurists.[47]

What do futurists think about? According to Marien, there is no real common agreement, although he states that there is a general emphasis on global topics and issues. He notes though that fourteen major categories of future studies are identified by the *Future Survey Annual*, which include:

- World Futures
- The Global Economy
- World Regions and Nations
- Defense and Disarmament
- Sustainability
- Environmental Issues
- Food and Agriculture
- Society and Politics (includes Crime)
- Economy and Cities (includes Work)
- Health
- Education
- Communication
- Science and Technology
- Methods to Shape the Future[48]

From a social perspective, Marien proposes that four relatively distinct "future cultures" have emerged. These cultures correspond to four main areas of study and concern in futurist thinking, and highlight two main differences in attitude and approach among futurists—optimism versus pessimism and short-term versus long-term thinking.

- Science and Technological Futurists tend to be optimistic and long-term.
- Business Futurists tend to be optimistic and short-term.
- Social Issue Futurists tend to be pessimistic and short-term.
- Green Futurists tend to be pessimistic and long-term.

As a final general point in the essay, Marien states that many futurists think that their primary function is forecasting or thinking about the most "probable" futures. However, the idea of many "possible" futures is gaining recognition due to the uncertainty of change, as is the idea about "preferable" futures, due to concern over how ethical values affect tomorrow. In his article, Marien states that future studies should look at all three—possible, probable, and preferable futures (though as noted above, Marien adds three additional "purposive categories" in a later article: examining present changes, panoramic views, and questioning).

A second way of categorizing the concerns of futurists is provided by research done by the members of the *Millennium Project* of the United Nations University. Based on a poll of futurists around the world in 1998, the Millennium Project compiled a list of main priorities for the future, which they described as **"global opportunities."**[49] Although this list identifies important goals for the future, rather than areas of futurist study, it overlaps in content areas with Marien's list. Futurists, obviously, are interested in studying those topics that are important priorities of action for the future. It also should be noted that this set of opportunities is a list of preferable futures—they are not just topics of study but valued and desirable directions or states of affairs. It is an agenda for the future. The fifteen most important "global opportunities" according to the Millennium Project survey are:

- Achieving sustainable development
- Increasing acceptance of global long-term perspectives in policy making
- Expanding potential for scientific and technological breakthroughs
- Transforming authoritarian regimes to democracies
- Encouraging diversity and shared ethical values
- Reducing the rate of population growth
- Emerging strategies for world peace and security
- Developing alternative sources of energy
- Globalizing the convergence of information and communication technologies
- Increasing advances in biotechnology

- Encouraging economic development through ethical market economies
- Increasing economic autonomy of women and other groups
- Promoting inquiry into new ideas and sometimes counter-intuitive ideas
- Pursuing promising space projects
- Improving institutions—evolving from hierarchy to network organizations—centralization to uncentralization

More recently, the list of opportunities was integrated with an earlier list of "**global issues**"[50] to form a list of central "**global challenges.**"[51] Many of the basic themes in this newer list are similar to or the same as those in the earlier list of opportunities, but there some items that are different in content or emphasis. These challenges are presented as questions rather than goals, but the questions still assume certain preferable futures.

- How can sustainable development be achieved for all?
- How can everyone have sufficient clean water without conflict?
- How can population growth and resources be brought into balance?
- How can genuine democracy emerge from authoritarian regimes?
- How can policymaking be made more sensitive to global long-term perspectives?
- How can the global convergence of information and communications technologies work for everyone?
- How can ethical market economies be encouraged to help reduce the gap between the rich and the poor?
- How can the threat of new and reemerging diseases and immune microorganisms be reduced?

- How can the capacity to decide be improved as the nature of work and institutions change?
- How can shared values and new security strategies reduce ethnic conflicts, terrorism, and the use of weapons of mass destruction?
- How can the changing status of women help improve the human condition?
- How can transnational organized crime networks be stopped from becoming more powerful and sophisticated global enterprises?
- How can growing energy demands be met safely and efficiently?
- How can scientific and technological breakthroughs be accelerated to improve the human condition?
- How can ethical considerations become more routinely incorporated into global decisions?

Synthesizing the above lists from Marien and *The Millennium Project*, it appears that the main concerns of future studies include the general categories of:

- Sustainable Development
- Science and Technology
- Energy
- Business and Economic Development
- Globalization and Global Issues
- Social and Political Issues
- Human Diversity, Democracy, Equality, and Human Rights
- Ecology, Environment, and Resources
- Human Health and Welfare
- Education and Communication
- Ethics and Values
- Work
- Religion[52]
- Urbanization and Regional Issues
- War and Crime

- Peace and Security
- Human Institutions
- Future Consciousness—Creativity, Decision Making, and Influencing the Future

This synthesized list is not intended to be definitive. In fact, it should be noted that there is no mention of human psychology, art, or the humanities in the list, which are glaring omissions. Many futurists have extensively studied and written on these topics. Nonetheless, the list provides a general map of the conceptual territory that futurists identify as most important within future studies.

It is interesting to compare this list with the list of areas of the future explored in science fiction. Although there is overlap, some noticeable differences can be seen: the science fiction list is more cosmic, highlights in more detail various technological possibilities, and addresses the topics of mental and spiritual evolution. The futurist list seems more earth-bound, highlights more economic, political, and organizational issues, and emphasizes a variety of human welfare concerns and priorities. Yet, these differences are more a question of emphasis than absolute distinctions, for futurists discuss all the topics on the science fiction list and science fiction writers address all the areas on the futurist list.

Turning from the subject matter of future studies to its methods, Alvin and Heidi Toffler in "Five Billion Futurists" contend that everyone is a futurist. We all think about and have assumptions about the future.[53] In support of this claim, such basic future focused processes as planning, goal setting, and foresight are integral to normal human psychology.[54] The Tofflers also argue that all civilizations have characteristic approaches to the future. Different modes of future consciousness, such as the rational, scientific, mystical, and narrative, have evolved throughout human history and different cultures and societies have created different belief systems, theories, archetypes, myths, visions, and values in conceptualizing the future.[55] According to the Tofflers, modern civilization, building on basic human psychology and traditions of history, has developed a new set of approaches or methods, which are fundamentally secular in nature. Note that the Tofflers include science fiction in their list.

- Utopian and Dystopian Political Literature
- Science Fiction
- Technological Forecasting
- Military Gaming
- Trend Extrapolations
- Corporate Strategic Analysis
- Central Planning in Governments

They also present the three-fold distinction of possible, probable and preferable futures and connect these three types of futures to different methods. Recall how Alvin Toffler described the way we approach each of these kinds of futures: "Visionary explorations of the possible, systematic investigation of the probable, and moral evaluation of the preferable."[56] They note, however, that the three forms overlap. Imagination, critical and rational thinking, and value judgments are not distinct psychological processes for they are interactive. According to the Tofflers, possible futures include science fiction, which they believe is an important contributing influence to future studies; probable futures are often quite systematic and scientific and frequently developed in government and business studies; and preferable futures often paint negative pictures of today and then offer solutions. However, preferable futures are often intended to be inspirational and psychologically uplifting, such as in Barbara Marx Hubbard's *Evolve: A Global Community Center for Conscious Evolution.*[57] Interestingly, the Tofflers contend that all types of futuristic approaches are art forms, involving subjective and assumptive elements and values, as well as scientific ideas and methods.

Contrary to Marien, the Tofflers do see some common areas of agreement among futurists, namely that most serious futurists agree that no one can predict the future, future consciousness is an essential survival trait, and we are in the midst of a fundamental transformation within our contemporary world.

Yet, as stated earlier, there is some clear disagreement on the issue of predictability of the future among futurists. Although Ed Cornish emphasizes possibilities and freedom of choice in thinking about the future, many futurists do make predictions. In identifying probable futures, futurists assume a degree of predictability regarding the future—it is simply that some range or set of different futures are being predicted. "Trend extrapolation"—a commonly listed method of futurists—is a form of prediction. It is simply probabilistic prediction. Cornish, in fact, acknowledges a limited degree of predictability to the future and the value of efforts to predict the future.[58]

In his article "Thinking about the Future," the science fiction writer Frederick Pohl adds some interesting twists to the "possible, probable, and preferable" conceptual scheme. He states that all along futurists have attempted to predict the future and have developed various methods for accomplishing this general goal. But he thinks that futurists should focus on the imaginative process of envisioning as many possible futures as they can and then distinguishing the most desirable possibilities and working toward realizing these highly positive possibilities. In fact, he suggests that futurists should be particularly concerned with low probability-high desirability futures and how they could be achieved. In general, Pohl highlights the process of **"normative forecasting,"** as he refers to it—identifying ideal futures and then attempting to "invent" or create these futures.[59] Although Pohl does not think that science fiction attempts to predict the future, his emphasis on imagining many possible futures aligns with one main function of science fiction.

Returning to his article on "Futurism," W. Warren Wagar believes that the methods of futurists range from the highly mathematical to the intuitive and speculative.[60] He does, however, list five popular methodologies presently used by futurists.[61]

- Trend Extrapolation
- Mathematical Modeling
- Delphi Technique (involving the polling experts on the future)
- Scenario Building (may include science fiction)
- Probabilistic Techniques

Wagar also notes that some futurists find inspiration in Marxist philosophy and ecological and spiritual thinking. Given his inclusion of science fiction, and intuitive and spiritual approaches in his list, Wagar clearly goes beyond defining futurist methods exclusively in terms of scientific and rationalist activities.

Michael Marien also takes a more liberal view on the methods of future studies. Aside from his list of six "purposive categories of futures-thinking," he also offers a twelve-category "continua for analyzing futures-thinkers." Each category includes opposing pairs of qualities, such as the category of disposition, which runs from optimistic to pessimistic and the category of breadth, which runs from generalist to specialist. Some of the other categories are style, rigor, culture, timeframe, and ideology. Given this list, Marien would include science fiction writers and "imaginative idealists" as futures thinkers.[62]

In the "Art of Forecasting," the World Futurist Society presents a list of the main methods used by modern futurists in thinking about the future.[63] This list of methods shows a strong emphasis on scientific research, data collection, empirical prediction, logical reasoning, and the use of technology. The list includes:

- Trend Projection
- Scenarios
- Consulting Others
- Models

- Simulations
- Computer Simulations

In his essay, "Futurists," Cornish also highlights scientific and rational methods.[64] He states that futurists use scientific and rational methods to understand the future and, generally, don't use mystical or supernatural means. He believes that futurists think the future is something that can be controlled, rather than being a matter of fate. Cornish connects the idea of fate with the mystical approaches to the future—he thinks that if someone is mystical then he or she believes in fate or destiny. Thus, Cornish supports the Enlightenment position that self-empowerment and social improvement is achieved through scientific and rational methods and, contrary to Wagar and Toffler, he wishes to restrict the discipline of future studies to scientific methodologies.

In his recent book, *Futuring*, Cornish continues to emphasize the rational and scientific qualities of futurist methods. He provides a synoptic list of the most common methods used:

- Scanning
- Trend Analysis
- Trend Monitoring
- Trend Projection
- Scenarios
- Polling
- Brainstorming
- Modeling and Simulations
- Gaming
- Historical Analysis
- Visioning

Furthermore, he provides extensive and very helpful descriptions with illustrative examples of all these methods, which, although "rational, empirical, and scientific," are usually refinements of common sense approaches to the future.[65]

What becomes clear through examining Cornish's description of futurist methods is that many of the methods are ways to make tentative predictions about the future. Cornish believes that we need to be cautious about predictions and often the best we can do is make probabilistic predictions, but he does not reject the role of prediction in future studies. In fact, although Cornish repeatedly states that the goal of future studies is not to predict the future but rather to create a positive future, he actually connects the goals of prediction and creation together. By thoughtfully considering what probable events may occur in the future, we can prepare for the future and consider whether such potential future events are desirable or undesirable. If we know what might happen, we can perhaps do something to improve the chances of positive futures and reduce the chances of negative futures. Cornish thinks, in the spirit of the Enlightenment, that knowledge is power. If we can understand nature, we can influence it.

The above lists of methods notwithstanding, Daniel Bell, in his "Reflections at the End of an Age," provides some necessary caution and balance on the rhetoric of future studies and its presumed scientific methods.[66] Bell believes that futurists are prone to hype and exaggeration in their views of the future. They often over generalize and oversimplify. For Bell, human society is too complex to reduce to some simple set of descriptive concepts. The futurists' use of scientific and mathematical concepts and tools is sometimes questionable. They may sound very objective and factual in their formulations, but it is often unclear exactly what is being measured. Further, futurists often make claims that cannot be measured or quantified.

Robert Nisbet, the historian, provides another balancing counter-point on futurist methodologies. Although not a futurist, he argues that futurism hasn't really changed much since its beginnings in the last century. In spite of all the apparent sophistications introduced into futurist methodology, futurists just identify trends and extrapolate on these trends into the future. Nisbet states that futurism also hasn't changed much because there continue to be, as there was a century ago, two basic groups of futurists—those that are optimistic and believe in progress and those that are pessimists and prophesize "doom and gloom."[67] Perhaps these two basic attitudes haven't changed much since ancient humans began thinking about the future.

A good way to conclude this section is to review a recent debate between Michael Marien and Wendell Bell on the nature of future studies. The debate illustrates some of major points of disagreement among futurists regarding the nature of the field and its methods. In general, Marien argues that future studies is not a coherent and clearly defined field of study, and is becoming increasingly more so—it is faltering and floundering. On the other hand, Bell believes the field is alive, thriving, relatively integrated, and its practitioners are in general agreement on a set of principles and beliefs. However, Bell does acknowledge that futurists are a "disputatious community," which is evidently illustrated through the debate between him and Marien.

Below is a list of seven myths about future studies identified by Marien followed by Marien and Bell's reaction to each:

1. *Future studies is a distinctive field or discipline.* Marien argues instead that at best it is a fragmented "very fuzzy multi-field" that blurs into other fields. Although Bell acknowledges that there is a degree of fragmentation and specialization in future studies, there is no more so than in other disciplines—in fact, perhaps less.
2. *Futurists are generalists.* Marien asserts that most futurists are specialists focusing on some relatively narrow area of study or consulting. Bell argues that futurists generally try to adopt a holistic perspective in their research and thinking.
3. *Most people involved in the study of the future see themselves primarily as futurists.* Instead, Marien contends that there are very few self-identified full-time futurists, and most people who write and do

research on the future are "secondary futurists." Bell acknowledges that most professionals have multiple professional identities.

4. *Future studies is a unique and distinctive field.* Marien argues that a lot of futures thinking occurs outside the discipline. Bell points out that no professional group has a distinct monopoly on making contributions to its area of study.
5. *Future studies is understood and appreciated by outsiders.* According to Marien, to the contrary, futurists are often criticized, misunderstood, and rarely consulted by mainstream publications on questions of the future. Bell points out that most academic and professional groups have problems with their public image and are misunderstood by the general public. Future studies is not unique in this regard.
6. *Future studies has a relatively stable identity.* Marien contends that futures thinking has gone through frequent and significant changes in its focus. Bell acknowledges that future studies has grown and changed throughout its history, but so what? Isn't this a good thing? Further, Bell counters that there are at least some central and enduring features to future studies.
7. *Future studies is a community.* Instead, Marien states that future studies is a plurality of communities that overlap with each other to some degree and communicate. It is a fragmented social reality rather than a unified one. Again, although Bell acknowledges that there is some level of fragmentation to the futurist community, it may be no more so than other academic communities, and there is clearly some level of connectivity and communication (both argumentative and mutually validating) among many of the major futurist individuals and organizations.[68]

Marien and Bell can be seen as representing pessimistic and optimistic viewpoints of future studies. They each acknowledge as much. Marien, the pessimist, uses a variety of statistical and factual indicators to illustrate his points and describes his view as "reality-based." Bell, the optimist, repeatedly counters Marien by arguing that future studies is no worse (perhaps better) than other academic or professional groups. Further, Bell uses other indicators to support his opposing position. They also interpret the same facts differently. Where Marien sees differences within the field as indicative of a lack of distinctive coherence, Bell sees differences of opinion among futurists as healthy.

Finally, Marien is more liberal in what he includes in futures research, study, and publications—thus he sees future studies as more diverse and hence fragmented. Bell limits what he includes as legitimate or acceptable futures thinking, and thus finds more commonality and unity in the discipline. Yet, it could be argued that although restricting future studies to acceptable scientific methodologies brings unity to the discipline, it also limits the discipline and may ignore important aspects of human reality.

Theories and Ideologies within Future Studies

At this point, let us look at how theories and ideologies of the future connect with future studies. A theory of the future is an explanation and description of the future, whereas an ideology is a set of values regarding a preferable future. A good place to begin this discussion is with Richard Slaughter's views on future studies. Slaughter provides a rich and multi-faceted description of future studies.[69] He has some important ideas regarding the scientific status and rigor of future studies and how to broaden future studies beyond traditional scientific methodologies. He also discusses the impact of theory, cultural world views, and values on future studies.

Slaughter clearly believes that future studies is an academic and scientific discipline, worthy of professional status, funding, and educational departments in colleges. He thinks that future studies has developed a core set of disciplinary concepts and terminology, a wealth of research methods, and numerous pedagogical activities and principles for teaching futurist thinking. For Slaughter, there is a set of key issues, themes, and applications defining the discipline.[70] Yet, he is an evolutionist and sees all of elements of future studies in a constant state of growth and redefinition. (Hence, total consensus among futurists is unrealistic and undesirable.)

In line with his evolutionary view of future studies, Slaughter traces four developmental phases in future studies: The American driven scientific and empiricist tradition (much of what was discussed in the previous section falls into this tradition); the European driven culturally oriented tradition (with more an emphasis on cultural values and how they impact futures research); the international and multicultural tradition, bringing in the ideas of non-Western thinkers; and finally, the Integral Futures movement, of which Slaughter is a principal advocate.[71]

Slaughter's conception of Integral Futures is based upon Ken Wilber's Integral Philosophy.[72] One key idea in Wilber's philosophy is the Four-Quadrant conceptual framework for describing the full breadth of reality and modes of understanding reality. Basically, the four quadrants consist of the inner singular consciousness perspective, the outer singular behavioral perspective, the inner plural cultural perspective, and the outer plural social and material perspective. The two inner perspectives are subjective ways of looking at reality, and the two outer perspectives are objective ways of looking at reality. In each case, we can focus on the individual or focus on the plurality of things. Wilber's contention is that a comprehensive view of reality and modes of human understanding must include all four quadrants. Slaughter takes this idea and applies it to future studies, arguing that a comprehensive understanding of the future must also look at all four quadrants. For example, we cannot simply focus on the collection of external objective facts as the sole basis for thinking about the future, but we must also look at the inner experiences of individuals and cultures and the sense they make out of understanding the future. In essence, reality, which includes the future, is more than the external physical world—it also includes the inner subjective experiences of people. Future studies must acknowledge the full breadth of human reality—both the inner conscious reality and the outer physical reality—and "integrate" both the subjective and the objective into its theoretical framework for understanding the future. Also, in line with the evolutionary perspective of both Slaughter and Wilber, all four quadrants of reality are seen as dynamical and developmental. Individual minds grow; cultures grow; the material world evolves, and these realities go through various stages in their growth and development.[73]

Based on this integral philosophy of the future, Richard Slaughter has extensively critiqued the Western bias toward scientism, rationalism, materialism, and objectivism and how these biases severely constrict the approaches to the future within future studies. For Slaughter, the methods of futurists presuppose theories or paradigms regarding the nature of reality, which are often influenced by cultural beliefs and values. The Western theory of reality, as defined and described through science, emphasizes materialism and the "outer reality" of the external physical world. Also, Western science focuses on rationality and empirical observations of the external world. Following Wilber, Slaughter argues that there are different standards of evaluation and different modes of inquiry and discovery associated with each of the four quadrants and that futures research shouldn't be limited to the methods of science which focus on the objective outer world. In general, Western science minimizes the importance of intuitive and a-rational methods and the "inner reality" of people. Further, the Western approach does not encourage or stimulate self-reflection and self-critique on its own assumptions and theories regarding reality.[74] In Slaughter's mind, we should broaden our approaches to the future to include the insights and methods of spiritual, mystical, introspective, and inner-directed traditions and philosophies.[75] Hence, Slaughter does not narrowly limit futurist methods to scientific methodologies, since he sees important roles for inspiration, heart, personal meaning, and "transcendent realities" in thinking about the future. Slaughter's vision is Apollonian and Dionysian; it is both empirical and observation based, as well as metaphysical and introspective.

It is Slaughter's contention that the Western scientific view of the future basically generates only two visions of the future: Techno-optimism and the pessimistic "Terminator 2" scenario. Technology will save us or technology will destroy us. The Western view sees the main drivers of change as science, technology, materialism, commercialism and greed; and these factors could either lead to more and more of the same thing or backfire and produce disaster. He even argues that much of modern science fiction buys into this either-or thinking. By not including other perspectives on reality or views of the future, our speculative imagination is severely restricted. As one example, he proposes that the pursuit of wisdom—of a wise society or culture, which brings into consideration the evolution of consciousness, inner meaning, and values—should be seriously considered as an alternative vision of the future to the Western obsession with technological power.[76]

It is important to see that Slaughter's main argument is for a comprehensive and balanced approach to future studies. Future studies should not be attached to any single perspective regarding reality and the future, but should be open to different cultural points of view, different theories of reality, and different methodologies. Yet there is a system, or as Slaughter puts it a "meta-paradigm," behind his "liberalism." Future studies should seek the broadest, most encompassing framework for understanding the future.[77] He believes that Wilber's four quadrants cover all of the basic perspectives one can take on reality and thus provides a framework for aligning and comparing different points of view; Wilber even claims he has provided a "theory of everything."

Slaughter believes that present Western culture, besides being excessively rational and materialistic, is too "present-focused" and that many of the problems of the modern world are due to this shortsightedness. For Slaughter, the secular philosophy and way of life is focused on the present without much concern for tomorrow. On this point, he concurs with other writers such as Howard Didsbury and Stephen Bertman, who also believe that modern life is focused too much on the present.[78] If we are to successfully address the problems and challenges of today, Slaughter thinks that future studies and the capacity of "foresight" must become more important and more evolved in human society. We need to become more future-focused. We also must transform our values and approach to life. He sees many purposes and values to future studies, including stimulating global dialogue on contemporary issues, providing essential ideas and methods for the further development of humanity, and in general, heightening future consciousness in all of us.

It is important to see that Slaughter's conception of future studies as an academic discipline has a value system and theory of the future behind it. He includes non-scientific methods in his description of future studies because of the theory of human reality, and the value system that he supports. He sees the contemporary world beset with a variety of problems, including the depletion of resources, ecological deterioration and pollution, and social conflicts, all problems he attributes to the materialistic and industrial way of life that emerged out of the Western philosophy of secular progress. For Slaughter, futurist thinking should involve seriously considering the limits of resources, the importance of sustainability of systems into the future, global and multi-cultural input into the creation of tomorrow, and a broadening of human values beyond technology and economic growth. He sees the value of humanistic, spiritual, and intuitive approaches in future studies because he believes that contemporary Western thinking has been too limited to materialistic, scientific, technological, and rational approaches to life. According to Slaughter, the secular approach is imbalanced and non-integral. He does not wish to exclude the humanistic and the intuitive in futurist thinking; rather, he thinks we need to broaden and deepen our understanding of life and the future to solve the problems of today and create a better tomorrow.

Futurists tend to develop both theories of the future (where we are, how we got here, and where we are going) and ideologies of the future (what we may be doing wrong and what we should do differently).[79] Futurists tend to develop theories that are descriptive and prescriptive—treating issues of both fact and value; theory and fact, and ideology and value are connected together. Slaughter's theory of human reality is that the modern West has an excessive (or lop-sided) industrial and materialistic culture and has created a "crisis" point in human history—potential disasters are imminent, due to our imbalanced mindset and manner in which we live. Therefore, (the ideology) we should expand our modes of thinking and values beyond this culture and limited mindset if we are to move successfully beyond this critical point in human history.

Numerous other theories of the future and ideologies or value systems of the future exist. For example, there are a significant number of mythological and religious theories of the future, and there are many theories of the future connected with the idea of secular progress.[80] Although I review an extensive set of theories of the future in the last chapter of this book, the following are some of the most influential and popular contemporary theories and ideologies:

- Globalization: The world is evolving into a global society along economic, political, and cultural dimensions. This is a good thing and we should embrace it.
- Accelerative Change: Change in almost all facets of human life is accelerating and will continue to do so in the future. We need to find ways to accommodate and thrive in this reality.
- The Technological Transformation of Humanity: Humanity and technology are integrating. Computer and communication technologies are becoming

increasingly embedded into our lives and our society. Biotechnology will transform humanity. All these developments are leading to increased self-empowerment and we should pursue them.
- The Adventure into Outer Space: Humanity will travel into and colonize outer space. This adventure is a positive expression of human curiosity and human evolution. We should support and embrace the adventure.
- Armageddon: We are heading toward the Apocalypse and Judgment Day. We should prepare ourselves for these events and the end of the world.

At times, a theory of the future is clearly stated;[81] while at other times it is implicit within the writings of the author. As a basic explanation and set of predictions for the future, a theory of the future provides a general and integrative overview of the future, often including a description and analysis of the present as well. Trends in the present are often connected with future events. Usually there is a dominant theme (e.g., the potential of technology, the emergence of a global society, or the transformation of humanity). The author of the theory invariably presents a variety of reasons, facts, and arguments to support the validity of the theory. In general, a theory of the future provides a guiding set of ideas, principles, or images for generating predictions and making sense out of the future.

Understanding the nature of future studies involves understanding the different key theories and ideologies within its domain. The discipline is not independent of theory and value. Many of the basic themes and concepts of futurist thinking derive from ideas in futurist theories and ideologies, e.g., "future shock," "progress," and "sustainable future." Further, competition and disagreement among the theories and ideologies create the central issues and debates of future studies. People get into arguments and conflicts over methods, subject matter, and courses of future action, to a great degree because they have different theories and value systems. Understanding a futurist's view of future studies (the nature of the discipline and its values) invariably involves exploring what theory and ideology of the future he or she supports. For example, Slaughter sees great potential for futurist thinking, futures education, and future studies in human society because of how he interprets the contemporary problems of the world and the remedies he believes are needed to set us on a better path. Other futurists might exclude non-scientific methods because of a different theory of the future; they may believe that superstition and irrational mindsets are excessive in our world and create rather than solve problems.

A good example of a theory and ideology of the future is contained in the writings of Alvin Toffler. Toffler has written three extremely popular books in which he has been progressively developing a particular view of the future. These books are *Future Shock, The Third Wave,* and *Power Shift: Knowledge, Wealth, and Violence at the Edge of the 21st Century.*[82] In essence, Toffler's theory is that modern human society is in a period of great transition, moving from an industrial, hierarchically organized, and standardized world to a global information society of network organizations, customization, and heightened individualism. Change is accelerating; diversity and complexity are magnifying; and flexibility and creativity are becoming central values. More recently, Toffler has collaborated with his wife, Heidi, on a series of additional publications that further develop themes in his earlier books.[83] Throughout these writings, Toffler and Toffler provide a detailed analysis of present trends, extrapolations into the future, and prescriptive value statements regarding how to guide and direct the future.

A theory of the future can be presented in a non-fictional format, such as in Toffler's books, Slaughter's publications, or H.G. Wells' visions of a better world. Theories can also be embodied within a fictional or novel form, such as Olaf Stapledon's *The Star Maker,* Arthur C. Clarke's *Childhood's End,* or Stephen Baxter's *Vacuum Diagrams.* Each of these novels tells a story, but each also contains a general theory of humanity's ultimate future and the forces directing it. At the other end of the methodological continuum, a book on the future may highlight statistical information on present trends and present a variety of mathematical and interpretive extrapolations on the future, such as in Meadow's *The Limits to Growth* and *Beyond the Limits*, Naisbitt's *Megatrends* and *Megatrends 2000*, and Moore and Simon's *It's Getting Better All the Time*. Although the statistics, numbers, tables, and graphs may create the appearance of pure scientific and factual objectivity, such books also contain both general theories and value systems. General explanations of the meaning of all the quantitative data are offered and proposals are made regarding values and desirable futures.[84]

Mixing fictional, metaphorical, mythological, theoretical, statistical, and even artistic methods are some of the ways to present a theory about the future. Two books on humanity's future, Dougall Dixon's *Man After Man: An Anthropology of the* Future and Peter Lorie and Sidd Murray-Clark's *History of the Future: A Chronology*, use art and visual graphics as a central mode of representation in communicating their ideas.[85] In fact, Kurian lists both these books among the hundred most influential books on the future in his *Encyclopedia of the Future*.[86] *Wired* magazine, one of the most popular publications on the future, also highlights visual imagery and graphics in communicating its ideas on the future.[87] Ian Pearson's *Atlas of the Future* combines maps, graphics, and statistics in a very readable and informative format.[88] Ancient mythological views of the future often contained key symbols, icons, images, and other visualizations integral to their understanding of the future.

Of contemporary significance is the fast growing use of computers, computer simulations and graphics, and virtual reality in visualizing possible futures; for example, *The Mind's Eye* video series of computer animation presents a mesmerizing array of futuristic scenes and scenarios.[89] The explosive growth of the World Wide Web (WWW) is supporting numerous efforts to create visual and multi-media images of the future; futuristic cities such as *Cybertown* on the WWW can be explored giving the viewer a dynamical and perceptually compelling sense of the future.[90]

The earliest theories and ideologies of the future were embodied in written histories, oral traditions, and mythological tales. Theories were conveyed mostly through stories, though perhaps we should also acknowledge the "mythograms" (cave paintings) of prehistoric humans as actually the first visions of the future.[91] With the emergence of the modern era, secular and scientific theories and value systems appeared; theories became abstractions and rational arguments. In the following century, science fiction literature—more visionary, concrete, and graphic than philosophical essays—came into being, and theories and ideologies of the future evolved that synthesized the dramatic and the scientific. Future studies continued the traditions of science and secular philosophy, combining reason and mathematics, in creating its theories of the future. Following Slaughter's ideas, future studies should also incorporate inner realities and a-rational methods.

With the advent of the modern media and computer technology, the future is being visualized like never before—both dynamically and interactively. The image is becoming as important as the word.[92] Numbers can be easily transformed into visualizations. As our capacities for representing the future have evolved, our theories and ideas about the future have found new media for their expression. Although it may not be completely true that the "medium is the message," our cognitive and theoretical understanding of the future is clearly dependent on the medium and perceptual modes of consciousness for representing the future. Resonant with Slaughter's developmental understanding of future studies, futurist thinking and futurist imagination have evolved and will continue to evolve in conjunction with the evolution of media and modes of representation for human understanding.

Bell's Comprehensive Overview of Future Studies

Having reviewed the history, methods, subject matter, and role of theories and values in future studies, it is worth looking at the ideas of Wendell Bell in some depth. Bell, in his two-volume *Foundations of Future Studies*, provides a relatively comprehensive and systematic overview of the discipline and, in particular, addresses how prediction, fact, value, and action can be integrated within the discipline. A good starting point is Bell's list of nine fundamental purposes of future studies and his list of basic assumptions in future studies. According to Bell, these assumptions give future studies a clear sense of focus and identity.[93]

Purposes of Future Studies

- Studies possible futures
- Studies probable futures
- Studies past images of the future—their causes and consequences
- Knowledge foundations—investigates in what sense can we have knowledge about the future
- Ethical foundations of future thinking—investigates the role and significance of values in futurist thinking
- Provides a basis for interpreting the past and present, or orienting to the present
- Integrates knowledge and values for designing social action
- Increases democratic participation in imagining and designing the future
- Communicates and advocates images of future—proposes actions, suggests possibilities and evaluates and advocates social actions

Assumptions of Future Studies

- Time is continuous, linear, unidirectional, and irreversible.

- The future contains novelty.
- Futures thinking is essential for "conscious or decisional" human action.
- Futures knowledge is the most useful knowledge.
- There are no future facts—there are future possibilities.
- The future is open—there are opportunities and freedom in directing the future.
- Humans make themselves.
- There is holism and interdependence within the world, which implies an interdisciplinary approach in the organization of knowledge for decision making and social action.
- There are better and worse futures.
- People are purposive and creative project pursuers.
- Society is a system of purposive beings and social results come from such purposive beings.
- There is an independent and objective external world.

To summarize Bell's general position on future studies, he clearly believes that thinking about the future and knowledge of the future has great value—it is essential for making decisions and engaging in purposeful action. He also thinks that the future is possibilities, probabilities, and even novelties rather than certainties, though he does believe that we can develop knowledge and understanding of these possibilities and probabilities. He sees future studies as interdisciplinary and involving the integration of fact and value. Contrary to Marien, he contends that future studies is a distinctive discipline with defining features and goals. In particular, Bell thinks that future studies is an "action science," where futurists attempt to predict or extrapolate possible futures based on present trends and other scientific data, consider these possibilities in light of human values, and propose plans and policies to realize those most desirable possibilities. One can see in the above list many items that revolve around the ideas that the future is possibilities, that we can evaluate these possibilities and make decisions regarding which are the most preferable, and that carried forth into plans and actions, our decisions create our future. Future studies empowers humans to determine their future.

It is important to see in Bell's views the integration of knowledge, value, and action. We make value judgments on informed predictive possibilities and guide our future actions toward the most desirable ends. According to Bell, futurists do attempt to predict the future (with degrees of possibility, probability, and the unavoidable contingency inherent in all human knowledge) based on whatever relevant factual evidence can be gathered. They make "presumptively true" predictions, e.g., if the following conditions and trends persist, the following future events will occur. These predictions constitute the scientific aspect of future studies and its knowledge base. But futurists don't simply make predictions; they evaluate (for desirability) the possibilities. To quote from Bell, "There are better or worse futures." Further, futurists don't simply think about these preferred directions; they attempt to encourage planning, policy, and action based on their predictions and value judgments.

Bell also thinks that values can be scientifically, objectively, and rationally assessed—contrary to the traditional distinction of value and fact that has developed in modern times. The factual consequences of adopting a value in the future can be (probabilistically) predicted and assessed. We can consider where a particular value will lead us.[94] Futurist thinking gives us a way to assess our values, as we attempt to ascertain the future consequences of our values.

Bell's integrative (or holistic) philosophy of future studies aligns with the contemporary scientist E.O. Wilson's ideas on the importance of inter-disciplinary thinking and his principle of consilience. Wilson believes that all the significant problems of our time require the input and integration of multiple disciplines, in particular, a pulling together and "consilience" of the sciences and the humanities.[95] Bell sees the future as a multi-faceted reality, involving technological, economic, social, and ethical issues. When we make predictions about the future, we need to consider all these dimensions of human reality and their interaction. Although there are differences between Slaughter's Integral Futures Studies and Bell's philosophy, there is a common agreement on the need for integration of different dimensions of human life in future studies.

Bell's integrative concept of future studies and his synthesis of knowledge, values, and action can be applied to our understanding of future consciousness. Future consciousness does not simply anticipate, but it judges or evaluates its creations and imaginative scenarios, which in turn fuels emotion and directs motivation toward action. Thought, imagination, motivation, emotion, and action are interactive and interdependent in future consciousness. Not only does Bell think that futurist thinking is essential for decision based action, but he also thinks that future knowledge (in the sense of probabilistic predictions) is the most useful form of knowledge. These points underscore the cognitive and practical values of thinking about the future. Finally, Bell believes that the various capacities and skills associated with futurist thinking can be further developed through education. I discuss all these points in depth in *The Evolution of Future Consciousness*.[96]

Another important idea to highlight from Bell's list of purposes and assumptions of futures studies is his belief that humans create their future. Bell's scheme of "thinking and action" is a framework for creating the future in an informed, methodical, and rational way. Basically, Bell concurs with Enlightenment philosophy. He believes that humans are responsible for their future, although humans often find ways to shirk their responsibility by blaming external forces, such as fate, destiny, or powers beyond their control.[97] His "action science" approach is a way to emphasize our responsibility for the creation of the future. Further, the general scheme of assessing facts and possibilities, evaluating the possibilities, and formulating and enacting plans to realize the most desirable possibilities provides a general outline for how to create the future.

Bell definitely believes that future studies is a science—in particular, an action science, for the scientific data collected is used as a basis for proposing and implementing informed action. He thinks future studies is scientific because it collects data and makes predictions based on data collection. Also, it studies facts, such as trends, patterns of change, and people's belief systems and values, and fundamentally strives for an understanding of the truth. I would add that future studies also contains various competing explanatory theories, and futurists often debate these theories using scientific and empirical data and evidence. Bell examines in significant depth the nature of science and scientific methodology, and attempts to demonstrate these scientific features in future studies thinking and research. For Bell, it is clearly important that our understanding of the future include relevant scientific principles and information—future studies and future consciousness must be factually informed as much as possible.

Yet, in attempting to demonstrate clearly the scientific quality of future studies, Bell excludes non-scientific aspects of future consciousness from his definition of future studies. Although Bell is critical of the idea that futures studies is more of an art than a science, he does acknowledge "artistic" elements in future studies. However, he is a purist in that, although he supports "methodological diversity" in future studies, he excludes science fiction and religious-metaphysical approaches from future studies. Marien and Slaughter's views of the discipline clearly seem broader. To exclude science fiction because it is more art and literature than science ignores the basic fact that science fiction has significantly contributed to futurist thinking and fuels the imagination of possibility thinking. Similarly, the basic archetypal concepts of ancient mythology and religion influenced modern thinking about the future and addressed the emotive and personal dimensions of future consciousness. Finally, it could be argued that introspective, intuitive, and mystical approaches to the future are excluded in Bell's system, yet all of these perspectives have value and provide a more complete view of human reality than objective science alone.

Also, Bell sees future studies as fundamentally a social science—its basic domain of study is the future of human society—but this definition also seems too narrow, for the study of the future also includes ecology, the environment, and the cosmos. Future studies must study the future of everything because humanity and society are contextually embedded within this reality; that is, the future of nature and the cosmos is relevant to the future of humanity.

Aside from his excessive and exclusionary emphasis on science in his depiction of future studies, Bell also presents a decidedly rational theory of future consciousness. He describes decision making about the future as a logical process. Facts are gathered and tentative conclusions and extrapolations into the future are derived. Values and consequently goals for the future are identified by considering the potential effects or results of following these values and goals. Plans are formulated and actions are initiated based on the thoughtful consideration of values and goals.

But future consciousness involves more than just scientific understanding and logical reasoning. In the creation of the future, people are influenced by inspirational visions and ideas. Images drive us as much as reasons. People are emotional and intuitive about the future as well as rational and factual. There is always an unavoidable element of faith in our beliefs and behaviors. Often, our purposeful actions feel more impulsive than methodical. The unfolding of our lives cannot be reduced to a set of calculations. Human life is personalized—it is more than just abstractions and impersonal facts. Humans live dramas and stories. Life is an adventure as much as a strategic plan. All these points were made by the Romanticists in their criticisms of Enlightenment philosophy.[98] The rationalist could respond that reason and evidence provide a better approach to the future than emotion, impulse, superstition, and faith, but the basic fact is that humans are psychologically multifaceted and richer in their mental reality than the rational theories of consciousness admit. Reason and evidence are critical and cannot be discarded in approaching the future, but one-sided rationalist theories of consciousness and action are too limiting.

On the positive side, Bell's concept of future studies as an action science does integrate two important dimensions of human reality. Science could be defined as the systematic search for factual knowledge and theoretical understanding about nature; i.e., it could be characterized as a descriptive discipline. Yet this view of science reflects the value-fact split of modern times. Our modern view of science states that science attempts to describe and understand the facts, and not to evaluate them or search for values. Within this view, science cannot make value judgments. But, the initial spirit of the Enlightenment was to integrate fact and value—to find a way for secular understanding to serve the ideals and goals of human society. Further, Enlightenment philosophy did support various values, such as freedom, wealth, and human happiness. Within Bell's scheme of future studies, he articulates a method for uniting fact and value. In thinking about the future, we should identify our basic values for tomorrow (our preferable futures), and through scientific data collection and theoretical understanding, we should assess the possible directions that our world is taking. We can then ascertain whether the world is moving in directions we think (evaluate) as desirable, i.e., aligning with our values. If trends are leading away from identified values, we can ask, "What can we do to redirect the process of change?" In comparing our predictions with our values, we can inform our decisions about what actions to take. Further, as noted earlier, for Bell, we can always subject our values to a futurist empirical test. We can make hypothetical predictions of the future factual consequences of our values. We can see if we really want to follow those values. To some degree, what Bell is describing here coincides with what humans normally do in developing goals and plans, though probably not with such systematic rigor.

Predicting the Future

"It is difficult to say what is impossible,
for yesterday's dream is today's
hope and tomorrow's reality."

Robert Goddard

"In 1901, I said to my brother Orville that man
would not fly for fifty years. Ever since I have
distrusted myself and avoided all predictions."

Wilbur Wright

Let's examine in more depth, two of the central issues in future studies and futurist thinking, the prediction and control of the future. Many religious and mythological views, as well as secular and scientific philosophies, assumed that the future could be predicted; however, the reasons for these convictions differed. Myth and religion often subscribed to destiny and fate; science supported causal determinism. The secular view of progress also assumed that the future could be controlled by humans, something that religious and mythical views often did not accept; for them deities and fate dictated the future. Following the scientific spirit of the secular view, Wendell Bell and other contemporary futurists have argued that, through scientific principles and empirical data collection, predictions about the future can be made with some degree of probability, and this information can be used to facilitate the control of the future.

We should recall that not all futurists state that they believe in the feasibility or desirability of predicting the future. Daniel Bell has questioned both the scientific rigor and actual validity of futurist predictions. Peter Russell argues that most of the significant changes in the last 100 years have been unanticipated because the changes went beyond the imagination of earlier forecasters.[99] Walter Anderson simply asserts that the future can't be predicted, yet he apparently does believe that certain general trends in our world, such as technological growth and globalization will continue, which are, in fact, general predictions.[100] Both Laura Lee and Graham May point out that futurists have made numerous mistakes in predicting the future.[101] May also states that talk about possible and probable futures could be seen as an excuse for incompetence. Pohl notes that futurists are good at predicting what "might" happen but not what will happen.[102]

Many futurists contend that future studies doesn't attempt to predict the future, but rather it presents alternative or possible futures.[103] Peter Bishop argues that it is better to think of the future as multiple possibilities rather than one clearly predictable eventuality.[104] Cornish, in *The 1990's and Beyond* and *Futuring*, contends that futurists don't try to predict the future.[105] Rather, futurists present different possibilities in order to lay out before us our choices for tomorrow. For Cornish, if the future could be predicted it would be determined (or predetermined) and, consequently, there would be no choices to make and no practical reason to think about the future. What would be the point, since our future actions would be determined as well? If the future is possibilities, including our own personal lives, then we have some options to consider, evaluate, and act upon. Hence, a very predictable future seems to preclude the possibility of choice—there is an apparent incompatibility between determinism and freedom of choice. But we have also seen that Cornish does support the idea that futurists engage in cautionary and probabilistic prediction, and in fact, many of the futurist methods he lists are of this type. Furthermore, it is only because we can thoughtfully anticipate the probable consequences of present trends and conditions that we can make informed choices about the future.

Ziauddin Sardar presents the argument that predicting the future is a way of controlling the future, to the point of destroying different possible futures that could have been realized. When futurists provide predictions they constrain the mindsets of individuals—they create anticipations and expectations that narrow our imagination and our actions. Predictions can become self-fulfilling prophecies; at the very least, they influence individuals to perceive and behave in certain ways. Predictions close the human mind. Hence, Sardar's argument is not that the future can't be predicted or that futurists don't attempt to predict -he believes that futurists do make many predictions—rather he believes that prediction destroys freedom and openness to the future.[106] If an individual firmly and unwaveringly believes in just one definite future, the person has closed his or her mind and stifled his or her imagination. In essence, the person becomes mentally rigid and blind to other avenues and possibilities for the future. Sardar's argument is that futurists, in making predictions, inhibit the minds of others and destroy their imagination and freedom.

Peter Russell makes an analogous argument. Uncertainty about the future has a positive effect. Uncertainty frees us from constraining expectations. If we are certain about the future, our mindsets are rigid; with uncertainty comes increasing flexibility.[107] Interestingly, Russell makes the general prediction, (based on the observable trend that change has been occurring faster and faster) that this accelerative trend will continue into the future, making it increasingly difficult to predict the future. We are approaching an "information" and "prediction horizon," in which the world will be so different from today that it will be next to impossible to understand it from our present mindset. In essence, Russell's prediction is that increasingly in the future we won't be able to predict the future at all.

On the other hand, as noted in the last section, Wendell Bell argues that even if futurists present a range of possibilities, they do engage in predicting the future. Predictions of the future are simply probabilistic rather than absolute. For Bell, predictions can be made with various degrees of certainty and precision, depending upon the complexity of the variables being examined and our scientific understanding of the variables—we can have degrees of **"warranted assertibility."**[108] Even if futurists predict a range of possibilities and their scientific rigor and evidence to warrant these predictions varies, they do anticipate and predict, and these informed predictions help us in thoughtfully directing the future.

Aside from the use of scientific inference based on empirical data, other futurists argue that history provides a foundation for predicting the future. Another well-known futurist, Graham Molitor, thinks that justifiable and accurate general predictions, based on past trends and historical developments, can be made far into the future. For Molitor, the future is part of a "seamless continuum," and "broadly speaking, there is nothing new under the sun."[109] Trend extrapolation is the best method for predicting the future, and according to Molitor, most discontinuities and surprises in the past could have been foreseen. Freeman Dyson also believes that the best way to predict the future is to study history.[110] Even Cornish acknowledges that one can predict the future from the past—in fact, the most reliable prediction for the future is that it "will be like today." Cornish identifies a set of "continuities of past and future," including continuities of existence, pattern, change, and causation, which can serve as a basis for extrapolating from the past to the future.[111]

The historian Robert Nisbet makes the important point that having a sense of the past is necessary for extrapolation into the future. He notes that in contemporary times, due largely to the influences of both Modernism and Postmodernism, the past has been disowned, rejected, and jettisoned. The faith of modernism that the future will be different and better than the past makes history seem either unimportant or negative. The fast pace of change leaves history in the dust—forgotten and abandoned. The Postmodern rejection of the Western ideal of some absolute and singular history of humankind opens the door to multiple versions of the past with no common foundation for interpreting the direction of time. Yet, if we reject the past, we have no way to identify patterns, directions, and continuities in time, and thus no basis for anticipating tomorrow.[112] If we destroy historical consciousness, cutting off one end of our extended temporal consciousness, we destroy the other end, eliminating our sense of the future. History and memory are clearly essential in anticipating and predicting the future.

As Toffler and Toffler state, there are various criticisms regarding the value and validity of predicting the future, including it is against the will of God; the future is governed by chance; predictions take the spontaneity out of life; and, historically, we have been mistaken in many of our predictions. The Tofflers believe though that humans have done a good job of forecasting and controlling the future as a survival and cognitive trait for a long time.[113] The capacity to anticipate the future—to extend our temporal horizons with some degree of accuracy—is an evolutionary development that has been going on throughout the history of humankind.[114]

If the future were simply chance, how could humans make relatively accurate predictions about the immediate future in everyday planning and behavior? In innumerable ways, even without the benefits of science or academic historical research, we correctly anticipate the future. The world around us possesses a great deal of order, continuity, and repeating pattern, as Cornish and Molitor note. It is not simply chaotic and unpredictable; our minds learn and absorb these regularities. Through the accumulation of memories, we develop cognitive schemes for dealing with the future. In normal human consciousness, a sense of the future based on memories of the past seems to be absolutely essential for basic psychological functioning. If the future were entirely due to chance, then our efforts to guide or control our lives would be pointless—a world of chance precludes control as much as a world of absolute determinism.

Futurists who provide possibilities rather than certainties are still constraining or limiting their visions of the future based on their assessment of reality. They do not believe in pure chance, or else why would they identify certain possibilities rather than the potentially infinite set of all possibilities of a universe of chance?

Reality seems to contain both: some degree of predictable order and some degree of openness and indeterminism. Within futurist writings, we find predictions and arguments for the value and validity of predictions, as well as arguments highlighting the openness and possibilities of the future. Sometimes, such as in Bell, we find both perspectives combined—futurists do make predictions but these predictions are of different possibilities or probabilities, rather than singular absolute eventualities. Even futurists who argue that the future can't be predicted, such as Cornish, often do make general predictions, albeit with cautionary notes about the contingency or probabilistic quality of their predictions.

Let us consider in more detail the relative elements of predictability and unpredictability regarding the future. A good place to begin our discussion is with the well known book *The Limits to Growth*, produced and published by the Club of Rome in the 1970s.[115] The book is a prime example of how mathematics, science, and computer simulations can be applied to the study of the future, yielding a variety of quantitative predictions about the future. The methods employed in *The Limits to Growth* were modeled on scientific principles of statistical and experimental research. *The Limits to Growth* was based on Jay Forrester's "World Dynamics" model of global society, economy, and ecology. Within the study, lists of key global variables (e.g., population resources, energy use, and environmental factors) were identified and quantified. Yearly statistics on these variables were collected, the rates of change were computed, and computer simulations were run, predicting future statistics on the different variables. It should be noted that the variables interacted with each other; hence, if industrial production went up, so would pollution, which in turn would affect climate and weather. The study predicted a variety of interaction effects among the different variables. Also within the study, aside from simply making predictions based on present trends, variables were hypothetically manipulated to see, for example, if world population growth were controlled, how that would affect future energy reserves.

The predictions in *The Limits to Growth* were highly pessimistic. Unless population and industrial growth are very quickly brought to zero, it was predicted that the world economy would collapse within a hundred years due to pollution and the exhaustion of essential resources. Although the study employed mathematical and scientific techniques, and its predictions were presented in quantitative form, the credibility and validity of its conclusions were hotly debated. The study has been criticized or questioned by other futurists and scientists and many of its short-term predictions have turned out to be in error.[116] What went wrong? Or should we say "What went right?" since many of the pessimistic predictions have—as of yet—not happened.

The Limits to Growth looks at statistical trends and makes predictions based on the observed trends. As noted, this is a common practice within future studies. Many other futurists also monitor and measure trends as a basis for predictions. This approach, in particular, has also been employed by John Naisbitt in his popular books *Megatrends* and *Megatrends 2000* and by Pearson in his *Atlas of the Future*.[117] Both of Naisbitt's books are full of statistics and facts concerning what variables are increasing or decreasing in economy, government, business, technology, and social life. Naisbitt makes predictions based on these trends and his predictions have had a considerable popular influence.[118]

Trend extrapolation is limited though in validity and applicability for a variety of reasons. First, change is not always linear. The term "linear" means moving in a straight line or single direction—it also means constant as opposed to wavering or fluctuating. Of course, many scientific laws of nature describe regular patterns of change that stay constant across time and place, but not all change is linear or constant. A trend may continue for many years and then either reverse or accelerate. Linear thinking and trend extrapolation assumes a steady rate of change. Natural history and the scientific study of change demonstrate that change is not always smooth and steady but sometimes sudden and dramatic.[119] Change can accelerate and can decelerate. Many phenomena in nature seem to show non-linear change.

Aside from the fact that natural change is not always constant, a second factor that undercuts linear extrapolation is "**wild cards.**" For example, global trends and variables are often significantly affected by individual and local events (e.g., an assassination, natural disaster, or technological discovery). History is filled with unique, unexpected singular occurrences that have global and holistic effects. It is a challenge to predict such individual events and how they will affect the total scheme of things. The futurist John Peterson describes "wild cards" as high impact/low probability events that may trigger consequences that are much more intense and pervasive than the original event—input does not equal output. Peterson thinks that most wild cards are presently being ignored, though there are usually indicators that point to them.[120] In his mind, it is possible, as well as desirable, to attempt to prepare for such events. To whatever degree we can prepare for such surprises, the fact remains that such "wild cards" contradict the simple linear model of change. An unexpected surprise can unsettle the whole system.

Russ Ray believes that humans can do better than chance in predicting wild card type events. In his article, "Catastrophe Futures," Ray contends that nobody can predict, except by random chance, the occurrence of catastrophes.[121] Yet, "futures contracts" are very good predictors of the future since people are investing money in expected prices, and, when money is on the line, people make excellent predictions. **Catastrophe futures** have developed as a new investment item where people invest money in the seasonal probabilities of catastrophes in different regions of the country. When investors collectively anticipate catastrophes, they do better than chance. Yet "catastrophes" are clear examples of "wild card" events that trigger non-linear change. Ray thinks that catastrophe futures could turn out to be the best predictor of catastrophes ever, since the investment of money and collective input seems to bring out the best in us as forecasters of the future.

Still another important factor relevant to the non-linearity of change is creativity and novelty in the future. Richard Fobes argues that we can use creativity to glimpse the future in a way that would be superior to the methods of trend analysis.[122] One simple way to see the connection of creativity and forecasting is to note that in the past new developments came through creative solutions to present problems. A negative trend is slowed, eliminated, or reversed by creative human problem solving. Change in human society is often nonlinear involving creative jumps in human ingenuity and invention. Hence, we could predict that the future will sometimes involve creative solutions to present problems, such as the global problems identified in *The Limits to Growth*.

The creative approach to the future assumes that problems will be solved and negative trends will not continue along the same path. Humans will not just sit idly by while the ship slowly sinks; they will do something, often with creativity.[123] Trend analysis is, therefore, too shortsighted and uncreative. I should note that there is some clear disagreement on this issue—humans do not always come up with ingenious solutions to negative trends, and thus, we should seriously consider the apparent long-term consequences of present trends and not just hope for some creative miracle.[124]

Over the last couple of decades, a new set of principles and techniques has emerged in science for understanding and predicting the nonlinear and fluctuating aspects of change. This new approach goes under different names including **"chaos theory," "open systems theory,"** and **"complexity theory."**[125] As introductory definitions, "chaos" means randomness and the lack of apparent pattern and order; "open systems" refers to the fact that systems in nature are open to each other (rather than closed) and mutually interact and influence each other; and "complexity" means degrees and levels of structure and differentiation within a system—the opposite of simplicity.[126] As it turns out, chaos, openness, and complexity are connected together within nature. The structure of open systems is described in terms of complexity theory. The interaction of open systems generates varying degrees of chaos and can produce non-linear effects and increasing complexity. Many of the ideas of these new scientific perspectives have been applied to future studies, futurist predictions, and computer simulations of the future.[127] These new ideas are helping to broaden our perspective on change and the prediction of the future—they take the prediction of the future beyond a simple linear model.

In his book, *Out of Control,* Kevin Kelly provides a good example of the newer ideas emerging in future studies that derive their inspiration from chaos and open systems thinking.[128] Kelly is particularly critical of the approach taken in *Limits to Growth.* The *Limits to Growth* model, Kelly argues, does not allow for the introduction of new significant variables that would alter the flow of events (e.g., a different energy source or innovative technology or industry). For Kelly, the linear model simply extrapolates on present conditions. Because the model adds nothing new, the world system in this model is not flexible or creative. But Kelly illustrates throughout his book that nature has often been quite creative and inventive in its evolutionary history. Numerous other scientists and philosophers of nature, such as Paul Davies and Ilya Prigogine have made similar arguments. Nature exhibits novelty and transcendence across time—it adds new structures and complexity to its organization.[129] Humans are part of this ongoing creative process. As Fobes has also noted, history demonstrates that the human species has repeatedly created new capacities, modes of behavior, physical structures and instruments, and novel ideas throughout its development.[130]

The model in *The Limits to Growth* does not acknowledge the central significance of learning. It portrays humans as incapable, if the need arises, of altering the direction of change. However, humans do learn and adapt, and they often come up with solutions to challenges and problems. Thus, the linear model of change is fundamentally stupid—it implies that humans will simply walk off the edge of the cliff rather than learn and avoid walking off into the abyss. Humans build bridges. In fact, the ultimate point in *The Limits to Growth* is that humans need to thoughtfully assess present trends and alter the direction of change or face social catastrophe. This is what humans have done throughout history. Humans can and do affect the direction of trends; they learn, adapt, and change.

Kelly also thinks that the linear model fails due to the unpredictable effects of multiple interacting variables. Although efforts were made in *The Limits of Growth* to connect and compute the interaction effects of trends, Kelly notes, following research in open systems theory, that when many variables are interacting, there can be significant fluctuations, chaos, and wild escalations. Similarly, Sally Goerner, who defines linear thinking as the belief that the effect (output) is proportional to the cause (input), points out that due to the complex interaction effects in natural systems, small changes in input can produce disproportional changes in output.[131] Recall Peterson's idea of wild cards. Goerner makes the general point that the combination of non-linear and interactive effects, which applies to most phenomena in nature, renders the linear model of change completely inadequate as a model for predicting the future.

Interactive effects within nature produce the phenomenon of "**unintended consequences.**" Because reality consists of open systems that interact with each other, any single event can produce effects that ripple out through the entire network of nature. Nature is an interactive whole, rather than a set of separate and localized realities. If, for example, some new technological device is introduced into modern society and this new device has a specific effect on human life, there could be a host of other effects produced that were not anticipated. The automobile was going to increase the speed of transportation, which it did; however, it also generated or stimulated the growth of congestion, pollution, suburbs, and a whole subculture of car enthusiasts. Because the effects of any new device or invention permeate through the whole network of human society and nature as well, its effects will not be localized but holistic. How can one predict all the possible consequences of an innovation throughout the whole system?

Steven Gillon, in his article, "Unintended Consequences: Why Our Plans Don't Go According to Plan," discusses some of the reasons why humans seem blind to the various consequences of their actions. He notes that complexity theory does imply a degree of unpredictability in the holistic effects of singular events, but he also points out that the normal human desire to achieve the planned results of our actions (the desire for control) will make us either oblivious or defensive to the possible ways something can go wrong with our plans.[132]

Edward Tenner has written an entire book on unintended consequences in technological innovation, aptly titled *Why Things Bite Back*. Tenner believes it is the human desire to control and subdue nature that causes all the problems. Instead of attempting to live with nature, humans try to dominate it, and nature reacts, so to speak, with a mind of its own. He even uses the expression "revenge effects" in referring to the counter-reactions of reality to our efforts to control it.[133] In using such an expression, Tenner may be anthropomorphizing both nature and technology, but nature clearly does react to our actions and these reactions can be holistic and very difficult to predict.

Gillon, in reviewing Tenner's book, does not think that we should give up in our efforts to improve our lives or control our reality. We should not despair but rather acquire humility. Humanity should be learning through the lessons of complexity theory a more realistic and valid sense of predictability and control.

Yet based on his understanding of interaction effects, Kelly is rather pessimistic regarding how far into the future we can predict. According to Kelly, chaos theory seems to imply that, in the short run, predictions have some level of accuracy, but for nonlinear systems, which include most systems in nature, predictions in the long-run drop to chance. Kelly believes that humans do very poorly with long range predictions. He argues that although there are times when long-term predictions are accurate, almost all long-range predictions are off the mark.[134]

At the other end of the continuum, the futurist Adrian Berry argues that predictions of the future get increasingly accurate as we move into the more distant future.[135] Berry's logic is simple: Whatever we may predict, eventually will happen given enough time. A variation on this idea would be that if in the future all things are possible, then eventually all possibilities are realized. Also, both Molitor and Wright have argued that long-term general trends in human history are not only quite apparent, but provide a basis for making general predictions into the future.[136] We might not be able to anticipate the specifics but the overall flow of history is predictable.

Further, I would add that chaos and open systems thinking actually provide new scientific ideas for increasing the sophistication and accuracy of our predictions. The linear model of change is too simple and consequently often inaccurate. Non-linear concepts enrich our scientific understanding of change and provide a more accurate depiction of reality. Within an open systems perspective of reality there is more fluctuation, novelty, and chaotic behavior, but these phenomena turn out to be real facts of nature that need to be acknowledged and incorporated into a theory of change. We can predict that there will be jumps in complexity, interaction effects, and a host of other types of change that would go unrecognized and unanticipated in linear predictions. We may not get the presumed (though mistaken) degree of accuracy of linear predictions with non-linear thinking, but we get a better (and more valid) idea of the general patterns of change that occur in nature.

To illustrate this point, a strong counterexample to Kelly's position, that is both contemporary and incorporates elements of chaos and open systems thinking, is the work of Theodore Modis, including "Life Cycles: Forecasting the Rise and Fall of Almost Anything," his book *Conquering Uncertainty*, and various publications that can be accessed on the WWW.[137] Modis contends that there is a general pattern to the growth and decline of natural systems. This pattern of the life cycle of systems in nature is nonlinear, but it is regular, showing the form of an **S-curve**. The growth of a system starts slowly, but goes through a process of positive acceleration and reaches a peak rate of growth halfway through its life cycle. Its rate of growth then begins to negatively accelerate and slow down, eventually coming to a halt. According to Modis, chaos within the system is at a maximum early in the system's history and late within its history. The beginning and the end are the times of greatest innovation, mutation, and risk taking. In its maximum growth period, during its middle age, the system is highly conservative, linear, and orderly. Modis, in fact, draws an analogy between the pattern of change in a system and the four seasons—a cyclical phenomenon. Spring is initial slow growth, summer is achieving maximum growth and extension, fall is a conservative streamlining and slowing down, and winter is decay, possible death, or conversely possible transformation into something creatively different.

Modis has applied this model to business companies, commercial products, industries, countries, and ecosystems, and he contends that his model fits all of these different phenomena very well. He proposes that through using this model, we can make more informed and successful decisions about guiding our future. A constant growth rate and proportionality of input and output are unrealistic ideas because most systems are non-linear;[138] they have a life cycle and they eventually fail, but for Modis, there is a pattern to non-linear change and consequently some basic features about the future can be predicted.

Another scientific theory that adds to our understanding of the predictability of nature is **quantum theory**. Quantum theory was developed in the early twentieth century, as a new way to understand the micro-structure of the physical world (atoms and sub-atomic particles and forces). It contradicted

Newton's strict deterministic physics. Within quantum theory, the behavior of sub-atomic reality is probabilistic rather than completely determined. One can predict a range of possibilities for states of sub-atomic particles very precisely, but not definite singular states.[139] Since the sub-atomic realm is the foundation of all physical reality, quantum theory seems to imply that there is an irreducible dimension of probability and uncertainty in the behavior of physical objects. Consequently, futurist predictions may be inherently limited to presenting a range of possibilities (or probabilities) due to the fundamental probabilistic nature of physical reality. Reality is not completely deterministic—the future is a set of forking paths.

The biologist Kenneth Miller contends that the indeterminism within quantum reality implies that the behavior of physical objects at the macro-level, which would include humans and all the familiar objects and systems of our world, contain a degree of indeterminism as well. The effects of indeterminism at the sub-atomic level generate indeterminism at the macro-level. For Miller quantum theory implies that the future is inherently uncertain. Miller carries this idea one step further: because reality is not entirely determined, humans can have freedom of choice. There would be no possibility of freedom in a totally determined world. We saw a similar argument made earlier by Cornish. If there are no real possibilities, there is no freedom.[140]

It is not clear though how indeterminism at the quantum level supports freedom of choice at the human level. Does quantum reality produce indeterminist effects at the human level? If so, wouldn't this produce a degree of chaos in the sequence of our thoughts and the consequences of our behavior, and how would this support freedom of choice? Freedom of choice is not the same thing as chaos.

As a general thesis, Miller emphasizes the element of chance or luck in the processes of nature. For Miller, quantum theory implies an element of chance at the sub-atomic level. But also, following the ideas of the evolutionary biologist Stephen J. Gould, Miller argues that chance has played a significant role at the macro-level in the evolution of life. Miller supports Gould's contention that history is contingent rather than deterministic, and if the history of life were replayed, it would not necessarily come out the same way. Luck or chance has played an important role in determining which species or ecosystems have survived and which have perished.[141] For example, dinosaurs, which were highly adapted and successful life forms, were victims of bad luck when a huge asteroid collided with the earth. This illustration, it should be noted, is an example of Peterson's wild card effect. One piece of rock, though rather large and moving very fast, changed the whole course of life on earth.

As with other theorists who espouse a belief in indeterminism and the uncertainty of the future, Miller does not always consistently follow his professed indeterminist philosophy. At times, he states that the present is a consequence of the past, which is clearly a deterministic viewpoint. Though he acknowledges the role of chance in evolution, he also invokes the Darwinian principle of natural selection—which is a law of nature—to explain the general pattern of increasing biological complexity through time.

As a general point, all futurists and scientists who argue that the future is not predictable will inevitably also present hypotheses and speculations regarding general patterns of change and general directions for the future. No one seems to be a pure indeterminist. No one believes that the universe is totally indeterminate through time. This is quite understandable since a purely chaotic vision of reality would be unintelligible. Everyone sees some degree of order or pattern across time in the universe.

To add some further support to the idea that the future, to some degree, can be predicted, *The Futurist* contains an interesting article on this topic, "What May Happen in the Next Hundred Years" by J. Watkins, reprinted from *The Ladies Home Journal*, December 1900.[142] Although many of Watkins' predictions are off the mark (e.g., he predicted that the letters C, X and Q would disappear from the alphabet), many hit the target. Almost one hundred years ago he predicted that telephones would circle the globe, autos would take the place of horses, kitchen appliances would become electric, photos could be telegraphed around the world, and planes, tanks, and submarines would be used in warfare. Such predictions do not encompass the totality of our present reality—many things have been surprises—but many specific yet important elements of the future were foreseen.

It can be argued that with so many people always trying to predict the future, some are bound to get predictions right just by chance. Yet, there seems to be more involved than simply chance. As noted earlier, H.G. Wells, who thought extensively about the future, made numerous, quite imaginative and accurate predictions. Recall that science fiction writers have made an incredible variety of accurate predictions. The great scientist, artist, and inventor, Leonardo da Vinci, also anticipated a number of modern technological developments. The list could go on. For example, see the predictions of the *Commission on the Year 2000* noted above, and also how Daniel Bell, the editor of the Commission report, in his later book, *The Coming of Post-Industrial Society*, accurately predicted the contemporary growing separation between the technological elite and the lower service class.[143] As Snyder notes, many futurists in the 1960s foretold the present traumatic changes of the Information Age.[144] It seems that people who are informed, think about the future, and possess high levels of imagination, creativity, and intuition are often quite successful in their predictions. Their capacity for foresight is enhanced. A key point to note, following Fobes, is that creativity or the ability to see beyond the constraints of simple linear change increases the power of human prediction.[145] Since the future is creative, it makes sense that thinking creatively will enhance one's foresight.

Although Laura Lee is one writer among many who points out the numerous examples of bad predictions by experts, she does not think predicting the future is a hopeless endeavor. She lists a number of considerations to keep in mind when making predictions. Lee argues that it is better to be bold and risky in one's predictions than cautious. The fear of being wrong inhibits making interesting predictions.[146]

If there are methods and ways to improve the quality of predictions, then not all predictions are equal, and there are ways to evaluate them. The following criteria proposed by the futurist Andy Hines are intended to provide some guidance (no guarantees) in evaluating predictions.[147] I have added comments in *italics* to further elaborate and explain Hines' points. He notes:

- Japanese forecasts tend to have high levels of accuracy because they are normative or prescriptive. The forecasts are actually plans to achieve something and become self-fulfilling prophecies. *This is a fitting example of the idea that the best way to predict the future is to create it. Normative predictions set goals that people attempt to achieve and thus fulfill the prediction.*
- We should check to see if the forecaster has an agenda. Hines says that having an agenda is fine if the writer is up front about it. Hidden agendas tend to work against predictions coming true. *This point notes that futurists have theories and ideologies, which clearly influence their predictions. As Wendell Bell points out, futurists often confuse their wishes for the future with predictions.*
- Methods for making predictions may be formal or informal. If the method is formal, though, it doesn't follow that the predictions are better; science fiction and intuitive hunches can be more accurate than statistical extrapolations. *The question is whether science fiction and other approaches, besides abstract and mathematical reasoning, should be included in the study of the future. Science fiction, intuition, and even mystical visioning may do more than inspire; these approaches may also inform.*

- Experts in an area are not necessarily better than non-experts. *This point highlights the importance of humility and contingency in making predictions.*
- Forecasts have underlying assumptions—technological and social. Forecasts often go wrong if assumptions are not clarified. *Again, it is important to understand and clarify the theoretical framework behind the predictions.*
- Putting specific time lines on forecasts makes them more exact, precise, and thoughtful. *Analytical and detailed thinking is important in futurist predictions.*
- We should ask what the trigger events are in a forecast—those events thought necessary to occur to lead to the forecast. *This point notes the importance of singular events in determining the direction of change.*
- We should ask what is missing in the forecast. Often it is the unique events, wild cards, and discontinuities that are missing. *These are all non-linear aspects to change.*
- Another common oversight is not taking into account the necessary resources for the predicted development. Who is going to pay for the innovation? (Resources can be psychological as well as physical, and the costs of a new development are often emotional and mental, as well as financial.) *Change requires energy and effort.*
- We should ask what the forecast means to us personally. What are the implications? Hines contends that there are not enough of the personal implications presented in forecasts. *Meaning and value are important in thinking about the future. Generally, it is the motivational, emotional, and personal-meaning features associated with a potential change that will drive it to realization. Following Slaughter, we should look at the*

subjective dimension of the future, as well as the objective.

- In evaluating forecasts, two of the most common errors regarding technology are the overestimation of speed of deployment and the underestimation of the magnitude of impact. *The second point highlights how significant technology is in understanding and predicting the future.*
- We shouldn't focus too much on what may be wrong or methodologically unsound in a forecast, but we should look for interesting ideas and possibilities. *This point again reasserts the importance of a balance of logical reasoning and evidence, and creativity and imagination in thinking about the future.*

Given Hines' comments on science fiction, intuition, and informal methods, let us compare future studies with science fiction regarding predicting the future. In his article, "A Funny Thing Happened on My Way to the Future, or The Hazards of Prophecy," W. Warren Wagar argues that instead of attempting to make one set of general statistical predictions on the future, the futurist should attempt to develop various alternative futures that are more concrete and specific in details.[148] The chances of being correct on any one detailed scenario are rather slim (the target has been narrowed), but the futurist creates many different possibilities rather than one. Also, it should be noted that it is the unique and colorful events in history that are often highly significant and interesting; general trends or conditions do not convey a complete picture of everything important. Creating detailed visions rather than general schemes captures this essential element of realism in predictions. Following Wagar's suggestions, science fiction writers who create various detailed and concrete stories of the future are on the right track. For Wagar, the predictive value of science fiction lies in its specificity of details and focus on the unique elements within any future.

A statistical or mathematical study on the future is clearly different in method and content from a science fiction novel on the future, but in both cases, a vision of the future is being created. Both forms of thinking present possible developments in the future, e.g., a generalized projection could be made regarding the future of weapons or warfare, or a science fiction story could be told involving various possible new weapons set in a hypothetical war. The science fiction novel may depend more on imagination, while the scientific study may depend more on analysis and computation. However, this difference at best is a matter of degree.

Both science fiction and future studies attempt to be realistic. Science fiction stories create more of an element of concrete realism. The realism in science fiction derives from the literary realism of detailed and plausible descriptions of characters, actions, and settings. Future studies attempts to present valid predictions and descriptions of the future, but its results are usually not framed within a personalized perspective with individual characters or specific scenes and settings. The resulting predictions are general conditions or facts, although specific examples of the projected future may be used to illustrate the general hypotheses. The realism of future studies derives from arguments, facts, evidence, scientific rigor, and logic. It is the type of realism and empirical validity created in support of a theory or hypothesis in science.

In actual practice and to their mutual benefit, the two forms of thinking and disciplines borrow greatly from each other. Futurists get ideas from science fiction stories and, conversely, science fiction writers get ideas for new stories by reading theories and projections about the future. Further, science fiction writers often do try to convince the reader through scientific or philosophical argument that the future described is plausible or possible. In general, our predictive capacities are enhanced through the combined strengths of scientific methods in future studies and the creative concrete imagining of science fiction. Relating back to the earlier debate on whether science fiction should be included within the domain of future studies, there are clearly some relative differences in approach, but the two approaches have been highly interactive and mutually enhancing activities throughout history. They have a common focus of concern: thinking about the future.

The themes of realism, prediction, and possibility thinking are also connected together when we consider the different cognitive levels of thinking that go into making predictions about the future. Predictions have degrees of cognitive complexity. A prediction can be and often is a simple straightforward extrapolation on some present trend. A prediction may simply identify some future event with a date for its occurrence. Such predictions convey a quality of certainty and provide people some level of security regarding the future. These types of predictions though often do not involve a high level of cognitive functioning. They are linear, single and, generally, not very realistic. Isolating one variable and drawing some straightforward conclusion about the future does not acknowledge the complexity of reality.

As open systems theory argues, reality is a network of interactive variables, and extrapolations into the future should consider the potential interaction effects that could occur among these variables. If one wishes to understand the future, one should study and consider the whole system in which future events will be unfolding. One should attempt to connect and relate different factors, such as both technological and social variables. This type of holistic thinking may not offer simple answers like linear and insulated thinking, but it does reflect a higher level of cognitive functioning. Juggling a host of variables in one's mind and considering different interaction effects is much more complex and challenging to the human mind than linear insulated thinking. As we have seen, futurists often attempt to understand the big picture and the interaction of multiple variables. It also should be noted that science fiction involves the creation of holistic scenarios, where many different aspects of life are considered and integrated together into a realistic and rich story.

When one examines a variety of different factors and their interaction with each other, it is quite understandable that predictions would be probabilistic and multiple, presenting a set of different potential scenarios. This is not so much a failure of futurist thinking as a reflection of the complexity of the reality being considered and the complexity of the thought processes involved in understanding the reality. Each variable in the equation could behave in different ways and the interactive results could vary as well. As we have seen, futurists attempt to think interactively and holistically about the future, and that makes singular and absolutist predictions unrealistic. Within science fiction, a multiplicity of futures is presented as well—one at a time. Different stories of the future are offered, each reflecting a particular perspective regarding how a host of different variables will interact and evolve in the future.

Prediction is clearly connected with understanding. Understanding, in fact, is often judged on the capacity to predict. In identifying natural laws, scientists provide a basis for both making predictions about changes in nature and understanding nature as well. Since laws describe general patterns in nature, they give the world a comprehensible order, and because they are about regularities of change, they allow for the prediction of specific changes in the future. Laws both describe nature and predict its behavior.

When we come to the complex interactions of multiple variables within human society and nature as a whole, the capacity to predict the future becomes probabilistic and conditional. Also, predictions are no longer singular but sets of different possibilities. But futurists, like natural scientists, attempt to understand and describe their subject matter. Futurists attempt to identify patterns of change and draw conclusions regarding the consequences of these patterns. Prediction and understanding are connected in futurist thinking. Understanding the future entails understanding a complex and interactive reality, and predictions invariably involve a range of possibilities because of the relative uncertainty of the effects of complex interactions.

In summary, we have seen that futurists do attempt to predict the future using a range of scientific methods and principles, e.g., trend extrapolation based on statistical and empirical data, historical research, "creativity" and "catastrophe" thinking, open systems and chaos concepts, and theoretical models of change (such as Modis). Also, futurist prediction is connected with futurists' efforts to make sense of the future—the complex array of variables and the general patterns and trends of change identified provide a basis for understanding, explaining, and predicting the future. Predictions of the future can also be made based on intuition and subjective hunches. We have seen that science fiction (futurist narrative) as a predictive tool expands the power and validity of futurist predictions. Science fiction provides multiple complex scenarios filled with specific detail and concrete realism. Some would contend that it is impossible to predict the future with any certainty or accuracy, yet the future clearly can be predicted within various ranges of accuracy and probability and all futurists engage in prediction.[149] Futurist predictions though, even if based on rigorous and informed scientific methods and concepts, are limited to ranges of possibility especially regarding the future of complex systems, such as human society and human technology. Yet for many futurists, this very limitation in predictability opens the door to human influence and control on the future—possibilities mean choices. Let us turn our focus more specifically to the issues of controlling the future and ways prediction and control are related.

The Control and Creation of the Future

"We are charting a land that is being created by the act of discovery... But to keep drawing that chart seems to be our self-appointed destiny."

J. T. Fraser

The activity of prediction (including anticipation and foresight) is intimately connected with other futurist activities and modes of consciousness, including the planning, creating, and controlling of the future. At the most general level, we develop predictions as a way to influence and control the future. Prediction is an effort to understand. We try to understand things better so we can have a greater and more effective influence on them. Knowledge is power. Foresight serves action.

From a psychological perspective, foresight is a perceptual understanding function, and planning and creating are action functions. In normal human psychology, perception and understanding, and planning and action form a complementary psychological pair—each process influences and guides the other. "Knowing that" and "knowing how" (to use a popular psychological expression), are coupled. We are continually guiding our behavior based upon our perception and understanding of what is going on around us, and our anticipations of what is to come. All choices and plans assume some level of foresight and understanding—they are based on beliefs about the effects of our actions and the behavior of the world. In turn, through feedback regarding our actions and interactions with the world, we revise and further develop our perceptual and conceptual knowledge of the world and our predictions about the future. Through feedback on our actions, our foresight grows and we learn to better anticipate the future. When futurists engage in prediction and articulate strategies and actions for the future, they are simply building upon a basic set of interconnected psychological capacities, including anticipation, perception, conceptual understanding, planning, and purposeful behavior.[150]

Clearly, the goals of futuristic thinking go beyond prediction and understanding, but also include the direction and control of the future.[151] This is a clear extension of normal human psychology and future consciousness. All humans attempt to control the future whenever they act purposefully. Humans develop plans based on their understanding and anticipations of the future and use these plans to guide their behavior and influence events.

The control and purposeful direction of the future is one central goal behind thinking about the future and one critical survival feature of future consciousness. Controlling the future is highly beneficial for survival. We need to anticipate change with some level of success if we are to survive, for reality doesn't stay put, and tomorrow is never exactly the same as today.[152]

Humans have always tried to see into tomorrow as a means to direct the future. What humans have been doing throughout history, from divination and revelation to reasoning, statistical extrapolation, computer simulations, and scientific thinking, is simply to build upon an existing adaptive ability in their biological make-up. We are always trying to get better at these abilities: understanding, prediction, and control. We think; we anticipate; we plan; we attempt to guide and direct events.

Specifically, what is the relationship of planning and prediction? Futuristic planning and predicting are not separate activities. With good planning, we take into account and anticipate (a predictive capacity) the challenges ahead of us; basically, a plan assumes an anticipated or predicted scenario in which it will be acted upon. We can anticipate with lesser or greater detail and this influences the level of planning. If we consider a range of possibilities in the future, we create flexible plans to reflect our uncertainty. The point of looking ahead is to narrow down and conceptualize the more probable scenarios as best as possible—to bring some order and focus to the chaos and ambiguity, so as to guide and inform planning and action.

What is the relation between the creation of the future and prediction? There is the hypothesis, mentioned earlier, that "the best way to predict the future is to create it." We can turn our prediction into a self-fulfilling prophecy by attempting to create the very thing we predicted. The reciprocal hypothesis also seems true that "the best way to create the future is to predict it." Since expectations influence not only the person who has the expectations, but also those who listen and believe, predictions can influence the future. (Recall our discussion of how science fiction has influenced human society.) Creative plans assume predictions about what will happen in the future. These reciprocal statements on creation and prediction demonstrate how prediction (a descriptive statement, "What will happen?") and calls to action (an evaluative statement, "What we should do?") are interconnected. Prediction informs and inspires action and creation, and we direct our actions to realize our future visions. This reciprocal relation between creation and prediction is reflective of the general complimentary relationship of understanding and action.

Prediction, planning, and control are fallible yet evolving processes. The above statements on creation and prediction assume a level of understanding and control over the course of events. Many predictions of the future have not come true, and efforts to create a particular future often fail. Obviously, if our predictions are based on inaccurate or incomplete assessments of factual evidence or patterns of change, our efforts to influence and control reality will suffer. Our predictions and efforts to influence the future must be realistic. Yet, our level of understanding is always in a process of trial and error and never complete. We are perpetually experimenting with the future. We cannot assume that somehow we will one day get it right and have it all figured out—that we will achieve omniscience or omnipotence.[153] But, even if our efforts at prediction and creation fail or only partially succeed, we keep trying to improve upon the process; our natural psychological inclination is to influence reality toward our envisioned and desired ends. We are purposeful beings that anticipate and desire. There are various ways in which planning and goal setting can be improved, and more generally, how futurist thinking as a skill can be developed. Humans throughout history have attempted to improve their capacities to predict and direct the future.

The expression "controlling the future" may seem too strong a phrase to describe human efforts to direct the future. "Control" may sound domineering, manipulative, and one-sided. One could argue that all human efforts to control reality involve the contribution, input, and even intrusion of external factors not within our control. At best, humans participate in the creation of the future, rather than pulling the strings from some detached position. Yet, humans clearly attempt to influence, direct, and create effects and results in the world. However we conceptualize and describe the process of control, humans are goal directed in their behavior, and continually work on developing more effective and efficient ways of achieving control over reality (and even themselves). Just as it is psychologically naïve to think that people, and especially futurists, don't engage in prediction since all humans (barring those with significant brain damage) anticipate and have expectations, it is also naïve to think that people shouldn't attempt to control reality since all humans show purposeful behavior directed toward affecting the world. It is just that there are different theories and interpretations of control.

Based on the psychological theory of human-environment reciprocity (or **reciprocal determinism**), I suggest that the most accurate way to describe human control and influence is as follows: Humans and the environment are interactive, each affecting the other. There is a loop of causality between human actions, environmental effects, and human reactions. Even if a series of reciprocal events begins in the environment, humans react and their actions have an effect back on the environment. That is, even in the act of adaptation or adjustment to environmental events, there is some purpose behind it and the action still produces some kind of effect within the environment. Within this context of mutual interaction, humans behave purposely, attempting to, in innumerable ways, manipulate or influence the environment. The results of human efforts will always involve an interaction effect, with both the environment and the human contributing to the effect. If there are two or more people participating in some event, each with their own purposes and goals, the result will be a combined effect of each of their purposeful actions, and subsequent reactions to the actions of the other(s). Control is never simply one way, with a human purposeful action producing an environmental effect; control and influence is always a two-way street, with action, reaction, and further actions and reactions. The bottom line is that because humans are purposeful and their purposes involve the realization of goals in the environment (or themselves), humans are attempting to create effects and alter conditions around them, regardless of whether their efforts fail or are only partially successful.[154]

There is also a strong connection between futurist theory and prediction and control. Depending on how we see and interpret reality, we will create alternative descriptions, make different predictions, and attempt to control reality in different ways. The theoretical framework of a futurist influences both predictions and actions. Some futurists interpret our present reality rather negatively, while others see the present more positively. Some emphasize technology more, and some emphasize humanistic elements. Futurists have different theories of change. These varying perceptions and theories, often highlighting different aspects of our complex and multifaceted contemporary times, create different predictions and attitudes. Depending upon the theory, we focus our attention toward different aspects of reality and become motivated to alternative courses of actions; thus, to some degree, creating differences in how the future unfolds. As the philosophers of science, Paul Feyerabend and Thomas Kuhn, have pointed out, the theoretical concepts in a scientific explanation of nature clearly influence and color the descriptions and predictions that a scientist makes about the world.[155]

Finally, values are a necessary dimension within our efforts to control the future. In thinking about the future, we invariably consider what we value.[156] Different values will lead to different evaluations—positive or negative— regarding the present as well as the anticipated future. We control things in an effort to achieve desirable or preferable ends—purposeful control is value and goal driven. Values define the relative desirability of the different choices and motivate and guide our planning and actions. One great attraction of religious and spiritual thinking on the future is that values are made quite explicit. The future is often seen as realizing or fulfilling some important cosmic or ethical value or values. The future studies concept of preferable futures also embodies the idea of values—everything may be possible, but what is desirable? Upon what prescriptive criteria do we make our choices?

Summary and Conclusion

Within the study of the future there is agreement and disagreement, as well as coherence and fragmentation. From the lists provided in this chapter, one can synthesize a relatively well-defined set of futurist topics of study and research. Based on the work of the Millennium Project, a consensually agreed upon list of major issues and challenges for the future can be identified. From Marien, Bell, Toffler, and other writers, it seems clear that futurists generally deal with the three central questions of the possible, probable, and preferable. In particular, I should note that in spite of statements made by some futurists that future studies doesn't attempt to predict the future, most futurists do engage in prediction, in one form or another. Also, even though many futurists wish to emphasize the scientific quality of the field, future studies deals with values and not just facts. (Bell, in fact, would contend that value judgments can be supported through reason and fact, thus connecting together scientific fact and value.) On a related note, although futurists engage in prediction, they also talk about choices for the future. In general, futurists have both theories of the future (what is going on and why) and ideologies for the future (where we should be heading).

Because futurists have different theories and ideologies of the future, areas of disagreement and fragmentation exist among futurists. Although there are networks of communication among futurists and geographically broad organizations, such as the World Future Society and the World Futures Study Federation that draw futurists together, there is no single unified community of futurists. Futurists tend to cluster around common mindsets and distinguish each other over fundamental differences of opinion and approach. Also, different futurists more narrowly or broadly define the nature of the field, some including literary, mythic, and even spiritual approaches. For example, Slaughter wants to include in futurist thinking and methodology other approaches and perspectives on reality besides science, taking into account inner consciousness and introspective and intuitive techniques. On these differences of opinion, I argue that although it is important to have scientific and rational standards in thinking about the future, the imagining of possible futures is not simply a logical process—it is visionary, intuitive, and creative. Excluding myth, art, narration, introspection, or science fiction as important contributions to the discipline of future studies misses the basic psychological fact that all these modes of consciousness significantly contribute to the imaginative process of visioning possible futures and tap into important dimensions of human experience. One should have standards for assessing beliefs and claims of knowledge, but one should also be open to the richness of the human experience of the future.

All in all, these disagreements can be seen as positive; disagreements reflect active thinking and freedom of thought within a discipline. Future studies is a relatively new area of study. For this reason, it is understandable and valuable for there to be different points of view regarding its nature and purpose. The disagreements drive its further evolution. Following Slaughter, it is best to describe future studies as evolutionary, dynamic, and growing.

References

[1] Wagar, W. Warren "Utopias, Futures, and H.G. Wells' Open Conspiracy" in Didsbury, Howard F. (Ed.) *Frontiers of the 21st Century: Prelude to the New Millennium*. Bethesda, Maryland: World Future Society, 1999; Wagar, W. Warren *H. G. Wells: Traversing Time*. Middletown, CT: Wesleyan University Press, 2004.
[2] Bell, Wendell "A Community of Futurists and the State of the Futures Field" *Futures*, Vol. 34, 2002a, 235 - 247; Bell, Wendell, "Advancing Future Studies: A Reply to Michael Marien" *Futures*, Vol. 34, 2002b, 435 - 447; Marien, Michael "Future Studies in the 21st Century: A Reality-Based View" *Futures*, Vol. 34, 2002a, 261 - 281; Marien, Michael "Rejoinder: My Differences with Wendell Bell" *Futures*, Vol.34, 2002b, 449 - 456.
[3] Bell, Wendell *Foundations of Future Studies: Human Science for a New Era*. Vol. I New Brunswick: Transactions Publishers, 1997.
[4] Anne Arundel Community College - Institute for the Future - http://www.aacc.edu/future/; Anticipating the Future - Future Studies Course - University of Arizona - http://ag.arizona.edu/futures/.
[5] Future Studies Masters Program - prtl.uhcl.edu/portal/page?_pageid=352,241210&_dad=portal&_schema=PORTALP.
[6] Hawaii Research Center for Future Studies - http://www.futures.hawaii.edu/.
[7] Australian Foresight Institute - http://www.swin.edu.au/afi/
[8] Tamkang University - Center for Futures Studies - http://foreign.tku.edu.tw/english/ctn11_TriObj.htm.
[9] World Future Society Home Page - http://www.wfs.org/wfs/.
[10] Didsbury, Howard F. (Ed.) *Communications and the Future: Prospects, Promises, and Problems*. Bethesda, Maryland: World Future Society, 1982; Didsbury, Howard F. (Ed.) *The Global Economy: Today, Tomorrow, and the Transition*. Bethesda, Maryland: World Society, 1985; Didsbury, Howard F. (Ed.) *Careers Tomorrow: The Outlook for Work in a Changing World*. Bethesda, Maryland: World Future Society, 1988; Didsbury, Howard F. (Ed.) *Frontiers of the 21st Century: Prelude to the New Millennium*. Bethesda, Maryland: World Future Society, 1999a; Didsbury, Howard F. (Ed.) *21st Century Opportunities and Challenges: An Age of Destruction or an Age of Transformation*. Bethesda, Maryland: World Future Society, 2003; Didsbury, Howard F. (Ed.) *Thinking Creatively in Turbulent Times*. Bethesda, Maryland: World Future Society, 2004; Wagner, C. (Ed.) *Foresight, Innovation, and Strategy*. World Future Society: Bethesda, Maryland, 2005.
[11] World Futures Study Federation - http://www.wfsf.org/.
[12] The Arlington Institute - Future Edition Newsletter - http://www.arlingtoninstitute.org/; Acceleration Studies Foundation - http://accelerating.org; Copenhagen Institute for Future Studies - http://

www.cifs.dk/en/omcifs.asp; Evolve - http://www.evolve.org/pub/doc/index2.html; Foundation for the Future - http://www.futurefoundation.com; Flynn, Elizabeth (Ed.) Center for Human Evolution: The Evolution of Human Intelligence. Bellevue, WA: Foundation for the Future, 2000; Transhumanist Resources and Alliance - http://www.aleph.se/Trans/index-2.html; The World Transhumanist Association - http://transhumanism.org/index.php/WTA/; The Extropy Institute - http://www.extropy.org/.

[13] The Odyssey of the Future - http://www.theodysseyofthefuture.net/web_bib.htm

[14] Bell, Wendell, Vol. I, 1997.

[15] Cornish, Edward *Futuring: The Exploration of the Future*. Bethesda, Maryland: World Future Society, 2004.

[16] Cornish, Edward, 2004.

[17] Bell, Wendell, Vol. I, 1997.

[18] Marien, Michael, 2002a.

[19] Slaughter, Richard (Ed.) *The Knowledge Base of Future Studies*. Volume II. Hawthorn, Victoria, Australia: DDM Media Group, 1996; Bell, Wendell, Vol. I, 1997; Cornish, Ed, 2004.

[20] Slaughter, Richard "The Knowledge Base of Futures Studies as an Evolving Process" *Futures*, Vol.28, No. 9, 1996b; Slaughter, Richard "Futures Studies as an Intellectual and Applied Discipline" *American Behavioral Scientist*, Vol.42, No.3, Nov./Dec., 1998; Foresight International - http://www.foresightinternational.com.au/menulinks/presentwork.php and http://www.foresightinternational.com.au/menulinks/earlywork.php .

[21] Slaughter, Richard "Futures Concepts" in Slaughter, Richard (Ed.) *The Knowledge Base of Future Studies*. Volume I. Hawthorn, Victoria, Australia: DDM Media Group, 1996c.

[22] Slaughter, Richard (Ed.) *The Knowledge Base of Future Studies*. Volumes I, II, and III. Hawthorn, Victoria, Australia: DDM Media Group, 1996.

[23] Kurian, George Thomas, and Molitor, Graham T.T. (Ed.) *Encyclopedia of the Future*. New York: Simon and Schuster Macmillan, 1996.

[24] Wagar, W. Warren "Futurism" in Kurian, George Thomas, and Molitor, Graham T.T. (Ed.) *Encyclopedia of the Future*. New York: Simon and Schuster Macmillan, 1996.

[25] See also Clarke, I.F. "Twentieth Century Futures Thinking: From Amateurs to Experts" in Slaughter, Richard (Ed.) *The Knowledge Base of Future Studies*. Volume I. Hawthorn, Victoria, Australia: DDM Media Group, 1996 for a discussion of Wells and early futurist thinking.

[26] Wagar, W. Warren, 1999.

[27] Lombardo, Thomas "Enlightenment and the Theory of Secular Progress" in Lombardo, Thomas *The Evolution of Future Consciousness*. Bloomington, Indiana: Author House, 2006.

[28] Toffler, Alvin, and Toffler, Heidi "Foreward: Five Billion Futurists" in Kurian, George Thomas, and Molitor, Graham T.T. (Ed.) *Encyclopedia of the Future*. New York: Simon and Schuster Macmillan, 1996.
[29] Bell, Wendell, Vol. I, 1997; Wilson, E.O. "Back from Chaos" *The Atlantic Monthly*, March, 1998.
[30] Bell, Wendell, Vol. II, 1997.
[31] Bell, Wendell, Vol. I, 1997.
[32] McGann, James "How Think Tanks Are Coping With the Future" *The Futurist*, November-December, 2000.
[33] Lombardo, Thomas "Enlightenment and the Theory of Secular Progress" in Lombardo, Thomas *The Evolution of Future Consciousness*. Bloomington, Indiana: Author House, 2006.
[34] Cornish, Ed, 2004, Chapters Thirteen and Fourteen.
[35] Bell, Daniel (Ed.) *Toward the Year 2000: Work in Progress*. Boston: Houghton-Mifflin, 1968.
[36] Kurian, George Thomas, and Molitor, Graham T.T. (Ed.) *Encyclopedia of the Future*. New York: Simon and Schuster Macmillan, 1996.
[37] Kahn, Hermann *On Thermonuclear War*. Princeton: Princeton University Press, 1960.
[38] Jouvenel, Bertrand de *The Art of Conjecture*. New York: Basic Books, 1964.
[39] Meadows, Dennis, Meadows, Donella, et al. *The Limits to Growth*. New York: Universe Books, 1972.
[40] Polak, Frederik *The Image of the Future*. Abridged Edition by Elise Boulding. Amsterdam: Elsevier Scientific Publishing Company, 1973.
[41] Toffler, Alvin *Future Shock*. New York: Bantam, 1971; Dahle, Kjell "55 Key Works: A Guide to Futures Literature", in Slaughter, Richard (Ed.) *The Knowledge Base of Future Studies*. Volume I. Hawthorn, Victoria, Australia: DDM Media Group, 1996; Kurian, George Thomas "One Hundred Most Influential Futurist Books" in Kurian, George Thomas, and Molitor, Graham T.T. (Ed.) *Encyclopedia of the Future*. Vol. II. New York: Simon and Schuster Macmillan, 1996.
[42] Marien, Michael "Best Books on the Future, 1996-2000: Future Survey's Super 70" *The Futurist*, May-June, 2001.
[43] Anderson, Walter Truett *Evolution Isn't What It Used To Be: The Augmented Animal and the Whole Wired World*. New York: W. H. Freeman and Company, 1996.
[44] Best Books 2005: The Future Survey "Top 30" - http://www.wfs.org/fstop30bks05.htm .
[45] Kurian, George Thomas, and Molitor, Graham T.T. , Vol. II, 1996.
[46] Marien, Michael "Future Studies" in Kurian, George Thomas, and Molitor, Graham T.T. (Ed.) *Encyclopedia of the Future*. Simon and Schuster Macmillan, 1996.
[47] Marien presents similar arguments in his later articles and debate with Wendell Bell. See Marien, Michael, 2002a and Marien, Michael, 2002b.

[48] Marien presents a slightly modified grouping of topics in his later article, 2002a. The only substantive difference is the inclusion of "religion" (which is grouped together with society) in the later list.
[49]Millennium Project - http://www.acunu.org/millennium/index.html; Millennium Project Opportunities - http://www.acunu.org/millennium/isandop.html.
[50] Millennium Project - Issues - http://www.acunu.org/millennium/isandop.html.
[51] Millennium Project - Challenges - http://www.acunu.org/millennium/challeng.html ; Glenn, Jerome and Gordon, Theodore *2004 State of the Future*. American Council for the United Nations University, 2004.
[52] As noted above, religion is included in Marien's 2002a list and he has included religion as a category in his *Future Survey* reviews. See also The World Network of Religious Futurists - http://www.wnrf.org/.
[53] Toffler, Alvin, and Toffler, Heidi, 1996.
[54] Lombardo, Thomas "The Value of Future Consciousness" in *Foresight, Innovation, and Strategy*. Wagner, C. (Ed.) World Future Society: Bethesda, Maryland, 2005; Lombardo, Thomas "Thinking Ahead: The Value of Future Consciousness", *The Futurist*, January-February, 2006; Lombardo, Thomas "The Psychology and Value of Future Consciousness" in Lombardo, Thomas *The Evolution of Future Consciousness*. Bloomington, Indiana: Author House, 2006; Reading, Anthony *Hope and Despair: How Perceptions of the Future Shape Human Behavior.* Baltimore, Maryland: The John Hopkins University Press, 2004.
[55] Lombardo, Thomas "The Origins of Future Consciousness" and "Ancient Myth, Religion, and Philosophy" in Lombardo, Thomas *The Evolution of Future Consciousness*. Bloomington, Indiana: Author House, 2006; Polak, Frederik, 1973.
[56] Bell, Wendell, Vol. I, 1997.
[57] Evolve - http://www.evolve.org/pub/doc/index2.html
[58] Cornish, Ed, 2004, Chapter Twelve.
[59] Pohl, Frederick "Thinking About the Future" *The Futurist*, September-October, 1996.
[60] Wagar, W. Warren, 1996.
[61] See Bell, Wendell, Vol. I, 1997 for detailed descriptions of these methodologies.
[62] Marien, Michael, 2002a.
[63] World Future Society. *The Art of Forecasting: A Brief Introduction to Thinking about the Future*. 1993.
[64] Cornish, Edward "Futurists" in Kurian, George Thomas, and Molitor, Graham T.T. (Ed.) *Encyclopedia of the Future*. New York: Simon and Schuster Macmillan, 1996.
[65] Cornish, Ed, 2004, Pages 78 - 79.

[66] Bell, Daniel "Introduction: Reflections at the End of an Age" in Kurian, George Thomas, and Molitor, Graham T.T. (Ed.) *Encyclopedia of the Future*. New York: Simon and Schuster Macmillan, 1996.
[67] Nisbet, Robert *History of the Idea of Progress*. New Brunswick: Transaction Publishers, 1994.
[68] Bell, Wendell, 2002a; Bell, Wendell, 2002b; Marien, Michael, 2002a; Marien, Michael, 2002b.
[69] Slaughter, Richard, 1996b; Slaughter, Richard, 1996c; Slaughter, Richard, 1998; Slaughter, Richard "Futures Beyond Dystopia" *Futures*, Vol.30, No.10, 1998b; Slaughter, Richard, "Integral Futures - A New Model for Futures Enquiry and Practice" in Foresight International- http://foresightinternational.com.au/catalogue/resources/Integral_Futures.pdf .
[70] Slaughter, Richard, 1996c.
[71] Slaughter, Richard, "Integral Futures - A New Model for Futures Enquiry and Practice" in Foresight International- http://foresightinternational.com.au/catalogue/resources/Integral_Futures.pdf .
[72] Wilber, Ken *Sex, Ecology, and Spirit: The Spirit of Evolution*. Boston: Shambhala, 1995; Wilber, Ken *A Brief History of Everything*. Boston: Shambhala, 1996.
[73] Slaughter, Richard, "Integral Futures - A New Model for Futures Enquiry and Practice" in Foresight International- http://foresightinternational.com.au/catalogue/resources/Integral_Futures.pdf .
[74] Slaughter, Richard, 1998c.
[75] Slaughter, Richard, "Integral Futures - A New Model for Futures Enquiry and Practice" in Foresight International- http://foresightinternational.com.au/catalogue/resources/Integral_Futures.pdf .
[76] Slaughter, Richard, 1998c.
[77]Voros, Joseph "Integral Future Studies: A Brief Outline" - Australian Foresight Institute - http://www.swin.edu.au/agse/courses/foresight/integral_futures.htm .
[78]Bertman, Stephen *Hyperculture: The Human Cost of Speed*. Westport, Connecticut: Praeger, 1998; Bertman, Stephen "Cultural Amnesia: A Threat to Our Future", *The Futurist*, January-February, 2001; Didsbury, Howard F. "The Death of the Future in a Hedonistic Society" in Didbury, Howard F. (Ed.) *Frontiers of the 21st Century: Prelude to the New Millennium*. Bethesda, Maryland: World Future Society, 1999 (b).
[79] Wagar, W. Warren "Past and Future" *American Behavioral Scientist*, Vol. 42, No.3, November-December, 1998.
[80] Lombardo, Thomas "Ancient Myth, Religion, and Philosophy" and "Enlightenment and the Theory of Secular Progress" in Lombardo, Thomas *The Evolution of Future Consciousness*. Bloomington, Indiana: Author House, 2006.
[81] Hines, Andy "A Checklist for Evaluating Forecasts" *The Futurist*, November-December, 1995.

[82] Toffler, Alvin, 1971; Toffler, Alvin *The Third Wave*. New York: Bantam, 1980; Toffler, Alvin *Power Shift: Knowledge, Wealth, and Violence at the Edge of the Twenty-First Century*. New York: Bantam, 1990; See also the following links on Alvin Toffler's futurist theory: http://www.wired.com/wired/archive/people/alvin_toffler/; http://www.skypoint.com/members/mfinley/toffler.htm; http://www.toffler.com/default.shtml; http://www.mfinley.com/list-toffler.htm.
[83] Toffler, Alvin, and Toffler, Heidi *War and Anti-War: Survival at the Dawn of the 21st Century*. Boston: Little, Brown, and Company, 1993; Toffler, Alvin, and Toffler, Heidi *Creating a New Civilization: The Politics of the Third Wave*. Atlanta: Turner Publishing, Inc., 1994; Toffler, Alvin, and Toffler, Heidi "Getting Set for the Coming Millennium" *The Futurist*, March/April, 1995.
[84] Meadows, Dennis, Meadows, Donella, et al., 1972; Meadows, Dennis, Meadows, Donella, and Randers, Jorgen *Beyond the Limits*. Toronto: McClelland & Stewart, 1992;Naisbitt, John *Megatrends: Ten New Directions Transforming Our Lives*. New York: Warner, 1982; Naisbitt, John and Aburdene, Patricia *Megatrends 2000*. New York: Avon Books, 1990; Moore, Stephen and Simon, Julian *It's Getting Better All the Time: 100 Greatest Trends of the Last 100 Years*. Washington, D.C.: Cato Institute, 2000.
[85] Lorie, Peter and Murray-Clark, Sidd, *History of the Future: A Chronology*. New York: Doubleday, 1989; Dixon, Dougall *Man After Man: An Anthropology of the Future*. New York: St. Martin's Press, 1990.
[86] Kurian, George Thomas "One Hundred Most Influential Futurist Books" in Kurian, George Thomas, and Molitor, Graham T.T. (Ed.) *Encyclopedia of the Future*. New York: Simon and Schuster Macmillan, 1996.
[87] Wired News - http://www.hotwired.com/
[88] Pearson, Ian (Ed.) *The Macmillan Atlas of the Future*. New York: Macmillan, 1998.
[89] 3-D Animation - The Mind's Eye and Other Videos - http://www.animationtrip.com/store/productlist.php?category=thematic
[90] Cybertown - http://www.cybertown.com/.
[91] Lombardo, Thomas "The Origins of Future Consciousness" in Lombardo, Thomas *The Evolution of Future Consciousness*. Bloomington, Indiana: Author House, 2006.
[92] Shlain, Leonard *The Alphabet Versus the Goddess: The Conflict Between Word and Image*. New York: Penguin Arkana, 1998, Chapter Thirty-Five.
[93] Bell, Wendell, Vol. I, 1997.
[94] Bell, Wendell, Vol. II, 1997.
[95] Wilson, E.O. *Consilience: The Unity of Knowledge*. New York: Alfred A. Knopf, 1998b.

[96] Lombardo, Thomas "The Psychology and Value of Future Consciousness" in Lombardo, Thomas *The Evolution of Future Consciousness*. Bloomington, Indiana: Author House, 2006.
[97] Bell, Wendell "Making People Responsible: The Possible, the Probable, and the Preferable", *American Behavioral Scientist*, Vol. 42, No.3, November-December, 1998.
[98] Lombardo, Thomas "Romanticism" in Lombardo, Thomas *The Evolution of Future Consciousness*. Bloomington, Indiana: Author House, 2006.
[99] Russell, Peter *The White Hole in Time: Our Future Evolution and the Meaning of Now*. New York: HarperCollins, 1992.
[100] Anderson, Walter Truett *All Connected Now: Life in the First Global Civilization*. Boulder; Westview Press, 2001.
[101] Lee, Laura "Forecasts That Missed By a Mile" *The Futurist*, September-October, 2000; May, Graham "After the Party is Over: Futures Studies and the Millennium" in Didsbury, Howard F. (Ed.) *Frontiers of the 21st Century: Prelude to the New Millennium*. World Future Society, 1999.
[102] Pohl, Frederick, 1996.
[103] Wagar, W. Warren, 1998; Bell, Wendell, Vol. I, 1997.
[104] Bishop, Peter "Thinking Like a Futurist: 15 Questions to Stretch Your Mind" *The Futurist*, June-July, 1998.
[105] Cornish, Edward *The 1990s and Beyond*. Bethesda, Maryland: World Future Society, 1990; Cornish, Ed, 2004.
[106] Sardar, Ziauddin "The Problem of Future Studies" in Sardar, Ziauddin (Ed.) *Rescuing All Our Futures: The Future of Future Studies*. Praeger, 1999.
[107] Russell, Peter, 1992.
[108] Bell, Wendell, Vol. I, 1997.
[109] Molitor, Graham T.T. "Millennial Perspectives: Strengths and Limitations of Long Term Forecasts" *World Future Society Conference*, 1998a; Molitor, Graham T.T. "Trends and Forecasts for the Next Millennium" *The Futurist*, August-September, 1998(b); Molitor, Graham T.T. "The Next 1000 Years: The "Big Five" Engines of Economic Growth" in Didbury, Howard F. (Ed.) *Frontiers of the 21st Century: Prelude to the New Millennium*. World Future Society, 1999.
[110] Dyson, Freeman *Imagined Worlds*. Cambridge, MS: Harvard University Press, 1997.
[111] Cornish, Edward "How We Can Anticipate the Future" *The Futurist*, July-August, 2001; Cornish, Edward, 2004, Chapter Eleven.
[112] Nisbet, Robert, 1994.
[113] Toffler, Alvin and Toffler, Heidi, 1996.
[114] Lombardo, Thomas "The Origins of Future Consciousness" in Lombardo, Thomas *The Evolution of Future Consciousness*. Bloomington, Indiana: Author House, 2006.
[115] Meadows, Dennis, Meadows, Donella, et al., 1972.
[116] Moore, Stephen and Simon, Julian, 2000.

[117] Naisbitt, John, 1982; Naisbitt, John and Aburdene, Patricia, 1990; Pearson, Ian, 1998.
[118] See also Barrett, David B. "Global Statistics" in Kurian, George Thomas, and Molitor, Graham T.T. (Ed.) *Encyclopedia of the Future*. New York: Simon and Schuster Macmillan, 1996 for an introductory survey of statistical sources and predictions on a host of variables from population and families to health, industrial production, and crime.
[119] Eldredge, Niles and Gould, Stephen "Punctuated Equilibria: An Alternative to Phyletic Gradualism" in Schopf, T. J. M. (Ed.) *Models in Paleobiology*. Freeman Cooper, 1972; Goerner, Sally *Chaos and the Evolving Ecological Universe*. Luxembourg: Gordon and Breach, 1994.
[120] Peterson, John "The Wild Cards in Our Future: Preparing for the Improbable" *The Futurist*, July-August, 1997; Cornish, Ed, 2004, Chapter Nine.
[121] Ray, Russ "Catastrophe Futures: Learning to Predict Natural Disasters" *The Futurist*, November-December, 1995.
[122] Fobes, Richard "Creative Problem Solving: A Way to Forecast and Create a Better Future" *The Futurist*, January-February, 1996.
[123] Zey, Michael G. *Seizing the Future: How the Coming Revolution in Science, Technology, and Industry Will Expand the Frontiers of Human Potential and Reshape the Planet*. Simon and Schuster, 1994; Zey, Michael G. "The Macroindustrial Era: The New Age of Abundance and Prosperity", *The Futurist*, March-April, 1997.
[124] Slaughter, Richard, 1996.
[125] Gleick, James *Chaos*. New York: Viking, 1987; Davies, Paul *The Cosmic Blueprint: New Discoveries in Nature's Creative Ability to Order the Universe*. New York: Simon and Schuster, 1988; Pagels, Heinz *The Dreams of Reason: The Computer and the Rise of the Sciences of Complexity*. New York: Simon and Schuster, 1988; Briggs, John *Fractals: The Patterns of Chaos*. New York: Simon and Schuster, 1992; Goerner, Sally, 1994; Goerner, Sally *After the Clockwork Universe: The Emerging Science and Culture of Integral Society*. Norwich, Great Britain: Floris Books, 1999; Santa Fe Institute - http://www.santafe.edu/.
[126] Gell-Mann, Murray *The Quark and the Jaguar: Adventures in the Simple and the Complex*. New York: W.H. Freeman and Company, 1994.
[127] Henderson, Hazel *Paradigms in Progress: Life Beyond Economics*. Berrett-Koehler Publishers, 1991; Henderson, Hazel *Building a Win-Win World: Life Beyond Global Economic Warfare*. Berrett-Koehler Publishers, 1996; Hubbard, Barbara Marx *Conscious Evolution: Awakening the Power of Our Social Potential*. Novato, CA: New World Library, 1998; Eisler, Riane "A Multilinear Theory of Cultural Evolution: Genes, Culture, and Technology" in Loye, David (Ed.) *The Great Adventure: Toward a Fully Human Theory of Evolution*. Albany, New York: State University of New York Press, 2004; Goerner, Sally "Creativity, Consciousness, and the Building of an Integral Society" in Loye, David (Ed.) *The Great Adventure:*

Toward a Fully Human Theory of Evolution. Albany, New York: State University of New York Press, 2004.
[128] Kelly, Kevin *Out of Control: The Rise of Neo-Biological Civilization*. Reading, MA: Addison - Wesley, 1994.
[129] Davies, Paul, 1988; Prigogine, Ilya *The End of Certainty: Time, Chaos, and the New Laws of Nature*. New York: The Free Press, 1997.
[130] Postman, Neil *Technopoly: The Surrender of Culture to Technology*. New York: Vintage Books, 1992.
[131] Goerner, Sally, 1994.
[132] Gillon, Steven "Unintended Consequences: Why Our Plans Don't Go According to Plan" *The Futurist*, March-April, 2001.
[133] Tenner, Edward *Why Things Bite Back: Technology and the Revenge of Unintended Consequences*. Vintage Books, 1996.
[134] Kelly, Kevin, 1994.
[135] Berry, Adrian *The Next 500 Years: Life in the Coming Millennium*. W. H. Freeman and Co., 1996.
[136] Molitor, Graham T.T., 1998a; Molitor, Graham T.T., 1998b; Molitor, Graham T.T., 1999; Wright, Robert *Nonzero: The Logic of Human Destiny*. New York: Pantheon Books, 2000.
[137] Modis, Theodore "Life Cycles: Forecasting the Rise and Fall of Almost Anything" *The Futurist*, September/October 1994; Modis, Theodore *Conquering Uncertainty: Understanding Corporate Cycles and Positioning Your Company to Survive the Changing Environment*. McGraw-Hill, 1998; Theodore Modis - Forecasting the Growth of Complexity and Change - http://ourworld.compuserve.com/homepages/tmodis/TedWEB.htm
[138] Goerner, Sally, 1994.
[139] Pagels, Heinz *The Cosmic Code: Quantum Physics as the Language of Nature*. Bantam, 1982; Fraser, J. T. *Time, the Familiar Stranger*. Redmond, Washington: Tempus, 1987; Gell-Mann, Murray, 1994.
[140] Miller, Kenneth *Finding Darwin's God: A Scientist's Search for Common Ground between God and Evolution*. New York: Perennial, 1999.
[141] Gould, Stephen Jay *Wonderful Life: The Burgess Shale and the Nature of History*. W. W. Norton, 1989.
[142] Watkins, John Elfreth "What May Happen in the Next Hundred Years" *The Futurist*, October, 1982.
[143] Bell, Daniel, 1968; Bell, Daniel *The Coming of Post-Industrial Society*. New York: Basic Books, 1973.
[144] Snyder, David Pearce "The Revolution in the Workplace: What's Happening to Our Jobs?" *The Futurist*, March-April, 1996.
[145] Fobes, Richard, 1996.
[146] Lee, Laura, 2000.
[147] Hines, Andy, 1995.
[148] Wagar, W. Warren "A Funny Thing Happened on My Way to the Future, or The Hazards of Prophecy" *The Futurist*, May-June, 1994.
[149] Bell, Wendell, Vol. I, 1997.

[150] Lombardo, Thomas "The Psychology and Value of Future Consciousness" in Lombardo, Thomas *The Evolution of Future Consciousness*. Bloomington, Indiana: Author House, 2006.
[151] Bell, Wendell, Vol. I, 1997.
[152] Toffler, Alvin and Toffler, Heidi, 1996.
[153] Postrel, Virginia *The Future and Its Enemies: The Growing Conflict Over Creativity, Enterprise, and Progress*. New York: Touchstone, 1999.
[154] Lombardo, Thomas "The Psychology and Value of Future Consciousness" in Lombardo, Thomas *The Evolution of Future Consciousness*. Bloomington, Indiana: Author House, 2006; Lombardo, Thomas *The Reciprocity of Perceiver and Environment: The Evolution of James J. Gibson's Ecological Psychology*. Hillsdale, NJ: Lawrence Erlbaum Associates, 1987; Hergenhahn, B.R. and Olson, Matthew *An Introduction to Theories of Personality.* 6th Edition. Upper Saddle River, NJ: Prentice Hall, 2003, Chapter Eleven.
[155] Feyerabend, Paul "Problems of Empiricism" in Robert Colodny (Ed.) *Beyond the Edge of Certainty*. Englewood Cliffs, N.J.: Prentice-Hall, 1965; Feyerabend, Paul "Problems of Empiricism II" in Robert Colodny (Ed.) *The Nature and Function of Scientific Theory*. London: University of Pittsburgh Press, 1969; Kuhn, Thomas *The Structure of Scientific Revolutions*. Chicago: University of Chicago Press, 1962.
[156] Bell, Wendell, Vol. I, 1997.

Chapter Three

Modern Times and the Contemporary Transformation

"It was the best of times, it was the worst of times, it was the age of wisdom, it was the age of foolishness, it was the epoch of belief, it was the epoch of incredulity, it was the season of light, it was the season of darkness, it was the spring of hope, it was the winter of despair, we had everything before us, we had nothing before us, we were all going direct to Heaven, we were all going direct the other way . . ."

Charles Dickens

"May you live in interesting times."

Confucius

Trends and Developments in the Twentieth Century

In this third chapter, I begin by describing the major events and trends as well as some of the central issues and problems that have emerged in the twentieth century. I highlight visions of the future and the impact, for better or worse, such visions have had on recent human history. Second, I examine the popular idea that we are in the midst of a significant contemporary world transformation in which humanity as a whole is moving from one way of life into a vastly different way of life. These two main sections of the chapter are connected, for the events and trends of the last century produced the conditions of our present world and set the stage for the contemporary transformation. Through looking at the immediate past, we gain an understanding and developmental perspective on the present and some sense of the direction in which we are heading in the future.

The twentieth century is often seen as the most eventful and unsettling era in human history.[1] The changes that have occurred, great and small, could be seen as a progressive continuation of developments of the previous centuries.[2] Alternately, the last century could be seen as a prelude or transition to something fundamentally different in human life in the centuries that lie ahead.[3] The answer is probably a combination of the two - of continuation and transformation. What was the twentieth century like? Where do we find ourselves upon entering the new Millennium?

In the West, the twentieth century began with a spirit of optimism and a general conviction in the promise of secular progress. The continuing growth of science, technology, and industry and an ever expanding economy stimulated innovation and invention in many spheres of human life. As an expression of humanity's intelligence, energy, and self-confidence, and based on new developments in architecture and engineering, skyscrapers in cities like New York and Chicago rose progressively higher in the early decades of the century, symbolizing humanity's high aspirations. Beginning with the telephone and automobile which emerged at the end of the previous century, the new century saw the creation of city-wide electrical lighting, the airplane, the movie, the phonograph player, plastics, and the assembly line, the last invention promising to bring the bounties of the modern world to the common person. Science and technology were fulfilling the promise of the Enlightenment and transforming the world.

Yet the very inventiveness and creativity of the early twentieth century produced a variety of disconcerting discoveries, new ideas, and innovations that would challenge the simple vision of progressive cumulative growth and progress. The development of relativity theory and quantum physics undercut the presumed certainty of Newtonian physics. Early twentieth-century art and music, through the great works of Stravinsky, Schoenberg, Picasso, Kandinsky, and others, transcended, if not contradicted, traditional aesthetic principles and standards of beauty. Contrary to the Enlightenment belief that reason could and should direct human destiny, at the beginning of the century Freud proposed that instinct and desire ruled the human soul and reason was, at best, a servant to the primitive unconscious. And finally, the philosophy of relativism, challenging the absolutism and sense of supremacy of Western civilization, became increasingly popular in anthropology, history, and cultural studies. As Watson titles the beginning of his intellectual history of the twentieth century, it was a time of "disturbing the peace."[4]

Moving into the next few decades, events and new ideas of the period 1915 through 1945 clearly disrupted and undermined the early optimism of the century. World War I, fought with the new creations of technology, was the most destructive and geographically expansive war ever fought. Clearly, humanity had realized great technological progress, but the new machines and technological systems were not used to improve the human condition, but rather to inflict death and destruction upon fellow humans. In the aftermath of World War I, belief in the inevitability of progress through reason, science, and technology declined. How could the enlightened and presumably superior European mind commit such atrocities of violence and destruction, unless there was something wrong or missing within the modernist image? Or perhaps Freud and other similar thinkers were right – reason and high ethical values did not rule the human soul - primordial instincts and passions controlled the psyche.

Coincident with the end of the war, Oswald Spengler published his famous prophetic book, *The Decline of the West*. Numerous other writers within the humanities, philosophy, and literature also found fault with modern Western society. Many believed, contrary to the image of progress, that humankind was degenerating. Popular critical works appeared which argued that liberal society, industrial capitalism, and bourgeois individualism were destroying a sense of shared culture and producing a world of alienation and excessive subjectivity. The materialist world of acquiring more and more possessions, of increasing consumption and production, was destroying the culture of knowledge. T. S. Eliot saw a spiritual and sexual sterility in the modern world and Kafka wrote of a world where the individual had lost control over his or her life. Once again, as in the nineteenth century with the critiques of romanticism, debate and conflict arose over the supremacy of reason.[5] As the sciences continued to progress and one intellectual stream of philosophy, Logical Positivism, argued for the central importance of reason and evidence, anti-rationalist movements also arose, such as the existential philosophy of Heidegger and surrealist art, the latter delving into the visionary symbolism of the unconscious.[6]

One of the strongest and most visible reactions against Western capitalism was the emergence of Communism and socialism in Russia. Inspired by the writings of Karl Marx[7] and fashioned into a twentieth-century social and intellectual movement through the leadership and ideas of Lenin and Trotsky, the Communist socialist state promised an end to social class and inequality - it provided a new and compelling vision of the future. Individualist capitalism, as well as the tyranny of royalty, was to be replaced by a centralized government and economy which would bring order, human equality, and a common purpose to human society. Yet as the decades of the 1920s and 1930s unfolded, what emerged in Russia was a new dictatorship and repressive authoritarian regime that, beginning in a violent revolution, continued its excessive carnage in subsequent decades as millions of people were imprisoned, tortured, and murdered in the name of the state. Communist Russia was one clear expression of the nineteenth-century idea that violence was justified in the name of the "the good of the people" - in this case, in the name the Communist state.

In fact, one could argue that both World War I and the Communist Revolution were instigated by elevated visions of the future that were used to justify war and violence. Throughout history, humans have frequently engaged in war and military conquest in the name of national and cultural ideals regarding the future.[8] In World War I, the German belief in cultural supremacy and their aspiration to become the dominant power in Europe were critical instigating causes behind the outbreak of war.[9] The Communist Revolution was propelled by Marx's vision of a better world and his support of social revolution as a necessary step to realize his vision.

The next dark chapter of the early twentieth century was the Great Depression. Beginning with the stock market crash of 1929, the Great Depression of the 1930s saw unemployment rates rise dramatically in the United States and Europe and thousands of banks fail worldwide. Inspired by the predictions of Marx, many writers saw post World War I inflation, increasing unemployment, and the economic depression of the 1930s as an indication that individualistic capitalism was failing. As Watson comments, the 1930s was "a grey menacing time." The Depression did not really ease up till the outbreak of World War II.[10]

World War I and the subsequent economic depression not only undermined faith in progress, they also contributed to a loss of faith in God. Hence, as Nietzsche had argued in the previous century, the central belief and value systems of the West, the religious and the scientific-secular, were both seen as unconvincing and in a state of decline. Not everyone experienced this generalized loss of faith, however, and the increasing spread of modernism, capitalism, science, and individualism in the early decades of the twentieth century also instigated a counter-reaction back to traditionalism and religious fundamentalism. While many were losing faith in everything, others were trying to find it again in the religious certainties of the past.[11] This fundamentalist counter-reaction to the flux and chaos of modernism would intensify in the decades ahead.

In the aftermath of World War I, an international effort was made to establish the beginnings of a world government with the Peace Treaty of Versailles. This new world organization was named the League of Nations. As writers such as H.G. Wells had argued, the ongoing violent conflicts among nations would be our undoing, and a world government needed to be created that would ensure peace and provide an international forum for nations to resolve their differences without having to go to war. With advances in transportation (intercontinental air flights), communication (radio), trade, and travel, humanity was increasingly becoming a global community, and the professed separatism and individual sovereignty of peoples and nations of the past was less and less a realistic vision of human life. A global vision of the future was needed - one founded on peace and cooperation rather than war and conquest. Though the League of Nations failed, the movement toward global governance and cooperation has continued to the present day, motivated by the dream of world peace and reinforced by the continued growth of global interdependency and a global culture.

Not everyone was satisfied with the Treaty of Versailles and the formation of the League of Nations. In particular, Germany, with its sense of cultural and racial supremacy, found its post World War I situation humiliating. In the midst of this national disgrace, and coupled with an economic depression following the war, Adolf Hitler and the Nazi party rose to power, promising the German people a renewed sense of national pride. Finding inspiration in his particular self-serving interpretation of human history and Germanic philosophy, and feeding on the discontent of the German people, Hitler instigated one of the most heinous and destructive series of events in the twentieth century. Hitler's futurist image of a "thousand year *Reich*" for the German nation catapulted humanity into World War II.

Hitler believed in the "great man theory of history" - that unique individuals possessing great vision and power have directed the course of human affairs throughout history. He saw individual leadership as a decisive factor in determining the future. Further, he thought that the father of all things was war and struggle (harkening back to Heraclitus and Hegel), and that Nietzsche's concept of the "superman" captured in many ways the new image he hoped to create for the German people and for German identity. He rejected the "trader" image of modern humanity, which he associated with Jewish people, the British, and the Americans, instead identifying with the "heroic" image of ancient myth. Further, he rejected both the hedonistic individualism of the modern West and the collective universalism of Communism, instead believing in the ethnic superiority and purity of the Arian race. It was the destiny of the German people to rule those inferior to them. He saw modern humanity as degenerative and regressive and he opposed traditional religion with its superstitions and emphasis on compassion and equality, replacing it with a new "religion of blood."[12]

Hitler and the German Nazi party were not the only ones who believed in a philosophy of supremacy and world domination. In alliance with both Mussolini and his Fascist government in Italy and the Japanese, who aspired, also with a sense of cultural and ethnic superiority, to rule Asia and the Pacific, Hitler and his Axis partners began a war of conquest in the late 1930s that spread across Europe, North Africa, the Pacific, and Eastern Asia. Again, I should point out that it was visions of the future, coupled with a highly aggressive philosophy for realizing these visions, that instigated great war and violence.

World War II was the bloodiest, most costly, geographically pervasive, and destructive war ever fought by humankind, even surpassing the carnage of World War I. Waged with the most technologically sophisticated weapons ever used, culminating with the atomic bomb, and fueled by the human failings of arrogance and pride, World War II was the great watershed point of the mid-twentieth century. Not only did tens of millions of soldiers die in battle, but at least as many civilians were killed, often by their own governments, with the infamous extermination of millions of Jews in Germany and other Nazi ruled countries being the most well known example of man's injustice to man during this great human disaster.

The eventual defeat of the Axis powers of Germany, Italy, and Japan, though at one level a great victory of human rights and democracy over violent tyranny and injustice, left a dark shadow on humanity that still haunts us today. The creation and subsequent mass production of atomic weapons left post World War II humanity in the ominous situation of having the capacity for total self-destruction and annihilation. We had walked up to the edge of the death of our species - a situation of our own doing - and there was no question that another global conflict, if waged with atomic weapons, would be the final bloody act in the history of human civilization. This cataclysmic image of the future was terrifying and had a great impact on subsequent world events.

Yet not everything during the period encompassing the two World Wars was doom and gloom. In great part due to the pressures of competition among antagonistic nations during the wars, science and technology continued to advance and even accelerate. Radar, submarines, jet airplanes, rocketry, and modern atomic physics were some of the noteworthy scientific and technological developments over this thirty year period. More within the domain of pure science, modern genetics and evolutionary theory emerged, combining the insights of Darwin with Mendel's work on genetic transmission.

Technology impacted the public and cultural world as well. "Talkies" (movies with sound) exploded on the scene in the 1920s and became immensely popular during the 1930s and 1940s. The American movie industry, centered in Hollywood, created one of the most significant cultural developments of the century, eventually sweeping across the entire globe. Movies became a primary source of public entertainment and "movie stars" were born, becoming cultural icons that millions upon millions of people idolized, if not worshipped. Because movies could be shown wherever there were movie theatres and were watched by so many people, the stars of Hollywood became highly visible figures in human society. Due to their heroic, larger than life presence on the screen, they increasingly became among the most admired people in modern culture. Additionally, movies became a great form of escapism from the darkness of the Great Depression and World War II.

Within the arts, the Art Nouveau and Art Deco movements brought innovative style and graceful beauty to painting, decorative art, the crafts, sculpture, and architecture. The modernist trends in painting and music, begun early in the century, continued to blossom and flourish. Chagall, Mondrian, and Dali, among others, would continue the challenge to classical artistic traditions begun early in the century. The great musical composers of the twentieth century Prokofiev and Rachmaninoff produced many of their best pieces during this time. Also breaking free of the constraints and norms of the past, in the realm of popular music, jazz and swing and the "Big Band" era provided new rhythms and exciting emotional tempos that freed the human spirit. Jazz and contemporary classical music would be integrated in the creative compositions of George Gershwin.

Supported by continuing advances in media, communications, technology, transportation, and mass production, and fueled by artistic and entrepreneurial creativity and initiative, in the first half of the twentieth century popular culture emerged as a highly influential social force and an ever growing part of the economy in many modern countries. The growth of the entertainment industry was a central force in this modern development, but so was the mass media which broadcasted and advertised the innumerable new products, ideas, icons, symbols, and possibilities of the emerging pop culture. Pop culture challenged the elitism of classical culture - it was for the masses - and not everyone was happy with this new development in modern human society. Hitler had waged a cultural war of purification against this "putrefaction" within human life and Communist Russia also saw Westernized pop culture as a form of degeneracy and conducted its own inquisition against it. Even in the West, various writers saw popular culture as shallow, fickle, hedonistic, low-brow, and threatening to the high ideals of human civilization. In spite of such criticisms and counter-reactions, the growth of the entertainment industry and popular culture, and the mass production and purchasing of all manner of new toys, gadgets, and cheap, commercial goods would further accelerate in the second half of the twentieth century.

Following World War II, the United States and the Soviet Union emerged as the two super-powers of the world, and although they had fought a common enemy in Nazi Germany, these two super-powers professed and practiced very different political and economic philosophies, namely democratic and individualistic capitalism versus a one-party, centrally controlled, collective Communist economy. Fueled by mutual distrust, post World War II - the "Cold War" - was a time of espionage and counter-espionage, competing cultural propaganda, and global paranoia, anxiety, and tension over a potential atomic war and perceived social and geographical threats. Two seemingly incompatible visions of the future competed for world dominance.

If World War I and the Great Depression had significantly damaged our belief in progress, World War II and the development of atomic weapons of mass destruction further undercut the optimism of the Enlightenment. With increasing tension between the United States and the Soviet Union and their thousands of nuclear missiles pointed at each other, perhaps, as many believed, hope for the future of humanity was naïve and unrealistic.

After the war, existentialism, inspired by the writings of Nietzsche, Kierkegaard, and Heidegger, emerged as an influential philosophy in France and other countries in the West. According to existentialists such as Jean Paul Sartre, humans are "condemned to be free" and can not find authentic meaning and purpose in any philosophical or religious ideology. Each human must create his or her own identity and values - without any abstract or absolutist foundation - and face a future that is uncertain and without any overall teleological purpose. There is no inevitable law of progress or universal spiritual destiny for humanity. As Nietzsche proclaimed, "God is dead" - in fact, all gods are dead. Sartre was particularly disdainful of a world overtaken with materialism, industrialism, standardization, and the growing Americanization of the culture. He hated the bourgeois and wished to free humankind of the tyranny of reason. For Sartre, it was the authentic self that should be pursued, liberated from the shackles of modernity and all types of false authority and security. This philosophy of authenticity and freedom would have a significant impact over the next few decades in the West.[13] The future was an individual and free act of creation.

Although critiques of Western culture and its capitalist and consumerist economy would continue, if not intensify, after World War II, the 1950s and 1960s was a time of rebuilding, increasing affluence, new discoveries and inventions in science and technology, and modernization spreading across the world. Capitalism and consumerism grew. After the war, Germany and Japan were rebuilt and the Japanese, in particular, embraced modernism and the technological and capitalist vision of the future. In the United States, the new pop culture and affluent lifestyle spread through the middle class. TV's, Walt Disney, suburbia, rock 'n' roll, supermarkets, shopping malls, and the building of a network of interstate highways all contributed to a renewed sense of optimism, adventure, and well being. As a consequence of efforts in World War II, when the first computers were created, continuing research and development into computer technology made steady advances in miniaturization and processing speed, laying the seeds for the information revolution in later decades. Even the Cold War with its tensions contributed to the upswing, for by the late 1950s, the United States and Russia found themselves in the "Space Race" as rockets, satellites, and eventually humans traveled upward into the heavens.

As Watson states, all efforts in social engineering in the twentieth century turned out to be great disasters. Although many nineteenth-century thinkers, such as Comte, Marx, and Saint-Simon, believed that science and the philosophy of secular progress could be applied to human society and its redesign[14], the authoritarian and totalitarian system of Nazi Germany had not only failed, but had caused great human misery and produced unbelievable human atrocities. After the war, as the Cold War intensified, there was the growing realization that the Communist system in the Soviet Union, under the ruthless and paranoid leadership of Stalin, was implicated in as much human torture, murder, and social repression of its citizens as the Nazis had been in Germany. In 1948, George Orwell published his fearful and scathing critique of totalitarian social control, *1984*. Social philosophers, such as Mannheim, von Hayek, and Popper, attacked the idea that there could be a science of human history and that its principles could be applied to the organization and direction of human society. Rejecting collectivist efforts, such as Stalinism in the Soviet Union, these writers argued for the need for spontaneous social order, flexibility, and liberty in the future evolution of human society. This philosophy resonated with the popular belief that World War II had been a victory of democracy over tyranny, as well as with the growing sense of individual freedom in the modernized West after the war. Indeed the 1950s and 1960s saw the philosophy of individual freedom and liberty become increasingly more powerful in many aspects of life in Western countries.[15]

As one significant example of this trend, the middle decades of the century witnessed the rise of feminism and the triumph of the sexual revolution. Through the feminist writings of

Simone de Beauvoir and Betty Friedan, the social and scientific research of Kinsey and Masters and Johnson, and the creation of the Pill, women were inspired to break free of the social and sexual constraints imposed upon them within traditional male-dominated Western society. Women had been a significant part of the workforce during World War II, but after the war, most returned to lives as mothers and stay-at-home wives. This stereotypical role was increasingly challenged in the 1960s as more women began to pursue professional careers and goals, increasingly sought higher education, and challenged the oppressive social norm of the submissive and subservient female. As "sexual consciousness" rose and came out from behind the psychological and social suppression of earlier times, both women and men openly broke free of the cultural ideal of life-long, religiously sanctioned monogamy, and engaged in pre-marital sex, frequently with different partners;, divorce rates also began to climb. As Kinsey discovered, there was actually more sexual freedom going on in the 1940s and 1950s than the general public realized, but with his publications, as well as many other noteworthy books in the next two decades, sexual freedom was more openly discussed and culturally reinforced and sanctioned.[16] The sexual revolution was an attack on tradition and a search for something new in the future regarding gender roles and human relationships.

Just as significant, during the 1950s and 1960s the Civil Rights movement became increasingly visible and influential, altering the social fabric and values of human society. Over the preceding two centuries, a frequent criticism of modern Western democracies had been that only a minority of the citizens of "democratic" countries really enjoyed the full benefits and opportunities of true freedom – there still existed in modern democracies a social stratification of power and wealth. Women, for example, couldn't vote in the United States until the twentieth century, and "Blacks" or African Americans were commonly treated by White European Americans as racially inferior and not deserving of the same respect and privileges as everyone else. The Civil Rights movement, led by Black Americans such as Martin Luther King, and increasingly involving other ethnic groups who had been ostracized in American culture, demanded equal treatment and equal opportunities. The Civil Rights movement, which would eventually spread through other countries in the world, was an attack on stereotypes, especially negative ones. It professed a philosophy of individualism, where each person should be judged on character and not the color of his or her skin or ethnicity. In modernized countries, the philosophy of liberty and equality, in many respects, was intensifying and spreading throughout all spheres of human life.[17] A new positive dream of the future that was more egalitarian and inclusive was emerging.

Also beginning in the post World War II era and continuing in the decades thereafter, the global phenomenon of decolonization spread across the world. In the Age of Exploration Europe had conquered and settled many large areas throughout the Americas, Africa, Asia, and Australia. Even up to the beginning of the twentieth century, this process of "colonization" had continued, with Great Britain, in an effort to find new economic markets, unsettling the Chinese Empire. As a consequence of this last significant expansionist effort, the great leading world economy of China fell apart. But the growing political wave of liberty and self-determination began to turn things around after World War II. India won its independence from Great Britain and steadily, in subsequent decades, colonies in Africa and Asia broke free of their European masters. Many nations and people around the world, no longer under the control and influence of Europe, could begin to create and implement their unique visions of the future. Coupled with decolonization, more and more countries throughout the world attempted to establish democratic governments, though not always very successfully. Still, the general political trend toward democracy and liberty has continued up to the present. As Francis Fukuyama has proclaimed, the modern political era can be summarized as the "Triumph of Democracy."[18]

In spite of such global movements toward democracy, liberty, and equality, the world struggle between capitalist democracies and Communist collectivist nations continued throughout most of the second half of the twentieth century; the political philosophies of collective totalitarianism and individualist democracy fought for control over the future of the world. From the 1950s to the 1980s the Soviet Union continued to aspire toward expanding its global leadership, in fierce competition with the United States and other Western democratic countries. Equally disturbing, Mao Zedong and his Communist party won control over mainland China in 1949 and created a new totalitarian government in the most populous country in the world. In the years ahead, the influence of Communist China would spread throughout Eastern Asia, and various other countries, such as North Korea and Vietnam adopted Communist governments as well. Wars were fought, again notably in Korea and Vietnam, between Communist and Western democratic forces over political and economic control of these lands. Ongoing internal conflict, in the name of Communist collectivist ideals, also occurred within Asian Communist countries, notably in the great Cultural Revolution in China, where presumed dissident voices against Maoist philosophy were "silenced" in vast numbers. Though worldwide there was a growing call for freedom and fair and humane treatment of all people, within Communist countries such as Russia, China, Cambodia, and Vietnam, tens of millions of citizens were killed by their authoritarian governments in the name of totalitarian ideals. In fact, such oppressive and violent actions by governments against their own people, if we also include Nazi Germany in the count, caused more deaths than combat fatalities in all the wars fought in the twentieth century.[19]

While the political and humanistic philosophy of freedom and equality made advances across different spheres of human life and across different parts of the globe, and while this philosophy and way of life wrestled with the "Communist Block" for control of the world, from within the confines of democratic and capitalist nations, critiques and counter-movements to the very system itself emerged and grew in power and influence. Life in the United States was not all it was cracked up to be. Notable writers such as David Riesman, in his book *The Lonely Crowd*, proposed that Americans, in spite of their professed individuality, were becoming increasingly other-directed (rather than inner-directed), and desperately in need of love and personal relationships. In a similar vein, the philosopher Hannah Arendt saw increasingly isolation and loneliness in an emerging mass society, which she feared was a step back toward totalitarianism. The psychoanalyst and social philosopher Erich Fromm, in his book *The Sane Society*, saw man in the modern capitalist world becoming a commodity - an identity to be sold - and expressing his Marxist philosophy, Fromm described modern work as dehumanizing, boring, and meaningless and leading to alienation. W. H. Whyte, in his very influential work *Organization Man*, argued that individuality was not on the rise but rather declining in middle class corporate America. We were becoming more conformist in our behavior, ruled by the dictates and expectations of modern society and business. The growth of homogenized and repetitious suburbia was one clear expression of this trend toward conformity. In *The Affluent Society*, John Galbraith expressed his concerns over the growth of a mass society, the advertising industry, and an excessive focus on the production and consumption of goods. Finally, Vance Packard, in his best selling series of books beginning with *The Hidden Persuaders*, argued that American consumers were being turned into "mindless zombies" who were being manipulated by the psychological techniques of the advertising industry.[20]

Such critiques of America and the modern capitalist and consumer society helped to lay the seeds of a cultural revolution that would blossom in the United States and spread into many other countries around the world in the 1960s and 1970s. The "Beat Culture" and the subsequent "Hippie Culture" presented an alternative philosophy of life and vision of the future that pulled together and expressed various existing trends and dissatisfactions within the modern world. It was a philosophy and lifestyle of extreme liberty and freedom - freedom of women, freedom of Blacks, and freedom to use illegal mind altering drugs and engage in premarital sex and "free love." The "beatniks" and "hippies" tossed off their middle class clothing in favor of jeans, flower shirts, and brightly colored beaded jewelry. Males let their hair grow long. Both young men and women strongly objected to the values of the "military-industrial complex," denouncing the materialism of the affluent West and condemning the incessant wars being fought by their countries around the world. The Peace movement, in fact, grew out of their efforts. The Environmental and Ecology movements were also strengthened through the hippie philosophy of "back to nature." Popular music, building upon the freedom of expression and boisterous quality of Rock 'n' Roll, became the soul of this new culture and popular singers such as Jimi Hendrix, the Beatles, Janis Joplin, and many others, became the central cultural icons of the movement. The music and lifestyle called for a breaking down of individual inhibitions and social taboos. In many ways the "Beat Culture" and "Hippie Culture" involved a re-assertion and further evolution of the philosophy of individualism (ironically a Western creation), with its emphasis on freedom and non-conformity, and was supported by the new popular psychologies of self-actualization, as espoused in the writings of Carl Rogers and Abraham Maslow. Instead of focusing on outer reality, there was an increased emphasis on inner reality, on spirituality, and the development of human potential. The cry and proclamation was to "Make love, not war!" This new counter-culture, as part of its rejection of modern Western society, also embraced many Eastern ideas, as well as the philosophies of indigenous peoples, such as the American Indians, and in opposition to the extreme rationalism of the West, turned to mysticism and a more emotional-romantic way of life. Personal expression and personal connection became more important than rationality, technology, and material gain.[21]

While the counter-culture of the beatniks and hippies was spreading across the United States and the West, a new intellectual and academic movement gained strength in Europe and eventually the United States that also contributed to the attack on the ideals of the modern West. Taking their inspiration from Nietzsche, Marx, and even Freud, the philosophers of Postmodernism and Deconstructionism achieved an increasingly powerful voice in the West in the 1970s and 1980s. Where the Enlightenment tradition had emphasized reason, science, and the ideal of secular progress, Postmodern thinkers, such as Foucault, Marcuse, and Derrida, argued that there was no objective and progressive direction to history, that all intellectual systems of thought were in actuality ideological systems for achieving and maintaining power, and that truth and value were historically and culturally relative. Neither science nor Western values had a privileged epistemological or ethical status. Philosophical relativism, in fact, has a long history going back at least as far as the ancient Greeks, but this new wave would have a strong impact on popular culture and thinking and clearly connected and was reinforced by many other resonant trends going on in the West.[22]

As expressed through the Beat culture, there was a growing distrust of tradition and authority, and a growing dissatisfaction with the creations of modern science and technology. Alienation, war, the destruction of the environment, consumerism, and poverty were all blamed on the modern Western system. The reasons given for justifying the status quo - big business, big government, and the smoothly running machine of modern life - appeared to be veiled attempts to maintain political and economic power. Twentieth-century anthropology seemed to reveal that non-Western cultures and people were not necessarily more primitive, but rather just different, and that the Western sense of superiority was egocentric and prejudicial. Respect for different points of view, lifestyles, and values, a consequence of individualist philosophy, seemed to imply that there was no superior way of either knowing or living. The Civil Rights and feminist movements and the process of decolonization all reinforced the value of pluralism and the ongoing critique of male-centered Euro-American dominance and superiority. According to philosophers of science, such as Thomas Kuhn and Paul Feyerabend, even the presumed objective and superior quality of scientific knowledge - basically a creation of the Western male - was also a chimera.[23]

Postmodern writers clearly agreed with those critics of American society and Western capitalism who had argued that the citizens of modern Western countries were not as free as they believed they were. The French writer Guy DeBord saw the modern world as a "society of spectacle" where the world had become a commodity and even consumption itself had become part of the show. Big business and advertising had deluded modern humans into thinking they were free, but in actuality humankind was enslaved. On a related front, the psychiatrist Ronald Laing accused modern psychiatry and psychotherapy, presumed avenues toward individual freedom and self-expression, as actually reinforcing the norms of the mass society and conformity.[24]

Philosophical relativism, as espoused in the writings of Postmodern thinkers, was especially critical of the idea of secular progress. If the beliefs and practices of different cultures and different times can not be comparatively evaluated relative to some absolute or objective standard, how can we say that there has been progress across time? Postmodernism reinforced the already growing loss of faith in the ideal of secular progress due to the two World Wars and the Great Depression. The Beat Generation also could not accept the standard view of progress, since it seemed that it was the very creations of advancing science, technology, and capitalism that were responsible for the most serious failings of the modern world.

The historian Robert Nisbet has chronicled the rise and fall of the idea of progress across the centuries and identifies a whole set of factors that has contributed to the loss of faith in progress in the twentieth century. For one thing, Westerners began to feel increasing guilt over their own affluence and success, becoming more and more aware that their material progress depended upon the enslavement and cultural destruction of non-Western people around the world (the results of colonialism). Nisbet also notes a steady loss of faith in the value of central institutions and the in the values such institutions presumably embody. As expressed through both the Beat culture and Postmodernism, there has been an ongoing revolt over the last half century against authority and traditional institutions such as government, business, religion, and the military. Enlightenment philosophers had made a strong connection between economic and social progress, yet in the second half of the twentieth century there has been a growing hostility toward the ultimate value of economic growth. Economic growth, based on a capitalist system, had generated increasing disparity between the rich and the poor, undermined the traditional values of family, community, friendship, tradition, and religion, and produced a society of materialist consumers. Presumably, moral constraints and principles had deteriorated in the West, in the name of self-interest and the pursuit of financial success.

Also, according to Nisbet, in the twentieth century there has been a loss of respect for knowledge, science, and scholarship. The question has been repeatedly asked: Has the growth of science and knowledge really benefited humanity? Throughout the century, there has been a draw back to nature, to simplicity, to some idealized, more humane and authentic reality of the past. Nisbet also sees the loss of faith in objectivity and reason, as a consequence of Postmodernism and other philosophical and cultural developments, as leading to increasing self-centeredness, subjectivity, and occult practices. Whereas the Enlightenment engendered hope that the principles of science could be applied to humanity and society, in the twentieth century, people lost faith in the social sciences, and arguing in a similar vein to Watson, Nisbet points out that many people have come to believe that the social sciences and social engineering in the last century have done much more harm than good.

Also, part of the problem, contends Nisbet, is due to the very success of modernization, for we have created a world of security, abundance, and leisure with no perceived need for pushing ahead. As Watson states, we are doing "too well to do good." Further, in the West, we have become complacent, if not bored with life. There are no visionary prophets of hope - no "Faustian" individuals expressing a "will to power." According to Nisbet, modern humans have escaped into hedonistic materialism, fanatical spirituality, and the violence and sensationalism of the mass media. We have lost our sense of true culture, of cultural heroes (these having been replaced by sports figures and media stars), of the sacred, and of something "beyond." In a sense, the future has died in a sea of momentary pleasures and complacency.

In the final analysis, in agreement with a similar view expressed by the historian of futurist thinking Fred Polak, Nisbet contends that a society, such as our modern world, that does not have a positive and progressive vision of the future will flounder and fail. Nisbet hopes for a renewal in the belief in progress.[25]

Every critique has its counter-critique; every movement has its counter-movement. After the anti-establishment movements of the 1960s and 1970s, a variety of counter-attacks and opposing movements have emerged. As one example, the Beat generation, in the eyes of many critics, was in actuality an extreme expression of self-indulgence, unbridled hedonism, and self-centeredness. There was nothing constructive about it. It was a "culture of narcissism," as the psychoanalyst Christopher Lasch argued. The free love (translated as non-committal sex), drugs, rock 'n' roll, and dropping out of society were simply a variety of pleasure seeking activities with no accompanying sense of responsibility, ethical constraints, or concern for others. "Doing your own thing" carried with it no sense of social sensitivity or responsibility. In the 1970s, the expression the "Me generation" became a commonly used negative label applied to the young adults of the era. The popular self-awareness sessions and encounter groups of the hippie period were seen by writers such as Lasch and the journalist Thomas Wolfe as simply opportunities for people to talk about their favorite subject - themselves. Although the expression "consciousness raising" was frequently used to describe one of the central goals of the Beat counter-culture, members of the movement were actually engaged in various forms of self-indulgence. Further, the whole movement had a strong regressive quality to it. It romanticized the past and rejected a great many of the achievements of the last few centuries. Moreover, if everyone was off "doing her or his own thing" and focusing on his or her own inner consciousness, who was minding the store, and where was the sense of a social vision and purpose for the future of humanity as a whole? The movement was excessively romantic and Dionysian, focusing on feeling, emotion, and personal reverie, and had lost touch with all the important values connected with the rational and Apollonian side of humanity. According to Lasch, the movement had failed to produce any real expansion in consciousness or real psychological growth.[26]

Yet the self-centeredness and hedonism of the 1960s and 1970s continued into later decades. One notable failing of the movement, as Daniel Quinn points out, was that it did not seriously consider how one goes about making a living. Underneath the veneer of freedom, members of the Beat and Hippie cultures were living off of the wealth and support of their parents; eventually the hippies had to "grow up" and find a way to support themselves. Interestingly, what happened was that many of the "hippies" turned into "yuppies," finding jobs in the corporate world (their professed enemy a decade earlier), and continuing their lives in middle-class affluence in the suburbs. But frequently they seemed to carry with them the self-centered narcissism of their youth, only now they had found a way to satisfy this psychological disposition through the making of money and the purchasing of goods, services, and high-tech toys. A frequent criticism of contemporary modern culture is that it is excessively sensationalistic, self-centered, and hedonistic, all qualities of the hippie culture as well.[27]

Another counter-attack of the 1980s and 1990s was directed against the relativism and liberalism of Postmodern philosophy and it came from a variety of fronts. Many philosophers and scientists disputed the subjectivist conclusions of Kuhn and Feyerabend, arguing instead that there was measurable objective growth in the sciences and that that there were cross-cultural standards for evaluating knowledge claims. The educator Allan Bloom, in his highly controversial book *The Closing of the American Mind*, argued that the shallowness and extreme liberalism of pop culture was destroying educational standards and that we were witnessing an intellectual deterioration not only of the academic curriculum but of modern civilization as well. According to Bloom, universities were giving into public opinion and the philosophy of political correctness and losing sight of the great repository of knowledge of the past. Again, the new freedom, as Lasch had argued, was superficial and without merit. In a sympathetic and supportive vein, Harold Bloom, an esteemed professor of literature, also challenged the relativism of contemporary culture and education and outlined a Western canon of great works of literature as foundational for higher education. The historian Gertrude Himmelfarb pushed the critique of Postmodernism even further. Not only was education being compromised and diluted, but our culture and our moral values were being threatened by Postmodernism. The liberalism of Postmodernism had become so extreme that it undermined "our duty to truth" and was actually subverting our liberty and freedom. Evil and human atrocity had become relative, and hence the actions of individuals like Hitler, which at one time were clearly seen as morally heinous and a global threat to the dignity and liberty of humankind everywhere, were now, through the eyes of Postmodernism, open to diverse and relativist interpretations.[28]

In response to the writings of Allan and Harold Bloom, there was a counter-defense. The two Blooms were accused of being stuck in the past and of being Euro-centric. According to the counter critics, Western academic education should acknowledge and include in its canon the ideas and writings of non-Western countries. One of the fundamental points of Postmodern philosophy has been that the Enlightenment embodies an extreme Western bias, elevating its culture, its values, and its body of knowledge to a position of absolute validity and superiority over all other cultures and belief systems. Therefore, in the name of fair-mindedness, contemporary education and cultural awareness need to go beyond the prejudices and biases of the West and acknowledge and appreciate the varied points of view and values of cultures from elsewhere around the world. Both our human heritage and our future should be conceived in pluralistic, rather than Euro-centric terms.

Yet, another development, strongly connected with recent research and thinking in the sciences, brought into question the factual validity of philosophical relativism. Early anthropological research in the twentieth century seemed to demonstrate great variability in cultures around the world and this presumed general finding seemed to support the idea that humans were very malleable and basically products of the particular culture and historical period in which they lived. Yet, more recent research done in the last few decades has not only uncovered serious methodological flaws and personal biases connected with the earlier studies, but has also revealed that there, in fact, are numerous universal features to human thinking, values, and practices all around the world. Cultural relativism, at least to a great degree, appears to be a factually invalid scientific theory. Underneath our apparent differences there is a great deal of commonality within the family of humankind. Further, with continued developments in the biological sciences over the last few decades, it appears that biology and genetics significantly constrain individual and cultural variation. As the contemporary biologist E. O. Wilson came to argue, biology and culture are not entirely separate, and biology - a biology all humans share as members of the same species - to a great degree determines culture.[29]

As Watson notes, many of the Postmodern and Deconstructionist writers remained relatively oblivious to the ongoing advances in science occurring in the second half of the twentieth century. While hippies were into doing their own thing and Postmodernists were proclaiming the relativity of all knowledge and values and critiquing the concept of progress, science steadily advanced in its understanding of the world and its technological creations continued to further transform the world. The growth of science and the increasing infusion of technology into human life have been two of the most powerful and influential trends of the twentieth century. According to Watson, one of the most significant developments of twentieth-century science has been the progressive synthesis of a comprehensive description of the evolution of nature and humankind. Derived from research across numerous disciplines from history, archeology, paleontology, and geology to physics, biology, anthropology, and ecology, a grand narrative of life, the universe, and humanity has emerged. While Postmodernists have attacked the idea of grand narratives, science has been producing one. While the Postmodernists talked about diverse points of view, science, involving the contribution of numerous disciplines and investigators from many cultures, found agreement and convergence from all these points of view on a general story and pattern of change. There are innumerable controversies and areas of mystery still within science, but a general picture of nature has emerged and this general picture has steadily grown in detail and comprehensiveness over the last century. This comprehensive view of nature is evolution.[30] Hence, while Postmodernists continue to dispute the concept of progress, and contemporary creationists and fundamentalists vehemently argue against the validity of the theory of evolution, and while modern popular culture wallows in a loss of faith in the idea of progress, science has been steadily uncovering across the whole panorama of natural and human history a fundamental and pervasive progressive process toward increasing complexity, organization, and intelligence.[31]

Watson identifies the three major intellectual forces of the twentieth century as science, free market economics, and the mass media. As we have seen there have been critics of modern capitalist and consumer culture throughout the twentieth century, and the attacks, in fact, go back to the eighteenth and nineteenth centuries. Yet a pervasive trend throughout the last century has been the growth and spread across the globe of modern capitalism and consumerist culture. Toward the end of the twentieth century, the Soviet Union and its communist government collapsed and though the process of transformation has been far from successful, the various countries of the former Soviet Union have moved toward the adoption of capitalist free-market economies. Though China still maintains a communist system of government, it has supported the development of a capitalist economy. While various countries and ethnic groups continue to oppose Western consumerist culture, people around the world increasingly embrace and participate in this way of life. In fact, perhaps the main force behind the ongoing globalization of the world is the progressive networking together of capitalist economies and consumerist cultures. For better or worse, production and consumption continue to rise around the world. Since the end of World War II, with its up's and down's along the way, there has been a great economic boom as production of goods has grown over six fold worldwide.

The media and science have also been prime targets of social criticism throughout the last century, yet again, there is no question but that the power of both these forces in modern life has grown immensely.[32] Watson, in fact, argues that the most significant trend across the twentieth century is the coming to terms with the advancements of science. Modern science has changed everything, including the way we think.

For Watson, the great global challenge facing humanity at the turn of the century is the non-West coming to terms with the West. The tension and conflict between the modernized West and Islam is particularly noticeable and pronounced, but Watson argues that Islam is still stuck in the past - in a philosophy of life that depends on ancient religion and faith. Writers such as Samuel Huntington see the cultural conflict between Islam and the West continuing, if not intensifying, in the decades ahead. Huntington also predicts escalating tension and conflict between the West and China, but other writers such as Thomas Friedman are more hopeful, believing that a compromise and balance can be struck between local cultures and the capitalist world economy sweeping across the world.[33]

David Christian, in his "big history" of the world, *Maps of Time*, provides an overview of the major trends of the twentieth century. According to Christian, it has only been in the twentieth century that the full significance and impact of modernism have become apparent. Along a variety of measures, in spite of all the "sound and fury," the twentieth century has shown phenomenal growth and change. World population has exploded from 1.6 billion at the beginning of the century to over 6 billion by 2000; of course, this may or may not be a good thing. Agricultural productivity has tripled in the same period of time. Energy usage per capita has almost quadrupled, as has the average global income. Perhaps, most astoundingly, the global economic output increased from 2 trillion dollars in 1900 to 5 trillion in 1950 and 39 trillion in 2000.

Other main trends in the twentieth century, according to Christian, are growing mass literacy, increasing longevity, the spread of democracy, widespread decolonization, an unprecedented economic boom since 1950, the death of peasantry and traditional tributary empires (such as those in China), the emergence of powerful transnational corporations, accelerated innovation, the breaking down of gender roles, and vast increases in wealth, productivity, and pollution. But, according to Christian, perhaps the most significant phenomenon of the last century has been the global transformation of the environment as industry, technology, government, and human society, in general, have altered, manipulated, and affected our planet in numerous ways. There have been great ecological changes in the last 50 years especially. According to many critics, humanity is quickly exhausting the resources of the earth and irretrievably polluting the environment. The question of where this ecological trend is heading has been the source of increasing controversy and debate over the last few decades.

Christian states that the continued growth of capitalism and the free market has been another major trend of the twentieth century. Eleven of the top thirty economic entities in the world are now business corporations, rather than nations. Capitalism has produced great wealth and abundance and the collapse of Communism in the Soviet Union was primarily due to the fact that it could not generate the productivity or innovation of capitalist economies. Yet, capitalism has not equitably spread its benefits to all people. The unequal distribution of wealth in the world continues to intensify. The gap between the rich and the poor over the last few decades has widened at an accelerated rate. According to Christian, the overall global effect of modernism and all its economic and industrial developments on those people and cultures in third world countries has been decidedly negative, producing more harm than good. Further, the continuing growth of capitalism and economic productivity has necessitated the reciprocal growth of a consumerist society. As critics of this system have repeatedly argued, economics in the modern world seems to direct life, rather than the other way around. We live in order to buy, rather than buy in order to live.

On a positive note though, Christian states that in the last century there has been a great increase in collective learning. Advances in communication, the spread of higher education, the growth of science, increasing travel, and most notably in the last couple of decades, the rise of information technologies have all contributed to increased knowledge sharing and consequently the creation of more new ideas and innovation. Referencing the contemporary social theorist, Manuel Castells, the flow and exchange of information around the world has now become the most significant sector of the capitalist economy. Supported by a high-tech global infrastructure and the spread of computers into business and the home, a new "information economy" has emerged that, according to Castells, is qualitatively different than the previous industrial economy. In this process, the power and significance of nations and national boundaries have significantly decreased and the wealth and power of multinational corporations has increased.

Christian argues that the acceleration of the pace and scale of change is both the most striking and equally most frightening feature of the twentieth century. By many indicators, more change has occurred in the last century than in all the rest of human history. As he notes, some fear that we are watching a "traffic accident in slow motion," as our industry and consumption continue to grow, unchecked by effective counter-measures to protect our environment and bring greater overall equality to all the people of the world.[34]

In summary and conclusion, the last hundred years has been a time of amazing and often unsettling changes. The physical and biological sciences have been revolutionized through relativity theory, quantum physics, big bang cosmology, contemporary evolutionary theory and research, astronomy and space exploration, incredible advances in archaeology and paleontology, and the discovery of DNA and genetic transmission. Our understanding of the laws of nature, life, the universe, and the vast extent and intricate detail of space and time has been totally revolutionized.[35] Technology has transformed our lives with the invention, mass production, and distribution of the automobile, airplane, telephone and cell phone, radio, phonograph/CD/DVD players, light bulb, laser, television, microwave, fax and copy machines, and computer. A hundred years ago there were no movies, movie stars, or home video machines. Basically, a hundred years ago we were still earthbound without satellites, rocket ships, and orbiting telescopes. In the twentieth century we landed on the moon, sent probes into the atmosphere of Jupiter, and mapped a billion galaxies in the heavens above. On a more down-to-earth level, in modernized countries, our kitchens, bathrooms, backyards, entertainment and living rooms, and homes, in general, have been transformed with a thousand different gadgets and modern conveniences, and our city streets, shopping areas, and business offices have become infused, illuminated, wired, animated, and automated with technology.[36] And finally, we have seen the beginnings of technology moving into our bodies, with artificial tissue and limbs, hearts, and other internal organs.[37] *Time* magazine identifies the advance of science and technology as the most significant defining trend of the twentieth century, nominating the preeminent scientist of the century - Albert Einstein - as the "Person of the Century."[38] The twentieth century has blown us away with its science and technology.

Yet in the last hundred years, we have been through two World Wars, a Global Cold War, and innumerable reshufflings of national boundaries and alliances involving hundreds if not thousands of local wars and skirmishes. In spite of (or maybe because of) our great nineteenth century dreams of human progress,[39] in this century we have fought, killed, tortured, and imprisoned our fellow humans over race, culture, religion, and politics in record numbers; the number of war deaths in the twentieth century quintupled over the number in the nineteenth century. We saw the rise and fall of Soviet Communism and the German Nazi state and suffered the historically unprecedented human atrocities created in their wake. *Time* magazine identifies the other two most important themes of the twentieth century as the struggle of democracy against totalitarianism and the struggle for human rights - two global battles that we are still fighting.[40]

Even the growth of science and technology in the last century seems to have been a mixed blessing. Through the rapid advances in our technology and science, we have threatened our own survival as a species. We created atomic weapons and the United States came very close to initiating an atomic war with the Soviet Union, and, though the Cold War has subsided, we have amassed an arsenal of nuclear bombs that easily could wipe out all humanity.[41] The production of weapons of mass destruction, including atomic, chemical, and biological weapons and all the techno-refinements associated with such weaponry, vastly exceeds our previous military capacities of a hundred years ago.[42] Our factories, transportation systems, escalating consumption of resources, and the offshoots of our techno-civilization - the increasing production of waste and garbage - have all significantly unsettled and polluted our earthly ecosystem.[43] And the cutting edge wave of new technologies, including robotics, genetic engineering, and nanotechnology, presents new global threats of an even more potentially devastating nature.[44]

There have been other important changes and developments in the last century - also often having both negative and positive features. As noted, the world population has skyrocketed and mega-cities of towering skyscrapers, incredible wealth, and overpowering slums have emerged across the globe.[45] Psychologically and socially, we have been transformed by pop culture, pop psychology, modern art and literature, rock music, and the rise of the feminist and Human Rights movements. But all these forms and self-expressions of freedom and the breaking of traditional constraints have instigated or created new conflicts, dangers, and uncertainties - for example, the loss of faith in progress and the rise of Postmodern relativism and unbridled self-indulgent hedonism. We discovered Freud and the dark secrets of the unconscious, Walt Disney, Mickey Mouse, and Jurassic Park, cyberspace and inner space, free love, mass divorce, credit cards, credit card debt, birth control pills, artificial sweeteners, and AIDS. As pop culture has circled the globe and communication and trade have increasingly networked humanity together, the local customs and cultures, which once provided simple and singular answers to the issues of life, have been replaced by a plethora, hodgepodge, and smorgasbord of choices, freedoms, ambiguities, and multiple visions of life.[46] Modern life has become filled with time saving conveniences yet is increasingly fast-paced and stressful.[47] We are more educated and psychologically self-aware, yet drug addiction and psychological disorders have reached epidemic proportions in modernized countries. And as we enter the new century, the Internet - the biggest machine and system for social, business, and recreational use ever constructed - seems to continue to grow with a will of its own, further networking the entire world together and further accelerating every dimension of human life. Yet as the computer and the Internet have come on the scene promising enhanced learning and interconnectedness, many people feel increasingly isolated and inadequate - stuck with the feeling that they are perpetually falling behind in what needs to be known.[48] The world today is in many ways a maze of contradictions.

The Contemporary Transformation

"Everywhere we look there is a tension between the past and the future, between a pessimism we cannot shake and an optimism we cannot believe in."

John Noble Wilford

"Living in the borderlands between the modern and the postmodern means negotiating constant conflicts between the old and the new and confronting perplexing and often disturbing change. Condemned to a seemingly unending state of transition, permanent tension and strife appears to be a defining modality of the between. Moreover, the discourses that strive to describe this condition are also in conflict and at odds with each other, condemning us to unending theory and culture wars with no truce in sight."

Steven Best and Douglas Kellner

"Somethin's happening here,
What it is ain't exactly clear...."

Buffalo Springfield

Watson states that we are in a "cross-over" or transition culture. We are moving from one way of thinking and living into a new way, yet there are many competing voices regarding the nature of the contemporary transformation, where it is heading, or where it should be heading. Watson identifies contemporary science, art, and business as three alternative ways of seeing the world, which are in competition with each other in our present times. But there are other levels and types of competition as well. Different cultures and people are at odds with one another across the globe, often engaging in violent conflict as a way to determine their destiny and future. Ideologies, social movements, and theories compete, as do corporations and businesses over who will dominate different areas of the economic market. All in all, according to Watson, we have entered the twenty-first century without a unifying grand narrative or common story of where we are and where we are heading. We have moved into a "Post - Postmodern" period.[49]

Although, according to Ed Cornish there seems to be a general consensus, as one defining feature of the contemporary transformation, that we are in a period of accelerative change, different groups of individuals interpret and react to this general trend in different ways.[50] Futurists such as Hazel Henderson, Robert Theobald, and Alvin Toffler point out that there are many who wish to deny, minimize, or openly resist the revolutionary nature of our times.[51] Instead of advocating for fundamental change in policies and philosophy, such conservative forces are attempting to continue along traditional paths, or even retreat to the past.[52] Yet we should keep in mind, as Mary Clark observes, that in the past, societies usually learned new ways of thinking only when they were pushed into it.[53] Rarely does a society initiate mental change - a society, in fact, will resist changing its worldview. Resistance to change is a basic human motive, due to the need for mental stability, security, and the need for a sense of identity.[54] And further, as Henderson notes, those individuals and organizations in power will especially resist change because they are benefiting the most from the status quo. Change is risky, costly, and time consuming. Yet as advocates for change argue, it may be more costly in the long run to deny and resist the transformational forces and trends around us.

The issue of stability versus change is one of the main theoretical and ideological conflicts of our times.[55] This conflict comes out in theories and paradigms of the future. Continuing the historical dichotomy between those views that saw time as cyclical or unchanging and those views that saw time as linear and progressive, one basic dividing line and argument among theories of the future is whether the future is going to be (or should be) significantly different or basically just more of the same.[56]

One good example of how change and the nature of the contemporary transformation can be interpreted differently derives from the sociological studies of Paul Ray and Sherry Anderson.[57] In a survey of American belief systems and values, Ray and Anderson identified three relatively distinct social-cultural groups: Those who emphasize a return to a simpler, less technological past (the "Heartlanders"); those who continue to embrace a modern secular and technological progress (the "Modernists"); and those who advocate a jump forward to a new way of thinking beyond either traditionalism or modernism (the "Cultural Creatives"). In a sense, all three groups might see themselves as advocates of change - it simply depends on where you are standing. Heartlanders (which would include traditionalists and religious fundamentalists) often see contemporary times as simply a continuation of secular and ungodly trends that go back centuries, if not thousands of years into the past. From their perspective, within the modern world there is really "nothing new under the sun." Yet, from outside this group, the Heartlanders are often viewed as attempting to retreat into an idealized past that, in fact, never really existed. Modernists, on the other hand, see the contemporary world as undergoing great change and progress and they see themselves as embracing this progressive movement forward. Yet, for the Cultural Creatives, the contemporary world is simply a continued expression of the eighteenth-century philosophy of material and secular progress - there is nothing qualitatively new, only much more of the same thing. The Cultural Creatives really want a change - not back to the idealized vision of the Heartlanders, but toward something really different. Many Cultural Creatives argue against the continued emphasis on unconstrained growth, calling for a "change" in direction and values toward a more sustainable society.[58] They want to move beyond the headlong rush and lack of foresight associated with modernism.

Keeping in mind these ambiguities and disagreements, it does seem, at least based upon a variety of indicators, that accelerative change is a defining feature of our era. To recall, as Christian documents, world population, industrial and agricultural productivity, wealth, innovation, the growth of knowledge, and ecological deterioration all show positively accelerated rates of growth over the last century. Other writers would add that humanity on a global scale is networking and connecting together at an accelerative rate as mass communication and the Internet are wiring us together into one "global brain."[59] And Ray Kurzweil and Hans Moravec, among others, would further add that the information processing and storage capacities of our computer technologies are also growing at an exponential rate, setting the stage for even greater changes in the coming decades.[60]

Accelerative change is intimately connected with increasing speed in innumerable aspects of human life. James Gleick, in his book *Faster: The Acceleration of Just About Everything*, chronicles in unsettling and often comical detail the growing frenzy of modern life.[61] For Gleick, increasing connectivity generates increasing speed in our lives. Within the business world, there is an increasing emphasis on efficiency and the modern worker feels busier than ever. There is a growing sense of information and choice overload and a never ending list of projects and tasks to complete. Information becomes obsolete more quickly and new products and developments have increasingly shorter life spans. There is a general expectation to be quicker and think faster. As Gleick puts it, we live in a state of "rat-race equilibrium" and are forced into multi-tasking as a way to get everything done. Yet, because new technologies, which supposedly were developed as ways to save time, actually create more to do, we never do catch up and feel guilty and uncomfortable doing nothing - we can not relax. We are a time conscious society, where time is money, and in fact, more valuable than money. Overall, the fast pace of our lives is changing us psychologically, such that old TV shows and movies seem too slow to keep our attention for very long. We live in an "unbearable state of distraction" and perhaps suffer from a collective case of attention deficit disorder.

Because many of the big changes of the last century, which include accelerative change and increasing speed, seem more negative than positive, it is not altogether obvious to many people whether the contemporary transformation is progressive or degenerative.[62] As Hazel Henderson has argued, in times of great upheaval, events will seem ambivalent.[63] We may be witnessing a collapse into a new age of darkness, or a rise to a higher plane of existence. Vaclav Havel states that we are presently in a transitional period where something is on the way out and something new is being painfully born.[64] During such times all existent value systems crumble and we enter into a time of confusion and uncertainty. As Zohar and Marshall put it, "Something new is in the air."[65] And as Buffalo Springfield would add, "What it is ain't exactly clear." Of special importance, we are experiencing the birth of a new millennium. This passage from the old millennium into the new is associated with the archetypal theme of death and rebirth.[66] It is seen as a turning point in the history of humanity - as a time for both hope and fear.[67] Barbara Marx Hubbard sees our contemporary world in a time of great crisis, but connects this threatening state with the opportunity for a new birth - the negative and the positive possibilities are inextricably linked together.[68]

From the perspective of open systems theory, the ambiguity of contemporary times is quite understandable. In open systems theory new order arises of out chaotic fluctuation. There are many who think that the world is going downhill because of all the apparent problems publicized in the news, as well as all of the local problems being experienced firsthand, e.g., stress, speed, terrorism, crime, inflation, pollution, traffic congestion, drug abuse, and divorce. David Pearce Snyder, a well-known futurist, acknowledges that in many ways things are indeed getting worse.[69] But as he notes, we shouldn't extrapolate, as many are doing, into the future from the immediate present. This is linear thinking. Snyder believes that we are in revolutionary times and that it will take two generations - 50 to 70 years - to assimilate the productive potential of the new technologies, ideas, and practices. During the first half of a revolutionary change, things will get worse. According to Snyder, the real benefits of a dramatic social change don't come until two-thirds of the way through the revolution. On a related note, Henderson states that status quo and conservative thinkers are trying to make sense out of the present transformation using industrial, Newtonian concepts and that this approach will not work.[70] For Henderson, the transition should be understood in terms of open systems and nonlinear, interactive, and creative concepts. From this more modern perspective, the problems and ambiguities of the present are seen as necessary events in a progressive and evolutionary jump forward.

Perhaps the optimists are right and the contemporary transformation is moving in a positive direction, but even many of them would agree that there is clearly a dimension of uncertainty associated with our present situation. As noted earlier, there are different interpretations of the nature and direction of events and trends in our modern world - often clearly at odds with each other. This multiplicity of points of view in fact contributes to the overall uncertainty of things.

There are various writers who do attempt to bring some order and coherence to the nature of the contemporary transformation. For example, Ed Cornish lists six fundamental "super-trends" in an attempt to capture the main features of what is happening around us.[71] These six super-trends are:

- Technological Progress
- Economic Growth
- Improving Health
- Increasing Mobility
- Environmental Decline
- Increasing Deculturation

Cornish sees these six super-trends as interconnected. Technological progress, as both a trend and a force, stimulates economic growth and economic growth not only feeds back on technological growth, but also contributes to improving health, increasing mobility, and on the negative side, environmental deterioration. Further, Cornish sees technological and cultural change as interconnected, so not only has technological change accelerated, but cultural change has accelerated as well. In the last century, both technology and human values changed drastically and continue to do so. Cornish notes that increasing mobility, due to increasing wealth and improved transportation, has been the primary cause of globalization and as globalization spreads, local cultures and ways of life are being homogenized into a world culture. Increasing deculturation involves the loss of traditions and values, which are replaced by one emerging international culture. As modernization spreads, many people are experiencing "culture shock."

In general, Cornish views our present period as revolutionary, involving a fundamental transformation in all spheres of human life. In a model similar to the one proposed by Alvin Toffler, Cornish argues that there have been three global revolutions in human society throughout history: The Agricultural, Industrial, and Cybernetic. (Toffler referred to the third revolution as the Information Revolution.)[72] We are still in the midst of the Cybernetic Revolution, which has involved the introduction of the computer and computerization into almost all aspects of our personal and professional lives. Cornish foresees a possible fourth revolution coming in the near future – the Biotechnological Revolution – which in its early stages right now promises the potential capacity to redesign life and in particular human nature. I would only add to this list a progenitive revolution – the Great Awakening of 40,000 years ago that saw the birth and flowering of a multitude of cultural and technological developments – perhaps the birth of language, abstract thought, and culture itself.[73]

If we place these four global revolutions on a time line, the most noticeable fact is that the rate of revolutionary change is accelerating. Each new revolution comes much faster/more quickly than the last. We have not even adapted to the present Cybernetic/Information Revolution and a new one, the Biotechnological, may already be emerging. Following the logic of Kurzweil, Moravec, and the science fiction writer Vernor Vinge, who first coined the expression, we seem to be accelerating toward a "technological singularity" where the pace of change and the exponential growth of computer intelligence will reach a level that will surpass human understanding and adaptability.[74] Hence, not only is the rate of change in our times accelerating, the rate of new waves of revolutionary change is accelerating as well.

One could argue that each new wave of revolutionary change brings with it a new set of technologies and social systems for organizing and disseminating information that, in fact, facilitates the speeding up of the rate of change. Computers, the Internet, improved transportation and communication, and an expanding global economy allow for increased collective learning and innovation, making our present social system much more powerful at producing change within the world. Further, each new wave also seems to reduce the time needed to move to the next wave of revolutionary change. If we follow the punctuated equilibria model of evolution proposed by Gould and Eldredge, the jumps in evolutionary development in human society come more and more quickly. Writers such as Kurzweil and the physicist Murray Gell-Mann, among others, see human cultural development as the latest and most powerful expression of the overall evolutionary process in nature; the evolutionary process itself evolves through increasingly more advanced stages that facilitate higher rates of progression and change. From this perspective, the accelerative rate of change in contemporary times is simply an expression of the mechanism of evolution evolving increasingly more powerful ways of generating evolution and change.[75]

The Postmodern interpretation of the contemporary transformation, as described in the writings of Stephen Best and Douglas Kellner, highlights some other significant features of the nature of change in our times.[76] As Best and Kellner argue, the Postmodern era is a decisive shift away from the previous period along many different dimensions of human life, including politics, philosophy, culture, and the arts. A critical feature of this transformation is the abandonment of a single unifying vision or narrative for humankind. Whereas people in the past generally adopted and lived by a single belief system and set of values, mass communication and globalization, among other factors, has exposed people around the world to multiple systems of thinking and philosophies of life. The Postmodern world is a reality of many points of view - it is a pluralistic rather than monistic culture. Of course, in the past, there were different cultures and sub-cultures across the globe, but these different cultures tended to be much more insulated, and usually each culture tended to see its beliefs and values as the best and truest. In today's Postmodern world, unless one defensively closes off the surrounding world, one lives in a reality of many different perspectives and lifestyles. It could be argued that such a pluralistic system of ideas and values, both socially and psychologically, opens the human mind to increasing flexibility, freedom, and creativity, but it also produces more uncertainty, ambiguity, and conflict in life. Thus the Postmodern era may be speeding up change, but also it is intensifying uncertainty as to where it all is heading.

There are numerous other efforts to describe the overall nature and significance of the contemporary transformation, and of course, this is a manifestation of Postmodern pluralism. I will describe many of these alternative viewpoints in the next chapter on theories and paradigms of the future, but I should mention now a few other comprehensive interpretations to provide a more complete picture of the main features of the contemporary transformation.

Ed Cornish does not identify globalization as a distinctive "super-trend" within the contemporary transformation, yet numerous other writers would include this trend as one of the most significant directions of change in our present world. Walter Truett Anderson, for example, in his book *All Connected Now*, focuses on globalization and highlights some of its most important features. In the last century, the number of international corporations has significantly increased and the financial and political power of these corporations has grown and spread across the globe, producing an ever expanding global economy. Global governance has also significantly grown, in part instigated by the need to monitor and regulate the expanding global economy. Hence, the growing global economy, global governance, and political system are tied together. There has also been a globalization of human rights and the development of principles of "World Law," as people around the world have come together to articulate a general set of moral and legal expectations regarding the treatment of fellow humans and basic living conditions and opportunities for everyone. According to Anderson, the emerging global society is a new kind of human society. It is responsible, to a great degree, for the increasing multiculturalism manifested in people's lives, for we are all being more exposed to different cultures and belief systems. (Hence, Anderson explains the pluralism of the Postmodern era as an effect of globalization.) Further, there is an increasing number of individuals who see themselves as cosmopolitan or international citizens, rather then identifying with a particular nation. Technology is also a significant factor in globalization. Our computer and communication technologies have become global systems facilitating the exchange of money, products, and information and allowing for the monitoring and control of numerous ecological conditions across the globe. The earth and humanity along with it is being "wired" together - technology is global and globalizing. Overall, Anderson sees the emerging global society as a more hopeful society than those of the past, a more "open society" with greater awareness of change and the future.[77]

Because of this general and pervasive trend toward globalization, a fundamental conflict has arisen in our times, and this conflict defines one of the key features of the contemporary transformation. The world has become divided between those who support globalization and those who do not and wish to preserve their local cultures and ways of life. Those who support globalization tend to advocate for a mixing of cultures and are pro-change. Those against globalization are against the mixing of cultures and tend to oppose change. Anderson describes this conflict as between "globalism" and "tribalism" while Thomas Friedman metaphorically describes it as "the Lexus versus the Olive Tree." Some writers, such as Samuel Huntington, see globalization as primarily a Western creation and foresee continued conflict between this Western vision of world unity and other cultures, such as Islam, which do not want to be enveloped in this spreading economic-cultural system. Finally, Benjamin Barber describes the conflict as "Jihad versus McWorld," where international corporations ("McWorld") in their attempt to create global economic markets are instigating frequent violent counter-reactions from local cultures ("Jihad") against global big business.[78] As Anderson states, globalization is becoming "the issue" of the global society.

Connected with the conflict over globalization is the ongoing disagreement between optimists and pessimists regarding the overall direction of change in contemporary times. This disagreement often takes the form of a dispute over whether the modern, technologically driven, expansionary capitalist-consumerist society is making the world a better place in which to live, or making things worse. Dinesh D'Souza describes the conflict as between the parties of "Yeh" and "Nah." As D'Souza states, the party of "Yeh" believes that material and moral progress come together and we should embrace the capitalist and free market way of life (This was the view of the Enlightenment as well.) The party of "Yeh" embraces technological advance and lives for the future, since, according to them, the past wasn't really that good and things are indeed getting better. The "Yeh" party believes in the promise of secular progress. (Basically they are Ray and Anderson's modernists.) The party of "Nah," in contrast, thinks that humanity's Golden Age was in the past and that the present embodies a deterioration of human society. For the party of "Nah," science, industry, and capitalism have produced a loss of community, a moral decline, and an end to the belief in a higher transcendent order (God).[79] Thus, in D'Souza's analysis, the pessimists have a regressive or backward looking perspective ("the good old days" syndrome), believing that the contemporary transformation is heading toward disaster; on the other hand, the optimists are forward looking and hopeful regarding where all the contemporary changes are leading. This debate, though, over the questionable effects of modernization goes back at least two hundred years to Romantic and earlier traditionalist critiques.

To add recent support to the arguments of the party of "Nah," Gregg Easterbrook in his book, *The Progress Paradox*, documents that although citizens of modernized countries have increasingly more material benefits and wealth, they do not necessarily feel any happier because of it. In fact, in the popular TV special, *Affluenza*, and accompanying book, a whole host of social, psychological, and moral problems are connected with the accelerative growth of production and consumption.[80]

Both D'Souza and Ray and Anderson provide a similar historical analysis of the underlying events that have led to this present conflict within human society. According to all of them, prior to the rise of modernism in Europe, the West was fundamentally a God-centered society. Values and purpose in life were defined by the Christian church. Revolutionaries of the modern era, though, perceived this Church-centered social order as exceedingly authoritarian and repressive. In the West, the rise of science, Enlightenment values such as reason, and democracy were various attempts to overturn the closed-minded and rigid social system of Medieval times. Yet what replaced the church and royalty was a market ideology and society of commerce. The West became a society that was money and time centered, held together by a shared selfishness, where greed was elevated to a moral virtue and imperative. This new modern society, though it created a whole set of benefits, produced increasing inequality, rootlessness, over-spending, alienation, and stress. The party of "Nah" could rightly ask, "Does this constitute progress?" Both tribal and fundamentalist critics of modernization frequently highlight the money-centered, un-godly way of life modernization has created.

There are other critical social and ideological movements that see a general trend toward decline, deterioration, and potential disaster in the present world. Postmodernists have critiqued the power hungry, supremacist behavior of the West, and environmentalists believe that in our rush toward technological advancement and economic growth, we are destroying our natural ecology beyond repair. All things considered, one defining feature of our contemporary world is a significant divide between those who see humanity heading in a positive direction and those who see us as spiraling downward. Many people, to recall the earlier discussion on the ambiguity of change in contemporary times, are mixed and uncertain in their assessment of where we are heading.

As Ed Cornish points out though, it is important to combat fatalism and nihilism about the future.[81] Even if pessimism has its value in highlighting problems that need to be faced and addressed, it can become a self-fulfilling prophecy.[82] It is important to have positive and realistic images of the future. Our contemporary world is faced, therefore, with a challenge, a challenge that has been with us probably since the beginning of human civilization: How do we create motivating and convincing positive images for the future. Clearly, there is presently a great divide among humanity regarding whether the future is something to fear, or something that engenders hope and optimism.

Leonard Shlain provides a different historical explanation of contemporary times and arrives at an optimistic conclusion regarding the overall future direction of humanity. For Shlain, one of the central conflicts within human history has been the emphasis placed on right-brain visual thinking versus left-brain linguistic thinking. He connects the former mode of thought with feminine values and behavior and the latter mode with male values and behavior. He sees male left-brain thinking as responsible for a great deal of the violence and injustice throughout history, whereas image-based, more feminine thinking tends to support peace and justice. Shlain believes that with the rise of visual media in the modern world, we are witnessing an "Iconic Revolution" and a more balanced mode of understanding and approaching life.

Shlain's analysis, in some ways, dovetails with the writings and ideas of the feminist philosopher Riane Eisler and the scientist Sally Goerner. Eisler sees most of recent human history as controlled by male authorities who organized human society in terms of dominance hierarchies. This system brought with it social injustice, inequality, and much war and violence. She does believe, though, that human society, especially in the last hundred years, is moving toward a more gender-balanced system. In her mind, the feminine approach to social organization is the establishment of partnerships, rather than top-down rule. Goerner, looking at the development of science over the last few centuries, argues that in science we are moving away from hierarchical and top-down models of natural order to network and connectivity models. Goerner believes that this new way of thinking in science is spreading outward into many different spheres of human life and will transform human society.[83]

Pulling together the ideas of Shlain, Eisler, and Goerner, there is perhaps a fundamental shift in human thinking, values, and modes of social organization occurring in the contemporary world. This general trend involves a set of inter-connected themes: The rise of women, the increasing importance of visual imagery, and a shift from dominance hierarchies to partnership networks. This general trend promises to create a more peaceful, balanced, and just world in the future. Yet, as Eisler, for one, is quick to point out, this new, more feminine mindset is a threat to the status quo, which is primarily ruled by males who think in terms of dominance hierarchies. Hence, the rise of women in contemporary times (which many writers would include as a fundamental trend of the contemporary transformation) has generated still another basic conflict within our world, as many individual cultures, that tend to be dominated by strong male rule, fight back against the global movement toward the equal rights of women.

The *Millennium Project*, introduced in the previous chapter, takes a somewhat different approach to describing the main features of the contemporary world. Instead of identifying basic trends as the key elements in its description, to recall, the *Millennium Project* identifies key challenges. This analysis provides a more action-oriented and dialectical view of our times; for each global challenge that is identified (there are fifteen) there will be a description of a desirable goal and of forces that are working both toward its realization and against it. The *Millennium Project* also describes provisional courses of actions, resources, and activist groups connected with addressing each of these key challenges. Included in their list are such challenges as bringing sustainable development to all people across the world; bringing more ethical considerations into the workings of the world economy; reducing the growing gap between the rich and the poor; using technology and science to improve the conditions of human life; and replacing authoritarian governments with democratic governments. The *Millennium Project*, coordinated by Jerome Glenn and Theodore Gordon, involves the systematic polling and analysis of judgments of experts around the world on key challenges and issues. Thus its assessment of the "state of the future" reflects a more global, balanced, and collective view than the ideas of any one individual or particular culture or ideology.[84]

The assessment of present times provided by the futurist Barbara Marx Hubbard also highlights the theme of challenges for the future. According to Hubbard, we are at a crisis point in the evolution of humanity. She sees two basic trends at work in our time: One trend is in the direction of self-transcendence and the other trend is in the direction of catastrophe. In her mind, which trend will dominate in the final analysis will depend upon the choices humanity makes; for Hubbard, the future is uncertain and can go either way. Hubbard believes though, as do Glenn and Gordon of the *Millennium Project*, that humanity possesses the resources and intelligence to meet the challenges facing us and to create a positive future. The question is whether we will pull together as a species and global society and constructively solve our problems. This is uncertain.

Hubbard presents her analysis of the present and the possibilities of the future in the context of human and natural history. She takes a "big picture" of time. According to her, the most fundamental pattern of change in history is evolution. The emergence and growth of human civilization is part of this overall cosmic process. As do many other writers, Hubbard believes that the evolutionary process is speeding up (or accelerating), but she adds a dialectical dimension to this process, arguing that evolutionary growth has a spiral form, where the evolutionary process passes through stages involving a crisis point or challenge at each stage that must be overcome if the growth process is to continue. (A similar model of growth was created by the psychologist Erik Erikson in describing the stages of individual psychological growth where a crisis at each developmental stage needed to be resolved if the person was to move on to the next stage of life.) Hubbard thinks that we are now at a crisis point in our evolution, but this crisis is not necessarily something to be depressed or fearful over, since evolutionary crises are the necessary preludes to advancement and jumps in evolution. As she puts it, "Our crisis is a birth." The nature of this fundamental crisis is best symbolized by the image of the mushroom cloud of an exploding atomic bomb, for this image represents humanity's recently created technological ability to self-destruct. We have come to the edge of the cliff and we can either fall into the abyss or take off and fly.

Many historical trends and developments, both positive and negative, have contributed to our present condition and world situation. On the positive side, Hubbard sees the growth of science and technology, democracy and individualism, and the recent feminist, human rights, environmental, and peace movements as all providing resources and ideas for successfully meeting the present world crisis. But there have also been negative trends, such as environmental deterioration, increasing military development, and continued conflict and animosity among the different cultures of the world, that threaten our very survival. Some trends, such as the growth of technology and economic productivity, are double-edged swords, providing resources for successful continued evolution, yet also creating negative and potentially very destructive effects in our world.

In Hubbard's mind, the key principle to our further development is "conscious evolution" where humanity purposefully guides its future evolution using the vast resources and knowledge base that we have created over the centuries. Further, she believes that it is critical that humanity comes together as a whole in this endeavor. We must move toward a "total resonance of all individuals" in a "co-creative" endeavor if we are to successfully meet our contemporary challenges. Hubbard sees the growing "social potential movement" - an outgrowth of the "human potential movement" - as the beginnings of such a collective effort. Thus, Hubbard stands with many other social theorists and writers throughout history, such as H.G. Wells, in basically arguing that "united we stand, divided we fall." In fact, Hubbard equates evil with separation and disconnectedness and good with creating an "Integral Culture."[85] Again, as in Goerner, we see the importance of the idea of connectedness.

Not everyone agrees that the solution to our contemporary problems lies in a coming together of humanity into a united whole. In fact, to recall, one of the central conflicts and issues of our time concerns the desirability of continued globalization. For many people, especially, but not exclusively outside of modernized countries, the movement toward globalization embodies an economic and political effort on the part of the West to dominate and assimilate all other cultures across the world. Many nationalists and cultural pluralists argue for the preservation of non-Western philosophies and ways of life, against the perceived continued colonialism and spread of Western concepts of progress across the globe.[86] One could argue that one of the most fundamental issues of our time is in actuality a continuation of an age old conflict between the forces toward social unity on one hand and those toward individuality and diversity on the other.[87] Each side of this coin, from an historical point of view, has had both positive and negative effects on the growth of human society; hence, there is no simple answer to this issue. Because this conflict has been so fundamental to the entire history of humanity, it is doubtful whether it will be resolved one way or the other in the immediate future. We will continue to struggle over unity versus diversity.

In two of his books on contemporary culture, Anderson distills two somewhat different classification schemes of the main "world views" or "stories" competing for our attention and allegiance in interpreting the meaning and direction of our times.[88] These maps of the ideological terrain provide a good conceptual bridge, summarizing a variety of the above ideas and views, and providing a lead into the more detailed review of theories and paradigms of the future in the next chapter. They present a perspective on the complex nature of our times.

In the first book, *Reality Isn't What It Used to Be*[89], Anderson lists and discusses the following six contemporary worldviews:

- The Western Myth of Progress
- Marxist Socialism
- Christian Fundamentalism
- Islamic Fundamentalism
- Green - Environmentalist Philosophy
- The "New Paradigm" Story

There are numerous theoretical and practical disagreements among these worldviews - disagreements that involve interpretations of our past, our present, and the future. Has the last century of human history been uphill, downhill, or more of the same? Are we creating, destroying, or are we stuck in a rut? Have we been going in one direction or many? Fundamentalism, either Christian or Islamic, emphasizes past traditions, in opposition to both "Progress" and "New Paradigm" views. (The "New Paradigm" view would include Ray's Cultural Creatives and writers such as Hazel Henderson and Barbara Marx Hubbard.) From the fundamentalist perspective, the last century has been morally degenerative. But Christian and Islamic fundamentalisms are clearly at odds with each other regarding many different cultural, political, and religious issues. Although both "Progress" and "New Paradigm" advocates may see themselves as leading the way into a "better world," fundamentalists, Greens, and Marxists, for different reasons, see Western progress as causing past problems and leading to new difficulties and perhaps worldwide disaster. Further, the "New Paradigm" group sees the Western progress view as stuck in the past and unwilling to change.

Anderson, though, wishes to add a seventh perspective, the Postmodern view, which he sees emerging out of this complex array of points of view. Postmodernism, according to Anderson, directly confronts the complexity of our times, acknowledging that there are multiple points of view, each created in a social-historical context containing valuable insights, yet no one view having some privileged access to "The Truth." In fact, the different points of view support flexibility in thinking, allowing us to look at things from more than one perspective and engage in dialogue, self-reflection, and further evolution. Following from this logic, one good reason for reviewing and discussing in depth the various theories and paradigms on the future is to broaden, enrich, and open up our view of today and tomorrow. But Postmodernism can also be criticized as generating too much ambiguity and chaos and a loss of standards without any sense of direction or purpose.[90]

Anderson in a later book, *The Truth About the Truth*,[91] a collection of essays by numerous contemporary writers, streamlines his classification scheme into four main views. In his article "Four Different Ways to Be Absolutely Right,"[92] he identifies and describes the following four world views:

- Postmodern-Ironist
- Scientific Rational
- Social Traditional
- Neo-Romantic

Once more, Anderson thinks that Postmodernism is the most flexible and "enlightened" way into tomorrow. He believes that all three other main views are to different degrees rooted and stuck in the past. The "Scientific Rationalist" supports secular Western progress, technology, and the supremacy of reason. The "Social Traditionalist," which would include fundamentalism and Ray's "Heartlanders," tend to ignore the cultural richness of humanity and oversimplify reality into a single vision. The Neo-Romanticist, which includes Green/ environmentalists and "New Agers," is, according to Anderson, most "strongly oriented to the past." From a technological or a scientific-rational perspective, to "Return to Nature" is both unrealistic and incredibly regressive. But all these views co-exist in our complex contemporary world, in opposition with each other over who possesses the "Truth."

Thinking through the above themes and points of debate, one is reminded of the opening lines of Charles Dickens's *A Tale of Two Cities*, quoted at the beginning of this chapter. Have the issues and concerns changed much in the last two hundred years since Dickens wrote the above words? In many ways they clearly have, but it is interesting to note that Dickens wrote these words during the last great social transformation as the Industrial Age spread across Europe and America. We are once again looking for the answer into the future - perhaps a new answer or some combination of old ones - and the confusion we feel is quite natural. As we have seen, there is a set of basic trends that provide a general picture of the nature of our times and the contemporary transformation. Yet, different theories interpret these trends differently. For example, is accelerative growth a positive or a negative trend, or is globalization good or bad? To add to the uncertainty of our journey into the future we find many answers - perhaps too many answers - often contradictory and conflicting - over the direction in which we are heading - vying for our attention and commitment on the horizon of tomorrow. At the very least, we can get an overall sense of the different views, something begun in the above discussion of Anderson's classification scheme, and that is the function of the next chapter.

References

[1] Glennon, Lorraine (Ed.) *Our Times: The Illustrated History of the 20th Century*. Atlanta: Turner Publishing, 1995; Christian, David *Maps of Time: An Introduction to Big History*. Berkeley, CA: University of California Press, 2004, Chapter Fourteen.

[2] Zey, Michael G. *Seizing the Future: How the Coming Revolution in Science, Technology, and Industry Will Expand the Frontiers of Human Potential and Reshape the Planet*. New York: Simon and Schuster, 1994; Wright, Robert *Nonzero: The Logic of Human Destiny*. New York: Pantheon Books, 2000.

[3] Toffler, Alvin *Future Shock*. New York: Bantam, 1971; Toffler, Alvin *The Third Wave*. New York: Bantam, 1980; Toffler, Alvin *Power Shift: Knowledge, Wealth, and Violence at the Edge of the Twenty-First Century*. New York: Bantam, 1990; Toffler, Alvin, and Toffler, Heidi *Creating a New Civilization: The Politics of the Third Wave*. Atlanta: Turner Publishing, Inc., 1994; Toffler, Alvin, and Toffler, Heidi "Getting Set for the Coming Millennium" *The Futurist*, March/April, 1995; Anderson, Walter Truett *Reality Isn't What It Used To Be*. New York: Harper, 1990; Anderson, Walter Truett *Evolution Isn't What It Used To Be: The Augmented Animal and the Whole Wired World*. New York: W. H. Freeman and Company, 1996.

[4] Watson, Peter *The Modern Mind: An Intellectual History of the 20th Century*. New York: HarperCollins Perennial, 2001, Chapters One, Four, and Six.

[5] Lombardo, Thomas "Romanticism" in Lombardo, Thomas *The Evolution of Future Consciousness*. Bloomington, Indiana: Author House, 2006.

[6] Watson, Peter, 2001, Chapters Ten to Thirteen.

[7] Lombardo, Thomas "Hegel, Marx, and the Dialectic" in Lombardo, Thomas *The Evolution of Future Consciousness*. Bloomington, Indiana: Author House, 2006.

[8] Bloom, Howard *The Lucifer Principle: A Scientific Expedition into the Forces of History*. New York: The Atlantic Monthly Press, 1995.

[9] Tuchman, Barbara *The Guns of August*. New York: Ballantine Books, 1962.

[10] Watson, Peter, 2001, Chapter Eighteen.

[11] Watson, Peter, 2001, Chapters Nine and Twelve.

[12] Watson, Peter, 2001, Chapters Thirteen, Seventeen,and Nineteen.

[13] Watson, Peter, 2001, Chapter Twenty-Three; Sartre, Jean Paul *Being and Nothingness*. New York: Washington Square Press, 1953.

[14] Lombardo, Thomas "Science, Enlightenment, Progress, and Evolution" in Lombardo, Thomas *The Evolution of Future Consciousness*. Bloomington, Indiana: Author House, 2006.

[15] Watson, Peter, 2001, Chapter Twenty-One.

[16] Watson, Peter, 2001, Chapter Twenty-Four.

[17] Watson, Peter, 2001, Chapters Seven and Thirty.

[18] Fukuyama, Francis *The End of History and the Last Man*. New York: The Free Press, 1992.

[19] Moore, Stephen and Simon, Julian *It's Getting Better All the Time: 100 Greatest Trends of the Last 100 Years*. Washington, D.C.: Cato Institute, 2000, page xi.

[20] Watson, Peter, 2001, Chapter Twenty-Five.

[21] Watson, Peter, 2001, Chapter Twenty-Six.

[22] Watson, Peter, 2001, Chapter Thirty-Five; Best, Steven and Kellner, Douglas *Postmodern Theory: Critical Interrogations*. New York: The Guilford Press, 1991; Best, Steven and Kellner, Douglas *The Postmodern Turn*. New York: The Guilford Press, 1997.

[23] Watson, Peter, 2001, Chapter Sixteen; Kuhn, Thomas *The Structure of Scientific Revolutions*. Chicago: University of Chicago Press, 1962; Feyerabend, Paul "Problems of Empiricism" in Robert Colodny (Ed.) *Beyond the Edge of Certainty*. Englewood Cliffs, N. J.: Prentice-Hall, 1965; Feyerabend, Paul "Problems of Empiricism II" in Robert Colodny (Ed.) *The Nature and Function of Scientific Theory*. London: University of Pittsburgh Press, 1969.

[24] Watson, Peter, 2001, Chapters Twenty-Eight and Thirty.

[25] Nisbet, Robert *History of the Idea of Progress*. New Brunswick: Transaction Publishers, 1994, Pages 317 - 357; Polak, Frederik *The Image of the Future*. Abridged Edition by Elise Boulding. Amsterdam: Elsevier Scientific Publishing Company, 1973.

[26] Watson, Peter, 2001, Chapter Thirty-Three; Lasch, Christopher *The Culture of Narcissism*. New York: Warner Books, 1979.

[27] Quinn, Daniel *Beyond Civilization: Humanity's Next Great Adventure*. New York: Three Rivers Press, 1999.

[28] Watson, Peter, 2001, Chapter Forty-One; Bloom, Allan *The Closing of the American Mind*. New York: Simon and Shuster, 1987.

[29] Brown, Donald *Human Universals*. New York: McGraw-Hill, 1991; Pinker, Steven *The Blank Slate: The Modern Denial of Human Nature*. New York: Penguin Books, 2002; Wilson, E. O. *Sociobiology: The New Synthesis*. Cambridge, MA: Harvard University Press, 1975.

[30] See Lombardo, Thomas "Darwin's Theory of Evolution" in Lombardo, Thomas *The Evolution of Future Consciousness*. Bloomington, Indiana:

Author House, 2006 for a history of the early development of the theory of evolution.

[31] Watson, Peter, 2001, Chapters Thirty-Four, Thirty-Nine, Forty-Two and Conclusion.

[32] Gitlin, Todd *Media Unlimited: How the Torrent of Images and Sounds Overwhelms Our Lives*. New York: Metropolitan Books, 2001.

[33] Watson, Peter, 2001, Conclusion; Huntington, Samuel *The Clash of Civilizations and the Remaking of World Order*. New York: Touchtone, 1996; Friedman, Thomas *The Lexus and the Olive Tree: Understanding Globalization*. New York: Farrar, Straus, and Giroux, 1999.

[34] Christian, David, 2004, Chapter Fourteen.

[35] Gribbin, John *Genesis: The Origins of Man and the Universe*. New York: Delta, 1981; Gribbin, John *In Search of the Big Bang: Quantum Physics and Cosmology*. New York: Bantam, 1986; Davies, Paul *God and the New Physics*. New York: Simon and Schuster, 1983; Davies, Paul *The Cosmic Blueprint: New Discoveries in Nature's Creative Ability to Order the Universe*. New York: Simon and Schuster, 1988; Davies, Paul *The Mind of God: The Scientific Basis for a Rational World*. New York: Simon and Schuster, 1992; Prigogine, Ilya and Stengers, Isabelle *Order out of Chaos: Man's New Dialogue with Nature*. New York: Bantam, 1984; Kaku, Michio *Visions: How Science will Revolutionize the 21st Century*. New York: Anchor Books, 1997; Adams, Fred and Laughlin, Greg *The Five Ages of the Universe: Inside the Physics of Eternity*. New York: The Free Press, 1999.

[36] Glennon, Lorraine, 1995.

[37] Kelly, Kevin *Out of Control: The Rise of Neo-Biological Civilization*. Reading, MA: Addison - Wesley, 1994; Anderson, Walter, 1996; Clark, Andy *Natural-Born Cyborgs: Minds, Technologies, and the Future of Human Intelligence*. Oxford: Oxford University Press, 2003.

[38] Isaacson, Walter "Time's Choice: Who Mattered - and Why" *Time*, Vol. 154, No. 27, December 31, 1999.

[39] Wilson, E.O. *Consilience: The Unity of Knowledge*. New York: Alfred A. Knopf, 1998; Wilson, E.O. "The Biological Basis of Morality" *The Atlantic Monthly*, April, 1998b.

[40] Fukuyama, Francis, 1992; Isaacson, Walter, 1999.

[41] Kahn, Hermann *On Thermonuclear War*. Princeton: Princeton University Press, 1960.

[42] Toffler, Alvin, and Toffler, Heidi *War and Anti-War: Survival at the Dawn of the 21st Century*. Boston: Little, Brown, and Company, 1993.

[43] Meadows, Dennis, Meadows, Donella, and Randers, Jorgen *Beyond the Limits*. Toronto: McClelland & Stewart, 1992.

[44] Joy, Bill "Why the Future Doesn't Need Us" *Wired*, April, 2000.

[45] Kauschke, Hans-Gerhard "Going Height Crazy: Super Skyscrapers of the Future" *The Futurist*, November-December, 1986; Barrett, David B. "Global Statistics" in Kurian, George Thomas, and Molitor, Graham T.T. (Ed.) *Encyclopedia of the Future*. New York: Simon and Schuster Macmillan, 1996.; Gappert, Gary "Cities" in Kurian, George Thomas, and Molitor, Graham T.T. (Ed.) *Encyclopedia of the Future*. New York: Simon and Schuster Macmillan, 1996.

[46] Anderson, Walter, 1990.

[47] Gleick, James *Faster: The Acceleration of Just About Everything*. New York: Pantheon Books, 1999.

[48] Sawyer, Deborah "The Pied Piper Goes Electronic" *The Futurist*, February, 1999.

[49] Watson, Peter, 2001, Conclusion.

[50] Cornish, Edward *Futuring:The Exploration of the Future*. Bethesda, Maryland: World Future Society, 2004, Chapter Two.

[51] Henderson, Hazel *Paradigms in Progress: Life Beyond Economics*. San Francisco: Berrett-Koehler Publishers, 1991; Theobald, Robert *Turning the Century*. Indianapolis: Knowledge Systems, Inc., 1992; Toffler, Alvin, 1990.

[52] Postrel, Virginia *The Future and Its Enemies: The Growing Conflict Over Creativity, Enterprise, and Progress*. New York: Touchstone, 1999.

[53] Clark, Mary E. "Mind: New Modes of Thinking" in Kurian, George Thomas, and Molitor, Graham T.T. (Ed.) *Encyclopedia of the Future*. New York: Simon and Schuster Macmillan, 1996.

[54] Lombardo, Thomas "The Psychology and Value of Future Consciousness" in Lombardo, Thomas *The Evolution of Future Consciousness*. Bloomington, Indiana: Author House, 2006.

[55] Postrel, Virginia, 1999, describes this conflict as the clash between "stasis" and "dynamism".

[56] Tipler, Frank *The Physics of Immortality: Modern Cosmology, God, and the Resurrection of the Dead*. New York: Doubleday, 1994.

[57] Ray, Paul "What Might Be the Next Stage in Cultural Evolution?" in Loye, David (Ed.) *The Evolutionary Outrider: The Impact of the Human Agent on Evolution*. Westport, Connecticut: Praeger, 1998; Ray, Paul and Anderson, Sherry *The Cultural Creatives: How 50 Million People are Changing the World*. New York: Three Rivers Press, 2000.

[58] Theobold, Robert, 1992; Meadows, Dennis, Meadows, Donella, and Randers, Jorgen, 1992; Slaughter, Richard "Futures Concepts" in Slaughter, Richard (Ed.) *The Knowledge Base of Future Studies*. Volume I. Hawthorn, Victoria, Australia: DDM Media Group, 1996.

[59] Stock, Gregory *Metaman: The Merging of Humans and Machines into a Global Superorganism*. New York: Simon and Schuster, 1993; Wright,

Robert, 2000; Bloom, Howard *Global Brain: The Evolution of Mass Mind from the Big Bang to the 21st Century.* New York: John Wiley and Sons, Inc., 2000.

[60] Kurzweil, Ray *The Age of Spiritual Machines: When Computers Exceed Human Intelligence.* New York: Penguin Books, 1999; Kurzweil, Ray *The Singularity is Near: When Humans Transcend Biology.* New York: Viking Press, 2005; Moravec, Hans *Robot: Mere Machine to Transcendent Mind.* Oxford: Oxford University Press, 1999.

[61] Gleick, James, 1999.

[62] Fukuyama, Francis *The Great Disruption: Human Nature and the Reconstitution of Social Order.* New York: The Free Press, 1999.

[63] Henderson, Hazel, 1991.

[64] Havel, Vaclav "The Need for Transcendence in the Postmodern World" *The Futurist*, July-August, 1995.

[65] Zohar, Danah and Marshall, Ian *The Quantum Society: Mind, Physics, and a New Social Vision.* New York: William Morrow and Co., Inc., 1994.

[66] Naisbitt, John and Aburdene, Patricia *Megatrends 2000.* New York: Avon Books, 1990.

[67] Capra, Fritjof *The Turning Point.* New York: Bantam, 1983.

[68] Hubbard, Barbara Marx *Conscious Evolution: Awakening the Power of Our Social Potential.* Novato, CA: New World Library, 1998.

[69] Snyder, David Pearce "The Revolution in the Workplace: What's Happening to Our Jobs?" *The Futurist*, March-April, 1996.

[70] Henderson, Hazel, 1991.

[71] Cornish, Edward, 2004, Pages 22 - 29.

[72] Cornish, Edward, 2004, Page 16.

[73] Lombardo, Thomas "The Great Awakening, Culture, and the Discovery of Death" in Lombardo, Thomas *The Evolution of Future Consciousness.* Bloomington, Indiana: Author House, 2006; White, Randall *Prehistoric Art: The Symbolic Journey of Humankind.* New York: Harry N. Abrams, 2003; Calvin, William *A Brief History of the Mind: From Apes to Intellect and Beyond.* New York: Oxford University Press, 2004; Diamond, Jared *The Third Chimpanzee: The Evolution and Future of the Human Animal.* New York: HarperPernnial, 1992.

[74] Vinge, Vernor "The Coming Technological Singularity: How to Survive in the Post-Human Era" *Vision-21: Interdisciplinary Science and Engineering in the Era of Cyberspace NASA-CP-10129*, 1993 - http://www-rohan.sdsu.edu/faculty/vinge/misc/singularity.html.

[75] Eldredge, Niles and Gould, Stephen "Punctuated Equilibria: An Alternative to Phyletic Gradualism" in Schopf, T. J. M. (Ed.) *Models in Paleobiology.* Freeman Cooper, 1972; Gell-Mann, Murray *The Quark and*

the Jaguar: Adventures in the Simple and the Complex. New York: W.H. Freeman and Company, 1994.

[76] Best, Steven and Kellner, Douglas, 1991; Best, Steven and Kellner, Douglas, 1997.

[77] Anderson, Walter Truett *All Connected Now: Life in the First Global Civilization*. Boulder; Westview Press, 2001.

[78] Huntington, Samuel, 1996; Freidman, Thomas, 1999; Barber, Benjamin *Jihad vs. McWorld*. New York: Ballantine Books, 1995, 2001.

[79] D'Souza Dinesh *The Virtue of Prosperity: Finding Values in an Age of Techno-Affluence*. New York: The Free Press, 2000.

[80] Easterbrook, Gregg *The Progress Paradox: How Life Gets Better While People Feel Worse*. New York: Random House, 2003; DeGraaf, John, Wann, David, and Naylor, Thomas *Affluenza: The All-Consuming Epidemic*. San Francisco: Berret-Koehler Publishers, Inc., 2001.

[81] Cornish, Edward, 2004, Chapter Fifteen.

[82] Seligman, Martin *Learned Optimism: How to Change Your Mind and Your Life*. New York: Pocket Books, 1998.

[83] Eisler, Riane *The Chalice and the Blade: Our History, Our Future*. San Francisco: Harper and Row, 1987; Eisler, Riane *Sacred Pleasure: Sex, Myth, and the Politics of the Body*. San Francisco: HarperCollins, 1995; Goerner, Sally *After the Clockwork Universe: The Emerging Science and Culture of Integral Society*. Norwich, Great Britain: Floris Books, 1999.

[84] Glenn, Jerome and Gordon, Theodore *2004 State of the Future*. American Council for the United Nations University, 2004; Millennium Project - http://www.acunu.org/millennium/challeng.html

[85] Hubbard, Barbara Marx, 1998; Hubbard, Barbara Marx *Emergence: The Shift from Ego to Essence*. Charlottesville, VA: Hampton Roads Publishing, 2001; Hergenhahn, B.R. and Olson, Matthew *An Introduction to Theories of Personality*. 6th Edition. Upper Saddle River, NJ: Prentice Hall, 2003, Chapter Six.

[86] Sardar, Ziauddin *Rescuing All Our Futures: The Future of Future Studies*. Westport, Connecticut: Praeger, 1999.

[87] Lombardo, Thomas "Science, Enlightenment, Progress, and Evolution' in Lombardo, Thomas *The Evolution of Future Consciousness*. Bloomington, Indiana: Author House, 2006.

[88] Anderson, Walter Truett, 1990; Anderson, Walter Truett (Ed.) *The Truth About the Truth: De-Confusing and Re-Constructing the Postmodern World*. New York: G.P. Putnam's Sons, 1995.

[89] Anderson, Walter Truett, 1990.

[90] Wilson, E.O., 1998; Wilson, E.O. "Back from Chaos" *The Atlantic Monthly*, March, 1998c; Watson, Peter, 2001, Conclusion.

[91] Anderson, Walter Truett, 1990.

[92] Anderson, Walter Truett "Four Different Ways to be Absolutely Right" in Anderson, Walter Truett (Ed.) *The Truth About the Truth: De-Confusing and Re-Constructing the Postmodern World.* New York: G.P. Putnam's Sons, 1995b.

CHAPTER FOUR

THEORIES AND PARADIGMS OF THE FUTURE

"It's all a question of story. We are in trouble just now because we do not have a good story. We are in between stories. The old story, the account of how the world came to be and how we fit into it, is no longer effective. Yet we have not learned the new story."

Thomas Berry

"If there is anything we have plenty of, it is belief systems."

Walter Truett Anderson

"...the future is usually a combination of all the stories you can construct to anticipate it."

Joel Garreau

Introduction

In this final chapter, I present an overview of the broad range of general theories and paradigms of the future. Theories and paradigms of the future often begin with the belief that humanity is in the midst of a pervasive world transformation and attempt to explain the transformation and where the big changes may be taking us. There are, though, significant differences among theories on these basic questions. I describe a variety of the debates and conflicts between these different points of view and conclude the chapter with a summary of key themes and issues.

I define a **theory** as an abstract description and explanation of some set of related facts or dimension of reality. A theory of the future both describes and explains present contemporary facts and trends and presents a variety of hypotheses regarding the future; it identifies the direction reality is taking into the future. Theories of the future often provide an interpretation of both the past and the present, suggesting trends and directions in our history that may continue into the future. A theory of the future may make specific predictions or describe some set of plausible possibilities for the future.

As noted earlier, theories of the future are prescriptive as well as descriptive. Theories do not simply explain the present and predict the future, but they also prescribe what we should and should not do in effecting the future course of events. A theory often proposes a general plan of action for tomorrow. Theories of the future are prescriptive because they are usually connected with values and ideologies; contemporary conditions and trends are evaluated as to their negative or positive qualities, and desirable or preferable future directions and goals are identified and argued for, all relative to a set of values. Further, a theory of the future provides a sense of inspiration also based on a set of values. Finally, the facts and predictions highlighted within a theory strongly influence the values prescribed and vice versa. Vision and value are connected.

Theories of the future are also often connected with stories - stories that describe in narrative and chronological form the great drama of time relating together events of the past, present, and hypothetical future. Such stories give a futurist theory psychological and social force, inspiring those who believe in the theory. The values of a theory are usually expressed within the story. In essence, a story is a theory put into dramatic form.

A **paradigm** can be described as a theory plus a way of behaving and living consistent with a theory. Living a paradigm not only involves understanding the world in a certain way, but attempting to practice the values and prescriptions for action within the core theory. Living a paradigm involves attempting to create the preferred sense of direction identified in the theory. Since a theory usually points out problems and challenges for the future, living the paradigm will involve ways to address and solve these problems and challenges. Following Thomas Kuhn, who popularized the concept of a "paradigm," following a paradigm will not just influence how we think and what we believe, but how we perceive the world and how we behave. Followers of different paradigms for the future often perceive the contemporary world around them very differently and because of such different perceptions respond differently to things.[1]

When a number of people live a paradigm for the future, a **social movement** may emerge. Within contemporary times there are a variety of social movements concerned with the future that profess different theories and express through their behavior different paradigms for the future. Within a social movement, believers in a theory collectively organize together, often with the intent of influencing others to adopt their ideas, values, and lifestyles. A social movement is a theory put into collective social action.

Our present world is multifaceted, with multiple meanings and dimensions - reverberating and evolving through our intricate communication systems as people debate and dialogue on different issues and points of view. Theories of the future attempt to provide explanations and synoptic descriptions of our present conditions and fundamental trends, but there are many different theories - each with its own slant on things - emphasizing one or more of the fundamental dimensions of change. The theories themselves are in a state of mutual amplification and development, both in competition and collaboration. There are many answers to the question of the meaning of our times - for we live in an era of multiple and complex theories and stories - and this kaleidoscopic array of views is in fact an essential feature of the answer to the question of the meaning of our times. We live in a pluralistic world - a competitive reality of different points of view and different social movements.[2]

We are in the midst of a fundamental world transformation, with a set of different theories and values systems of the future attempting to define and guide the direction for tomorrow. The future, in fact, will be greatly influenced by these theories and paradigms. Clearly though, there are fundamental disagreements, perhaps some of which can be reconciled and some of which probably can not. Further, social movements, as expressions of different paradigms, will clash, often violently, in this competitive and creative process. As noted in the previous chapter, history is filled with wars over opposing theories of the future and will probably continue to be so in the years to come. Theoretical conflicts can be healthy though - promoting dialogue and an ongoing evolution of our understanding of where we are and where we are going. The disagreements among those holding these theories highlight the different possibilities for the future and the different values for the future.

Theories of the future often focus on some particular theme or themes considered of central importance in the nature and organization of future events. I have grouped theories and paradigms according to a set of basic themes. Some theories, because they are broad and comprehensive and integrate several themes, could be listed under different themes. The theories are organized according to the following general themes:

- Theories that Highlight Time and Change
- Theories that Highlight Science, Technology, and Rationalism

- Theories that Highlight Ecology and Nature
- Theories that Highlight Psychology and Human Relations
- Theories that Highlight Society, Culture, and Morals
- Theories that Highlight the Spiritual, Religious, and Mystical
- Integrative Theories

The descriptions of theories included below provides a general overview of contemporary futurist thought; I have attempted to identify what I think are the most important, influential, and distinctive theories. The list is not intended to be definitive or complete - there are simply too many theories. The list does, however, provide a broad sampling of different approaches and areas of focus. Each theory usually teaches something important, adding to the richness and wonder of reality and the future. Further, the theories presented are not all mutually exclusive; various writers and futurists often combine various theories in their particular view of the future.

Theories of Time and Change

Stasis, Traditionalism, and the Return to the Past: Included within this theoretical perspective would be viewpoints that see modernism and continued growth and change as negative and destructive to humanity and which promote a move backward in time to a simpler, more stable and uniform way of life. Ray, in his study of American sub-cultures and values, identifies the traditionalist group in America as "Heartlanders," accounting for approximately thirty percent of the total population. Heartlanders tend to believe in the literal truth of the *Bible* and believe that the *Bible* provides a moral foundation for assessing the present and a moral compass for guiding the future. In general, they emphasize the importance of virtue in life. In particular, Heartlanders support small town, rural values, and believe that modern urban life is, in many ways, immoral. They tend to be anti-science, anti-technology, and pro-naturalistic. They are also often patriotic and antagonistic to foreigners. Aside from being culturally and religiously conservative, "Heartlanders" also show a strong belief in traditional domestic roles for women.[3] Because Heartlanders attempt to practice what they believe and form a relatively big organized group, they constitute a strong social movement in the United States.

Along similar lines, Virginia Postrel describes the fundamental conflict of our times in terms of a clash between "stasis" and "dynamist" points of view. According to Postrel, advocates of stasis believe that the world has gone terribly wrong and that someone needs to take control of the situation to set things right. Postrel argues that supporters of stasis fear the unpredictability of the future and uncontrollable change. Hence, the stasis viewpoint emphasizes the need to control the future, usually by maintaining the status quo or returning to some idealized and better way of life in the past. Postrel would include both conservative cultural reactionaries and technocrats in the stasis group.[4]

The themes of stasis, tradition, and returning to the past are also strong features of various cultural and nationalist movements around the world. Again the perceived threat is the fast changing modern world and the forces of globalization, but from the point of view of many non-Western people, it is the West, and in particular the United States, that threatens time honored ways of life. Hence, interestingly, from within the United States, traditionalists tend to be patriotic and pro-United States, whereas from the outside, traditionalists are often anti-United States. Both Barber and Friedman describe various anti-Western, antiglobalization reactions of local indigenous cultures that are attempting, in opposition to the forces of Western modernism, to preserve their traditions and ways of life.[5]

Fundamentalism: Fundamentalism is any theory of the future that adheres to a strict and literal interpretation of some traditional religious text and doctrine, e.g. Christian or Islamic fundamentalism, and rejects as invalid and morally incorrect any deviation from these ideas. Fundamentalism assumes an absolutist theory of truth and value - arguing that its particular belief system is unquestionably true and all other belief systems are wrong insofar as they conflict with it. It is usually based on a singular cultural framework and tends to be grounded in past ideas and values.[6]

Walter Truett Anderson describes one of the central conflicts of our time as between relativists and absolutists. He argues that twentieth-century developments in science and culture destroyed many "classical truths," yet with the "triumph of science" old religious belief systems did not go away. In fact, the rise of fundamentalism with its emphasis on absolutist thinking emerged in counter-reaction to the growth of science and liberal, relativist thought. Fundamentalism is therefore a recent development in reaction to the threat of modernism. Its appeal is that it provides a sense of security and stability in a world of ambiguity and flux and clearly and cleanly divides reality into good and evil forces.[7]

In his book, *Powerful Times*, Eamon Kelly identifies "the sacred and the secular" as one of the fundamental dynamic tensions in our contemporary world and highlights religious fundamentalism in his discussion of the sacred perspective on human life.[8] What Kelly emphasizes about fundamentalism is its divisive quality. In reaction to the perceived threatening nature of modern secularism, fundamentalist social movements (such as Christian and Islamic) are often openly antagonistic and at times even violent in their counter-reactions to secularism. For example, terrorism coming out of the Middle East is strongly connected with Islamic fundamentalist groups. Among its followers, fundamentalism generates strong feelings of separateness and exclusivity and a "tribal" mentality, and the future is literally conceptualized as a war between believers and non-believers. This warlike mentality is understandable since at least for Christian and Islamic religious traditions, the future is described as involving a great final conflict between good and evil. Fundamentalists believe that they are on the side of good and that the secular way of life embodies all that is evil in the world.[9]

Although fundamentalism is a recent development, its vision of the future is unequivocally rooted in the past. For both Christian and Islamic fundamentalism, the truth about the meaning and purpose of life was revealed long ago and recorded in the sacred texts of their religion. The struggles and problems of modern times and the eventual resolution of our present challenges are likewise interpreted in terms of fundamentalist doctrine, presumably anticipated by their great religious prophets in ancient or earlier times.

Creationism: Christian fundamentalists tend to support a "creationist" view of nature and the origin of humankind, arguing that God purposefully created the universe and humans, as explained in the Book of *Genesis*. Creationists believe in the possibility of a literal and singular, unequivocal reading of the Bible; they believe that this literal reading provides the "Truth," and they believe they know what that correct reading is; hence, they are absolutists. Strictly speaking this theory concerns the past rather than the present or future, but adopting this viewpoint carries with it certain implications regarding the future. Because creationists vehemently oppose the theory of evolution, they believe that life and humanity in the future will not be determined by natural evolutionary processes - humankind will not transform biologically. Creationists see different forces than evolution at work in history and the future. Generally, they are skeptical or antagonistic to science, believing science and evolution, in particular, do not provide any answer to the meaning and purpose of life and do not provide any moral guidance. Creationists find purpose in life (a sense of direction for the future) through God. They object to evolution because evolutionary theory seems to imply that chance plays a significant role in the origin of humanity; creationists believe that there is a divine purpose behind the existence of humanity and a divine purpose to the future.

They see humankind as a special creation, rather than just part of the overall evolutionary process. They are also dualists, objecting to a materialist view of reality, and instead believe that there is an "immaterial" soul within each person as well. Creationists associate evolution and science with materialism. They also associate evolution with a competitive, violent vision of life, which undermines, in their view, a moral vision of the universe. Though creationists would argue that they base their position on religious tradition and the principles of a time-honored ancient text, the modern rise of creationism, as with fundamentalism, is based on a strong counter-reaction to recent trends in thinking in science and culture. Creationists often support the "**Intelligent Design**" argument for humanity and the universe, but the "Intelligent Design" argument does not necessarily imply creationism and can be formulated in ways that involve neither a Christian God nor any kind of God whatsoever.[10]

Dynamism and Rapid Change: As noted above, Postrel believes that the most basic clash within approaches to the future is between those who emphasize change and those who emphasize stability; as she puts it, the conflict of "dynamism" versus "stasis."[11] Postrel supports a dynamist view and literally believes, as indicated in the title of her book, that those who support a stasis view are the "enemies" of the future. According to Postrel, instead of trying to control and direct the future from a top-down authoritarian position, we should allow free-flowing creativity and diverse points of view to generate the future. Postrel, following the popular view in contemporary open systems thinking, argues that new order arises out of chaos.[12] Freedom, creativity, and diversity all stimulate growth and change which, in her mind, is the preferable type of future. She sees an infinite and never-ending array of forthcoming inventions, new ideas, and achievements if we embrace a "dynamist" approach to the future.

Most social and technological observers would agree that we live in a world of rapid change. Yet, I think Postrel is correct in her assessment that there is a divide between those who believe change is good and those who believe that change is bad or at the very least excessive in our times. Toffler offers a related distinction in comparing the conflict between "the fast" and "the slow."[13] There are many individuals, as well as cultures, that seem to embrace change and want to move full speed ahead, and other individuals and groups who want to move much more slowly, if move at all. Within the United States, following Ray and Anderson's research, modernists by and large find change a positive thing, whereas Heartlanders do not. In the following review of theories that highlight change in some manner or form, we will find some views that emphasize the positive features of contemporary change and some views that stress the negative aspects of change.

Future Shock: The Industrial Era was marked by a significant increase in the rate of change as measured by population growth and economic, technological, and scientific developments. Change proceeded much more slowly during the earlier Agricultural Era.[14] Toffler, in his first book *Future Shock*, proposed that a new jump in the rate of change is occurring in present times. We are living within an unsurpassed technological and informational explosion that is coupled with a social reality of increased transience and turnover in products, friends, spouses, neighborhoods, and vocations. The first main section of *Future Shock* is titled "The Death of Permanence" - as Heraclitus argued, everything is flux - nothing stands still. We live in an era where change and growth are hot topics of study, discussion, debate, and concern.[15] Media, government, business, and pop culture all play up the importance and value of change.[16]

According to Toffler, things in our lives are changing so fast that we are being pushed to the limits of our abilities to adapt. We are experiencing future shock. We live in an era of stress and overload. The world around us is changing faster than humans -- psychologically, physically, and culturally -- can keep up. Future shock does not simply impact modernized countries where technological advancements are more pervasive and readily embraced, it also impacts less-developed countries where old ways of life are quickly being disrupted, if not destroyed, as new ideas, products, and technologies move into these regions. Toffler contends in *Future Shock* that a balance needs to be achieved between stability and change. Accelerative change needs to be recognized and controlled.

In his later book *Power Shift*, Toffler points out that the fast rate of change in modern life has instigated various counter-reactions that emphasize the importance of tradition and conservative values.[17] Such traditionalist movements can be seen as a way to cope with future shock - in this case to retreat into the past. Yet, this negative reaction to change provokes the question of whether rapid change is inevitable or not. If the future is a matter of choice, then perhaps we have some control over the pace of change. But if rapid change is inevitable, it makes more sense to try to understand it and make the best of it, rather than resist it. Many evolutionary theorists, such as Gregory Stock and Ray Kurzweil, think that evolution is inevitable and that an increasing rate of change in our world is a necessary consequence of this natural law.[18] And even regarding Toffler's distinction between "the fast" and "the slow," which, among other things, contrasts the pace of life in modernized versus less-developed countries, the wave of rapid change (the transformation of cultures from the "slow" to the "fast") is spreading into more and more regions of the world, seemingly engulfing all of us, and sending everyone into future shock.

Hyperculture: As discussed in both Stephen Bertman's *Hyperculture* and James Gleick's *Faster*, modernized human society is increasingly obsessed with speed, efficiency, and short-term goals.[19] Not only is everything moving faster, but also we increasingly attempt to squeeze more information, tasks, activities, and experiences into the now. We pay less attention to both the past and the future, as our attention span shrinks to ever-smaller, more-compressed units of time.

As Gleick notes, our attention jumps quickly from one thing to the next, continuously distracted and immersed in the "buzz," and as a population living in such a maniacal reality, we collectively suffer from a lack of sustained focus, a social attention deficit disorder. Bertman is particularly concerned that we are losing our sense of history; that we are suffering from a form of cultural amnesia. And if we lose our sense of history, then we will lose our sense of the future. Due to the hyperculture in which we live, we have become lost in an overpowering present.[20]

Howard Didsbury, adding to Bertman's ideas, argues that the speed and efficiency of high technology increasingly supports the immediate gratification of our needs. Technology speeds up communication, production, and delivery. More and more of our wants and desires are taken care of through our gadgetry. We need neither patience nor foresight if our every want is quickly provided for us, and thus, according to Didsbury, we are seeing a "death of the future" in our modern world.[21]

Accelerative Change: There are many theorists who highlight the idea of accelerative change in their writings. Scientists-inventors, such as Ray Kurzweil and Hans Moravec, present the two-fold argument that the information processing speed and information storage capacity of systems in nature have exponentially grown throughout the history of the universe and that in the last century the speed and storage capacities of computing machines, in accordance with Moore's Law, have also increased at an exponential rate.[22] The second accelerative trend in computing power is simply an extension and expression of the first accelerative trend in nature as a whole. Extrapolating on these accelerative trends, they predict continued exponential growth in computational complexity in the future. In the near future, following from Moore's Law, computer intelligence will take the lead over humans in computational complexity and create a world, involving both artificial intelligence and technologically amplified humans, of continual and exponentially increasing complexity and mentality.[23]

Without particularly focusing on the exponential growth in computer capacities, other writers point out that along many diverse parameters of human life there has been accelerative growth and change and that this trend is especially noticeable within the last century. As noted earlier, Christian cites accelerative growth in human productivity and energy usage, Moore and Simon cite the measured accelerative growth in patented inventions and innovations in the last century, and Gleick, in his detailed commentary of contemporary life, describes the "acceleration of just about everything."[24] John Smart and his *Acceleration Studies Foundation* focuses on the study of accelerative change, why it appears to be occurring in the world around us, how it connects with technological growth, as well as with cosmic evolutionary and developmental processes, and what we can do to selectively guide the process.[25]

Peter Russell, in *The White Hole in Time*, presents a general theory of accelerative growth and change that connects human history with cosmic evolution.[26] Russell traces the development of intelligence and complexity from the beginnings of the universe to the present time, and draws the general conclusion that evolution is speeding up. There is an increasing level of novelty and complexity being created within the universe and the growth of human intelligence and culture is the latest expression of this overall cosmic process. (In this regard he is in general agreement with writers such as Kurzweil and Moravec.)[27] Russell sees the accelerative growth of intelligence and complexity connected with increasing freedom and the capacity to control nature. He predicts, based on this historical analysis, that accelerative change will continue into the future, but because change is coming faster and faster, we will become increasingly uncertain regarding the direction of events; we will approach an "information horizon" pass which we will not be able to see. Yet, in spite of this general trend toward increasing change, according to Russell, humans tend to resist change, fearing the uncertainty of the future, and hence they attempt to control it. Russell believes that if humans can let go of this need to control themselves and the world around them, they will realize a much more flexible and liberated state. He thinks that if humans can move beyond "ego-centric," time-bound, and materialist thinking - present modes of thought that are leading to our extinction - and embrace an inner-focused evolution of consciousness and the mind, the accelerative rate of change within our lives will continue forward at an even faster rate. Mental evolution will accelerate more quickly than our present materialistically focused evolution. As our evolution speeds up, we will approach an asymptote of the evolutionary process, "a white hole in time" or "omega point," where human consciousness will achieve a sense of eternity and cosmic oneness.

The General Theory of Progress: The general idea of progress - that life will improve in the future - extends back in time to ancient Greece, Zoroastrianism, the salvation theology of Judaism, and early Christianity, especially as espoused in the writings of St. Augustine.[28] The modern theory of progress derives from thinking in the Age of Enlightenment in Western Europe during the seventeenth and eighteenth centuries. As described by Nisbet in his history of the idea of progress, the modern view of progress is that human history exhibits an overall direction of advancement along numerous parameters (social, technological, ethical, and epistemic) and that humanity can realize further improvement in the future, especially through the use of reason, science, and modern economic and political principles, such as capitalism and democracy. In the nineteenth century, progress was often seen as an inevitable law of nature - that is, it was proposed by writers, such as Spencer, Comte, and Marx, that there was a "law of progress" within the universe and the historical growth of human civilization was simply a manifestation of this cosmic law. Also, in the nineteenth century, the idea of progress became connected with Darwin's theory of evolution.

It was popularly believed that evolution provided the natural mechanism by means of which progress occurred. Evolution was a progressive process. In general, although early in its history the idea of progress was frequently grounded in metaphysical and religious visions, such as in Zoroastrianism and Augustine's Christian theology, in modern times, progress took on a secular and naturalistic emphasis. However interpreted, the theory of progress is an optimistic view of the future based on the idea that the present is better than the past and the future will be better than the present.[29]

There are innumerable debates concerning the idea of progress. Does human history actually show progress, or is it just "one damn thing after another"? Isn't progress basically just a Western idea - a culturally biased grand narrative of the past and future? Is evolution in any objective sense really progressive? And if we were to define progress in secular terms, as many of the philosophers of the Enlightenment did, have our lives really improved in the last two centuries while we have pursued this ideal? Are things really "getting better all the time"? Is it that the old secular concept of progress is too narrow, missing what is really important in life, and needs to be revised?[30]

Contemporary Evolutionary Theory: The general idea of evolution is usually associated with a naturalistic or scientific perspective of reality and the future, but the concept of evolution has been applied to numerous dimensions of reality including natural forms, technology, civilization, mind, ethics, spirit, and the cosmos.[31] Still, following from Darwin's original formulation of the theory, the process of evolution is usually seen as due to forces and causes at work within nature, rather than being imposed or guided from some supernatural source toward some predetermined end - hence evolution is often contrasted with teleologism. Evolution is seen as a fundamental law of all nature and consequently the future will be a result of this pervasive natural process. Evolution is often connected with a progressive direction to time and nature as a whole, but there is ongoing debate regarding whether evolution necessarily implies progress.[32]

Although Darwin's theory of biological evolution through natural selection is usually seen as the modern starting point for contemporary evolutionary thought, the theory of evolution has continued to evolve over the last century and a half.[33] As Walter Truett Anderson has aptly titled one of his books, "*Evolution isn't what it used to be*," or at the very least, it is much, much more.

As mentioned above, the theory of evolution has been applied to the cosmos as a whole. According to advocates of this point of view, the entire history of the universe can be described as an evolutionary process, with the continued and ongoing "emergence" of new natural forms arising out of simpler forms; even the laws of nature have presumably evolved. The idea has even been suggested, by the contemporary physicist and cosmologist, Lee Smolin, that the creation of our universe is due to an evolutionary process.[34] In general, the cosmic theory of evolution proposes that evolution produces increasing complexity across time and this overall trend defines a progressive direction to the process.

Gould and Eldredge have hypothesized that evolution (at least in biology) is not a smooth and steady process, as Darwin believed, but occurs in spurts or relatively rapid transformations "punctuated" by periods of relative stability.[35] This idea has been applied to physical and social evolution as well.

Scientists such as Ilya Prigogine, Lee Smolin, and Stuart Kaufmann have attempted to demonstrate empirically that evolution involves the creation of new order out of chaos, and in fact, in general, chaos, accident, and chance play a significant role in contemporary evolutionary thinking.[36] Evolution appears to involve an interplay or reciprocity of order and chaos.

Kaufmann, following the lead of Prigogine, has argued that along with natural selection, a second fundamental principle is at work in natural evolution, usually referred to as the principle of "self-organization." Whereas natural selection involves competition among individuals (and perhaps also groups), self-organization involves an integrative or synthetic process where natural entities come together to form more complex wholes. Writers, such as Lyn Margulis, John Stewart, Robert Wright, and Howard Bloom, argue along similar lines that evolution not only involves competition but cooperation and symbiotic merging. In general, evolution throughout nature seems to produce the complementary and reciprocal trends of increasing integration and differentiation.[37]

One final noteworthy contemporary development within evolutionary theory is Richard Dawkins's argument that ideas or units of meaning actually evolve according to the principle of natural selection. Units of meaning or "memes" (the term Dawkins coined) compete with each other within the minds of humans and reproduce by passing from one mind to another. Through this process of competition and reproduction, memes evolve.[38] For Dawkins, evolution through natural selection can explain the dynamics and development of both life and the growth of ideas.

Fraser's Evolution of Time: The philosopher-physicist J.T. Fraser, founding president of *The International Society for the Study of Time*, has proposed, in a series of books on time, the general theory that time itself evolves in complexity through an ascending hierarchy of levels from the primitive to the more advanced, and this cosmic process is open-ended and never-ending.[39] For Fraser, this means that there are levels of complexity to time, beginning in the simplest form within light and subatomic particles and becoming more intricate and complex as we move upward in evolution through the chemical, biological, and psychosocial realms. According to Fraser, time is not an absolute or universal constant across the entire panorama of existence, as in Newton, but a variable of the universe that undergoes transformation. At each new level of time, new patterns and structure emerge; for example, according to Fraser, only humans and animals exist in a reality of a past and future. At the primordial level of light there is neither past nor future, only a timeless present. Hence, a direction to time is an evolutionary development - there is no direction to time at the most basic level of the cosmos.

In essence, Fraser turns most classical metaphysics and religious doctrines (for example, Platonic, Christian, and Hindu philosophies) on their head. Instead of thinking that the most advanced or highest level of reality is the timeless realm of eternity, Fraser believes that eternity is the most primitive level of reality and that through the evolution of time we are moving away from what is most primitive - the eternal. The evolution of time embodies an ascent to higher levels of reality.

Since Fraser believes that the evolution of complexity in the universe adds increasing structure to the cosmos, there is a sense in which the universe becomes more deterministic as it evolves. As the universe has evolved structure, it has evolved new types of forms and processes, enriching its temporal complexity. Each level of evolution has also brought with it an additional range of possibilities of behavior. Therefore, though things have become increasingly more structured through the evolution of time, the universe realizes more possibilities of behavior. Increased determinism has brought with it more freedom. The future will contain possibilities that would seem incomprehensible or impossible relative to the present. For Fraser, evolution is truly creative.

A key feature within Fraser's evolutionary thinking is his theory of order and chaos. In his book, *Time as Conflict*, he argues that the basic driving force behind the evolution of time is the irreconcilable conflict of order and chaos. Each level of time is a uneasy synthesis of these two basic cosmic forces - yielding dimensions of both stability (derived from order) and change (derived from chaos) - that eventually spills over into a jump forward to a new level of time and increased complexity. Yet each new level of time carries with it the conflictual synthesis of order and chaos thus perpetuating the process of evolution without end. In a basic sense there is time because of this irreconcilable conflict at the heart of existence.

Fraser believes that our contemporary world is in a period of transition from one level of time to a new level. This new level of time that is struggling to emerge is a global "sociotemporality." As our world increasingly integrates into a global society, time is being further standardized and controlled, so that all the people of the world can function as a coordinated whole. Computerization and the networking of computers into a global system are contributing to the creation of a "time compact world." Various forces though, such as governments, international corporations, and local cultures, are in competition to define a unified sense of humanity. Further, there is a struggle over defining a singular history for all humanity - a collectively agreed-upon sense of the past. Presently, we have multiple stories of our past competing with each other. Fraser believes that only if we can arrive at some unifying vision of our past, can we create a clear sense of our future. The future of the sociotemporal world is uncertain.[40]

In summary, within Fraser's theory, evolution is the basic form of time and the fundamental law of the universe - reality lawfully changes by evolving. Time moves from the simple and undifferentiated to increasingly higher levels of complexity - hence time is progressive. The process of evolution is generated by an ongoing conflict of order and chaos, which cannot be resolved - consequently it is never-ending. Further, the evolution of time leads to increasing degrees of freedom and potentiality. Each new level cannot be understood from lower levels of reality. Because of this upward and expansive motion, it is in principle impossible to predict and understand the future in its entirety.

Wright's Cultural Evolution: Robert Wright, in his book, *Nonzero: The Logic of Human Destiny*, argues that there is a discernable progressive general direction to all of human history.[41] He believes that a scientifically valid conception of cultural and historical evolution exists that can be applied to all human societies. This directionality provides a basis for making general predictions about the future evolution of humanity, society, and culture.

According to Wright, the general evolutionary direction in human history is toward increasing social complexity based on the development of mutually beneficial relationships or transactions among people, individually and collectively. Although we all possess a basic need to serve our own individual interests, we repeatedly find that we can establish "win-win" transactions among us that also benefit us individually. These arrangements, both supported and fueled by technological innovations, add to the complexity of our societies. In essence, although Wright does not highlight the term, social complexity grows through the evolution of "reciprocities" that mutually support the individual lives of those involved. Other futurist theorists such as Henderson and Eisler[42] use a similar concept of "win-win" relationships in articulating a new philosophy for human society - in contrast to "win-lose" relationships that they argue have presumably dominated Western thinking in the past - but Wright's argument is that non-zero or win-win transactions have been evolving throughout human history, as well as even prehistory.

Wright contends that if we look at human history we find that cultures do not remain static, but at different rates, invariably move in the direction of establishing more and more "win-win" relationships among the members, and consequently move in the direction of increasing complexity. There are, of course, many cases where individual societies collapse or disintegrate, e.g. the Roman Empire, the Egyptian Empire, and the civilizations of the Aztecs, Mayans, and Incas, but the overall direction across the entire globe has been increasing complexity and "win-win" relationships. Wright contends that when cultures collapse, the accomplishments and advancements of the culture are not totally lost - they are discovered and passed on, to different degrees, to emerging cultures. The overall effect is growth, even in the repeated and frequent demise of civilizations. Even the most primitive cultures show a history of progress - while they remain viable - and our increasing understanding of prehistory demonstrates that progress was occurring throughout our evolution running back millions of years.[43] Although he is aware that universal histories and grand narratives of progress fell into disfavor during the twentieth century, with the rise of cultural relativism and Postmodernism, Wright contends that this basic pattern of evolution or progress applies to not only all individual human cultures, but also to the total global panorama of humanity. As a social species, our world is more complex and filled with "win-win" reciprocities than in the past, and we are increasingly integrated and tied together via these transactions and arrangements.

As a final point regarding Wright's theory of cultural evolution, he clearly sees the importance of technological and material advance in understanding human history and progress. Throughout our history, new technologies have repeatedly served the function of improved connectedness among us (e.g. writing, the printing press, radio, TV, and the Internet). Technologies have also served the function of destruction, but in so far as evolution moves toward the enhancement of mutually beneficial reciprocities, then technology will be pushed in this direction as well.

Stewart's Evolution of Evolvability and Cooperation: John Stewart, in his book, *Evolution's Arrow: The Direction of Evolution and the Future of Humanity*, argues that the fundamental pattern of change throughout the history of life (if not the cosmos) has been evolution.[44] Over the last couple of centuries, discovering this general form and the details of this pattern has been a highly momentous advance in the evolution of life. Life has become conscious of its own direction and the *modus operandi* of its existence. For Stewart, this discovery required the emergence of conscious thought and the capacity to abstractly model reality. In articulating various concepts, models, and theories of both the past and the future throughout history, earlier humans laid the groundwork for this relatively recent scientific achievement.

Coming to understand evolution has been part of a general process in the history of life away from a narrow sensitivity and adaptability to the "immediate here and now" toward an expanding awareness of past and future Stewart argues that becoming increasingly aware of both past and future, of the pattern of evolution and how humans fit into the grand evolutionary saga, amplifies human freedom. For Stewart, humans should use their knowledge of evolution as a foundation to guiding their development into the future.

One particular future development Stewart foresees will be the conscious and intentional enhancement of "evolvability." In tracing the evolution of life he notes that increasingly more complex living systems find ways to evolve more quickly and efficiently. Their evolvability evolves. For example, the capacity to model reality in thought and imagination provided a mechanism for consciously considering the different possible consequences of changes in behavior. This was an advance over genetic mechanisms that operated through trial and error across generations. The development of culture provided a mechanism for passing on knowledge acquired within one generation to the next generation. Each stage in evolution brought with it some new more enhanced mechanism for evolving further. Understanding the process of evolution itself and consciously using this knowledge to evolve is, in Stewart's mind, a quantum leap forward in evolvability.

Using the concept of evolution to guide our actions will further move the human mind from short-term goals to longer-term goals. The human self will become less present-focused and less self-centered. Stewart believes that the general goal for future humans should be to contribute to the overall evolutionary success of the species; in fact, extending the sphere of awareness further outward, future humanity should guide its thinking and behavior toward contributing to the overall evolution of the cosmos. This shift to cosmic dimensions within the human mind is a natural extension of the overall evolutionary direction of moving from the immediate here and now to the broadest expanses of space and time and the general trend from short-term evolutionary success to increasingly longer-term evolutionary success.

Another key feature of Stewart's thinking is his emphasis on the principle of cooperation in understanding the dynamics and direction of evolution. Although Stewart acknowledges the role of competition in evolution, he believes that the overall direction of evolution ("evolution's arrow") is toward increasing cooperation. Cooperatives are more adaptable and capable than individuals and can successfully deal with more complex, larger scale environmental challenges. Yet, individuals will not cooperate unless it serves their own self-interest; hence the evolution of cooperation always involves ways that benefit the individual. Cooperation is only naturally selected for when it helps individuals.

Stewart argues for the emergence of a planetary civilization where the benefits of cooperation are shared by all individuals. He believes that a planetary civilization organized in this way would show a high level of innovation and evolvability. In fact, Stewart thinks that evolvability (the capacity for progressive change) should be the central focus or value of future human societies. Societies should have a transformative or dynamic self-image instead of a static one. Similarly, individuals within such transformative societies should guide their lives toward making themselves more evolvable. In essence, the connected goals should be to make human societies and individuals more flexible, innovative, cooperative, and focused on continual long term success.

Purposeful and Conscious Evolution: For scientists like Stewart and many others, especially those involved in the area of biotechnology, humanity will soon gain control over evolution and learn to guide it. We will probably be able to amplify, improve upon, and accelerate evolution. Although part of this new level of control will be within the biological sphere, in a more general vein there are futurists who think that evolution and its purposeful control should be the guiding principle in all aspects of human life. Two contemporary writers who take this position are Barbara Marx Hubbard and Mihaly Csikszentmihalyi.[45] They think that evolution should be the guiding theme in the development of the human mind and human society. Csikszentmihalyi, a world famous psychologist who has extensively studied creativity and the experience of "flow" in individuals, believes that becoming active, conscious agents in the evolutionary process is the best way to find meaning and happiness in our lives. He contends that the central task of the new millennium is to control the direction of evolution.

Introduced in the last chapter, Hubbard, who uses the expression "conscious evolution," has developed a network of writers and social activists sympathetic with her cause - in essence, she is spearheading a paradigm and social movement resonant with her theories.[46] According to Hubbard, humanity should purposefully guide its evolution, using both scientific and technological knowledge and spiritual and humanistic values. She also believes that a new creation story is needed to inspire and unite us. Hubbard believes that the old stories of our origins don't work very well any more - they have lost credibility. As noted earlier, she feels we are at a crisis point and in a confused state in human history; we need to define and create a new direction and story for tomorrow. She proposes a core set of values necessary for guiding our future ethical evolution: higher consciousness, freedom, love for each other, unitive aspirations, reverence for what is higher, and synergistic thinking.

Hubbard sees herself as facilitating a new social architecture for implementing the cultural and philosophical changes she sees ahead of us. She is presently working on the development of futurist communities that will network and participate in the contemporary social transformation. Toward this end, she has constructed an "innovation wheel" where innovations from around the world are collected together, providing a central source for cross-fertilization, mutual inspiration, and dialogue. Her hope is that this social and organizational effort will facilitate the purposeful evolution of humanity. The innovation wheel is divided into the following categories:

- Government and Law
- Education
- Economy and Business
- Health/Relationships/Personal
- Science and Technology
- Spirituality and Religion
- Environment
- Culture

Within each category, Hubbard highlights certain key themes and values. For example, under "Government and Law" she lists participatory democracy, equality, and justice as important ideas being addressed. Under "Health/Relations/Personal" she lists personal harmony, growth, and partnership as basic values and goals. Although Hubbard presents her ideas as both innovative and future focused it should be noted that most of her main values and goals are reflective of long-standing ideals in human history. Perhaps these ideals haven't been realized - perhaps in fact, their opposites have held sway, but most of the basic goals of Hubbard's philosophy can be found among the values of the major traditional religions and those of Enlightenment philosophy. In part, this is not surprising, for Hubbard draws a good deal of her inspiration from what she considers the positive contributions of past traditions, both secular and spiritual.

Aside from Csikszentmihalyi and Hubbard, Frank Tipler should also be mentioned. Tipler presents a theory of purposeful evolution that is cosmic in scope, encompassing the total future of the universe.[47] In essence, Tipler thinks that through the use of advancing technologies and the spread of humanity throughout space we will intentionally direct evolution toward the creation of a cosmic mind. Evolution has an ultimate purpose and we will be involved in its realization.

Finally, the connected social network of The Evolution Research Group, The Club of Budapest, and The Darwin Project should be included. Under the leadership of Ervin Laszlo and David Loye, among others, this network advocates for an evolutionary view of tomorrow, involving the conscious and purposeful guidance of humans in the process. They have published a variety of books on this theme, including *Evolution: The Grand Synthesis* by Erwin Laszlo and *The Evolutionary Outrider: The Impact of the Human Agent on Evolution* by David Loye in which they discuss evolutionary theory in depth and examine its implications for the future of humanity.[48] More recently, Loye has published a collection of essays titled *The Great Adventure: Toward a Fully Human Theory of Evolution* where the argument for purposeful, ethical evolution is developed and connected with Darwin's original thinking on the nature of evolution.[49]

Sustainability: Versions of this theory of the future are often presented as critiques of the modern concepts of progress, growth, and accelerative change. The general emphasis within this point of view is on developing ways of life that will ensure for a continual high quality of life for all humanity. The fear is that the present modern pace of life is leading to an exhaustion of our natural resources and environmental catastrophe - we are moving too fast, motivated by the greed of the moment, and we are going to collapse or crash. As Dennis and Donella Meadows and The Club of Rome have argued, we have exceeded the resource capacities of the planet earth in our drive toward continued industrial and economic growth[50]. Richard Slaughter, who is an advocate of this perspective, believes that our modern high production - high consumption world is too present-focused and that seriously dealing with the issues and challenges of sustainability is a much more future-focused mode of thinking and living than our present general mindset and behavior.

Many ecological thinkers believe that we are exhausting the resource base of the earth with our massive and ever growing industrial and technological developments.[51] There is a call to return to a simpler way of life - to cut back on our modern obsessions with economic growth, technological advance, and extreme consumerism.[52] Of special note, sustainability theorists highlight that we need to consider much more the type of world we will leave to our descendents and children. The argument for a sustainable human society can be stated with various degrees of emphasis. Sometimes it sounds like what is being called for is a regression to a more primitive way of life; sometimes the position simply calls for a slowing down of growth. At its most basic level, the sustainability argument emphasizes "foresight" regarding our present ways of life and the possible negative consequences of overextending our environment and ourselves.[53] We need to think ahead to ensure a quality existence for humanity in the future.

Catastrophe, Decay, and Extinction - The End of World Civilization - A New Dark Age - The End of Humanity - The End of Everything: This perspective on the future, which actually includes a whole variety of different theories, encompasses all those views that predict a decline or extinction of human civilization, if not humanity as a whole. Included within this perspective would be theories that predict biological or ecological catastrophes, scientific and technological disasters, social and economic collapse, moral decline, and spiritual failure.

At the opposite extreme of progressive and optimistic theories of the future is the general attitude of impending doom. This pessimistic perspective on the future can take different forms. One possibility involves a great and cataclysmic Third World War.[54] As discussed earlier, the threat of a global nuclear war over the last half-century spawned a wave of science fiction novels, stories, and movies depicting an age of barbarism or a new Dark Ages after the Third World War.[55] Wagar provides a review of governmental and scientific projections regarding a third world war and the possible consequences.[56]

A third world war could conceivably spell ecological disaster, wiping out many forms of life, including humanity.

Aside from a third world war, there are many other negative possibilities for the future. Many of these possibilities have been captured in popular science fiction or disaster movies: 1) Humanity irrevocably pollutes or unsettles the environment and/or uses up most of the earth's natural resources - people die by the millions or billions, and our present world civilization collapses - *The Day After Tomorrow* or *The Core* scenario; 2) A great plague sweeps over humanity, perhaps due to biological experiments gone out of control - the *Twelve Monkeys* scenario; 3) Humanity is invaded, conquered, and enslaved by some powerful alien life form - the *War of the Worlds* scenario; 4) A powerful dictator gains control over the world and humanity is enslaved under a sinister totalitarian government - the *1984* scenario; 5) Governments lose control over the growing crime, violence, terrorism, and disorder sweeping the globe, and global humanity collapses into a "dog-eat-dog" lawless existence - something akin to the *Snow Crash* scenario; 6) The earth is hit by a huge comet or meteor and we go the same direction as the dinosaurs - the *Armageddon* or *Deep Impact* scenario. We are victims of bad luck; 7) Humanity creates a technological intelligence that decides to wipe out or enslave our species - the *Terminator* or *Matrix* scenario.[57]

Richard Moran, in his comprehensive overview *Doomsday: End-of-the-World Scenarios*, identifies and discusses ten basic types of possible catastrophic future events:

- Disasters due to Weapons of Mass Destruction
- Asteroid Impacts on Earth
- Massive Volcanic Eruptions
- The Greenhouse Effect and Global Warming
- A New Ice Age
- Mega-Tsunamis
- Plagues
- Cyber-Terrorism
- Insect Invasions
- Bioengineering Mistakes

According to Moran, all of these types of doomsday scenarios are scientifically plausible and could possibly occur. Such events could spell the end of humanity and conceivably even wipe out all life on earth.[58]

The cosmological scientist, Sir Martin Rees, in his book, *Our Final Hour*, focuses on scientific and technological disasters that could wipe out humanity. It may not be natural disasters that constitute the worst threat to our continued existence; humanity itself may be the greatest threat. He includes, as potential disasters, the emergence of super-intelligent machines; biotechnological terrorism and bioengineering error; nanotechnologies that get out of control; scientific experiments at the sub-atomic level that accidentally create black holes or expanding disruptions in the structure of space; cascading environmental disasters due to our technology and industry; and the continuing threat of nuclear weapons, especially if such weapons fall into the hands of terrorists. Rees sees the potential annihilation of humans as having cosmic significance if it turns out that humanity is the only intelligent life in the universe. If humanity is indeed the only intelligent life form in the universe, a view supported by a number of scientists, our destruction would cut off the chance of intelligence spreading throughout the cosmos. With our death, the life of the mind in the cosmos dies as well.[59]

The extreme cosmic expression of a pessimistic or doomsday view is the "Heat Death of the Universe" scenario. Inspired by the discovery of the three laws of thermodynamics in the nineteenth century, many scientists came to the conclusion that disorder in the universe is going to continually increase in the future. Eventually the universe will run out of available energy and consequently the capacity for creating new order, and all existing order and complexity will crumble. Along similar lines of thinking, perhaps the universe will continue to expand, spreading matter and energy ever more thinly so that the universe will die a slow death in a "Big Chill." Or, if there is enough matter and gravitational force to reverse the expansion sometime in the distant future, the universe will start to collapse, shrinking back to a pin-point, which would destroy all structure, life, and intelligence. Everything would end in a "Big Crunch."[60]

Destiny and Determinism: As a general perspective on the direction of the future, many religious and scientific thinkers have argued that the future is destined or determined. There is no element of possibility in the future. Either the flow of time is being orchestrated and controlled by God or supernatural beings, or all the events in nature occur in a predictable sequence due to the laws of nature and the deterministic connection of causes and effects. The destiny or "teleological" view of the future can be found in both Western and Eastern religions, such as the Christian theory that God has a divine plan for the universe and there are a set of pre-determined events that will inexorably come to pass, and the Hindu theory that the universe is a dream in the mind of *Vishnu* and that the God *Shiva*, in his dance of creation and destruction, will bring the universe to an end in a great cosmic conflagration. From the scientific end, writers such as Pierre-Simon Laplace in the eighteenth century popularized the view that given a complete knowledge of the laws of nature and a complete description of the present conditions of the universe, the entire future history of the universe down to the smallest details could be predicted.

Possibility, Freedom, and Uncertainty: As noted earlier, after World War II the philosophy of existentialism, especially as espoused in the writings of Jean Paul Sartre, became a popular and influential way of thinking, first in Europe but eventually in the United States as well.[61] A central tenet of existentialism is that humans are free agents and that our future is a matter of choice, rather than due to destiny or determinism. According to Sartre, to anchor one's identity, one's values, one's life, or one's future to some fixed, pre-determined conceptualization, especially involving authoritarian doctrine or authority figures, is to live in "bad faith" and abdicate one's self-responsibility and self-determination. For Sartre, we freely create our future and freely create ourselves.

The philosophy of determinism has also been criticized from within science, especially with the development of quantum physics. According to contemporary interpretations of quantum physics, for example in the writings of Murray Gell-Mann, the future of the universe is probabilistic rather than entirely determined. The principle of indeterminism, a key concept within quantum physics, seems to imply that at the sub-atomic level the behavior of particles can not be completely predicted and this lack of predictability is not a limitation in our knowledge but an inherent property within quantum reality. Many contemporary writers, such as Kenneth Miller, argue that the indeterminism at the quantum level opens the door for human freedom, though not everyone agrees with this conclusion. But as a general theory of the future, quantum physics implies that the future of the universe (which of course includes human existence) is open to different possibilities.[62]

A common position held by many members of the World Future Society is that the future is a set of possibilities rather than one definite trajectory. Because the future is possibilities, humans have a choice in what future will be realized. Most futurists in fact talk as if they believe that the decisions made today will influence what our future will be like. We are not passive victims of supernatural destiny or natural laws.[63]

Finally, I should mention those contemporary "cultural edge" writers who emphasize chaos, creativity, and freedom in describing the nature of our modern world and how we should best approach the future. For example, in the essays of the anthology, *Mondo 2000*, arguments are made that nature is filled with chaos and creativity and that humans should embrace these principles in their lives.[64] Such views resonate with Postrel's argument that the best path to the future involves free creativity and a rejection of top-down efforts to control the world – the future is (or should be) open.

Theories of Science, Technology, and Rationalism

The Second Scientific Revolution: Twentieth century science has introduced many fundamental changes regarding the way scientists think about reality.[65] Among the most noteworthy scientific developments have been relativity theory, quantum physics, cosmological evolution, open systems, chaos and complexity theory, and string theory. Over the last century, there has been a Second Scientific Revolution. These changes in thought, as well as our very perception of reality, are beginning to have an impact on how people view human society, psychology, business, and other areas of life. It may be that our total mindset is in for a fundamental transformation in ideas, attitudes, and values as a consequence of this second revolution of science.

A popular argument is that the Industrial Age was based on Newtonian science and, consequently, with the transcendence of this earlier scientific perspective, a new age will emerge based on the newer concepts of science. Some classic examples of this argument are Zohar and Marshall's *The Quantum Self* and *The Quantum Society*, Fritjof Capra's *The Turning Point*, and Hazel Henderson's *Paradigms in Progress*.[66] Zohar and Marshall propose that we reorganize our society and psychology in terms of the principles of quantum physics; both Henderson and Capra suggest an open systems approach to society and social planning. More recently, Sally Goerner, in her book, *After the Clockwork Universe*, has provided a systematic comparison between Newtonian and contemporary science and the social implications of each system of thinking. Central to her position is the thesis that contemporary science emphasizes the theme of connectivity and networking in nature, which is a fundamental shift away from Newtonian science with its emphasis on linear thinking.[67]

In general, this theory of the future is that due to a re-conceptualization and re-organization within the scientific community regarding the nature of reality, science is going to drastically alter the collective mind of all human society. This transformation will not just involve technological changes;

the revolution will be psychological, philosophical, and social as well.[68] As we will see below, many theories of the future that incorporate contemporary scientific ideas also draw the implication that the human mind and society will fundamentally change as various new ideas of science increasingly impact everyday human life.

Wilson's Neo-Enlightenment Philosophy: The socio-biologist E. O. Wilson supports the basic principles of the Western Enlightenment in his vision of the future. His theory of progress rests on a belief in the powers of reason and science and a materialistic philosophy of reality. For Wilson, humanity is empowered through reason, science, and a belief in progress. Wilson believes that the basic tenets of the Enlightenment were that the universe was lawful and could be understood through science; that all human knowledge could be united through a set of fundamental scientific laws - laws that gave order to nature; and that through understanding and applying these laws of nature the potential for infinite progress in humanity could be realized.[69] Further, he believes that nature ultimately can be understood in materialistic or physical concepts, and his theory of ethics and values is naturalistic rather than supernatural or absolutist; he conceives of morality as having evolved in a social and natural context.[70] Wilson thinks that the great goal of the Age of Enlightenment and the West's greatest contribution to the world was the idea that secular knowledge (science and rational philosophy) could facilitate and drive the evolution of human rights, ethical and moral advancement, social development, and human progress. Although Wilson's theory of reality is materialistic, he does think that it is important to address humanistic issues and values and connect them to materialistic science. He believes that the evolution of human culture and human psychology can be understood and explained within a biological framework.[71]

The Technological Revolution and Technological Evolution: The technological perspective on the future includes those theories and paradigms that emphasize computer technology, robotics, energy and resource development, technological super-structures and macro-industrial projects, micro-technology (nanotechnology), transportation, biotechnology, and space technology.[72] Technology is seen as the central driving force in the evolution of the future and, in particular, the accelerative rate of change in present times is viewed as primarily caused by technological innovation and development.

According to Michael Zey, although humanity in the last three centuries has made great progress in technological and industrial development, perhaps this is simply a prelude to what lies ahead in the coming centuries. Perhaps, as Zey suggests, we are on the verge of a "Macroindustrial Revolution"[73] that will dwarf in scope, complexity, and size the physical inventions of the last few centuries. Some strong possibilities in the very near future include mile-high buildings housing hundreds of thousands of people, tiny machines and motors the size of single molecules, intelligent houses and automobiles (that drive themselves), global mega-projects including transcontinental highway systems, and a multitudinous variety of robots performing a thousand different types of tasks for humanity.[74] Michael Zey, a strong spokesman for the future promises of technology, argues that the recent exclusive emphasis on information technology has been both misleading and counterproductive. Industry and manufacturing are not lessening up, but rather accelerating in complexity, scope, and efficiency. We are moving into an era of hyper-progress in all types of technology and industry.

The technological revolution is multifaceted and the innumerable areas of development cross-stimulate each other. Michio Kaku identifies three fundamental "scientific revolutions" in the twentieth century critical to the growth of new technologies in the world around us. These three revolutions are the quantum, computer, and biomolecular. According to Kaku, these three areas of research and study are mutually reinforcing, and technological developments based on scientific ideas in one area frequently instigate technological innovations coming out of another area. Walter Anderson particularly emphasizes how the growth of computer technology and biotechnology are not independent of each other; without computer technology, progress would have been much slower in identifying the human genome, and the study of biological systems has provided a host of ideas in the development of computer processing systems. Many of the newest innovations in biotechnology involve computer information processing systems, often implanted in or connected with biological organisms.[75]

Still, with the possible wonders of technology comes a sense of anxiety and apprehension over where it is all leading. There is the fear that we will lose control of our biotechnological or nanotechnological creations.[76] Perhaps we will literally integrate and merge with our technology?[77] More and more parts of our bodies are becoming replaceable with technological devices.[78] More and more of the activities within our lives are becoming dependent on the use of technology. As Naisbitt notes, at least in the United States, we are becoming "technologically intoxicated."[79] Perhaps worse still, at a more insidious and deeper level, our values and our purpose in life are being re-defined and ultimately controlled by the technologies we are creating.[80] Beginning with the Industrial Revolution, humanity has expressed a fear and concern over technology and machines. Are we becoming slaves to our machines? Are we being taken over by them? Yet, will our relationship perhaps turn out to be more symbiotic than submissive?[81]

The Information - Computer Revolution: In the last few decades, the emergence and spread of computer technology into many spheres of human life has both stimulated our sense of fascination and excitement, as well as our fear and apprehension over modern technology. The rich and varied set of interdependencies between computers and human life continues to escalate. The ever-growing pervasiveness, complexity, and capacities of computer technology can be seen as a real threat to our sense of power and superiority.

According to some predictions, computers will become more intelligent than humans in two or three decades. Yet, equally, computer technology promises to vastly increase our abilities and power.[82] Never has there been such an intelligent and complex machine; never has a machine become so integral to so many different features of human life so quickly. Are we going to become totally dependent upon and swallowed up within an exploding technology of computers? Or are we going to emerge empowered in ways presently beyond our imagination and grasp?[83]

The computer unites two different paradigms for the future. First, the creation and development of computers is an extension of the manufacturing and industrial mindset. Computers are machines and, in the last couple of decades, mass-produced machines, that embody scientific and technological advances that have been taking place over the last three centuries. But the computer is a unique type of machine in that it stores, represents, processes, and communicates information. It is a higher-order machine, one step removed from earlier machines that harnessed physical forces and energy and manipulated, molded, and moved physical matter. As the steam engine was a central icon of the Industrial Age, the computer has emerged as the central icon of the Information Age. More and more of our professions focus on the manipulation of information, rather than the manipulation of physical matter, and the computer is the primary tool in these information professions. Thus, according to many writers, the computer ushers in a new type of human society - an Information Society. And computers are even transforming our Industrial Age machines. As computer technology works its way into more of our instruments, artifacts, and machines, our manufacturing becomes more intelligent and information-driven.[84] All told, computer technology is becoming the integrative "nervous system" for all our other technologies.

One very popular view regarding the contemporary transformation is that humanity is moving from an industrial society into an information society. Alvin Toffler, particularly in his books *The Third Wave* and *Power Shift*, subscribes to this theoretical view.[85] To a great degree, so does John Naisbitt in *Megatrends* and *Megatrends 2000.*[86] Of particular, more recent note, Manuel Castells, in his three-volume work *The Information Society: Economy, Society, and Culture*, highlights the unique features of the Information Age and how it embodies a qualitatively different way of life than previous eras in human history.[87] Although neither Peter Drucker nor Daniel Bell, two well-known and influential theorists of contemporary times, explicitly advocate for an "Information Age" theory of the future, both strongly emphasize the passing of the Industrial Age and the growing significance of information technology, information professions and disciplines, and knowledge as wealth and power.[88] Basically, the Information Age theory states that humanity is moving from a society ruled by Newtonian machines and philosophy to a society ruled by knowledge and information. The information revolution is being supported by computer and communication technologies which store, process, and transmit information.

In discussing computers and the Information Age theory of the future, the sudden and ubiquitous expansion of the Internet, the World Wide Web, and other communication technologies must be included as an essential part of the big picture of things. The Internet, the largest machine humans have ever built, globally connects the ever growing population of computers worldwide. The Internet supports collective thinking and discourse, increased collaboration among the people of the world, and ties all the world's computers into a global brain and mind. It has been just as critical in the development of the Information Age as the computer. It facilitates financial, economic, scientific, and personal transactions, at a whole new scale, among the myriad businesses, research institutes, social organizations, and people across the globe. The World Wide Web provides an electronic platform for creating a global visual presence. Coupled with the Internet, other communication technologies such as cell phones and wireless transmission are intensifying the exchange of ideas, information, and an immense torrent of small talk. In the Information Age, information is being broadcasted and received at a level that dwarfs all previous eras in human history. Projections are that this trend of "information overload," for both better and worse, will continue to accelerate in the future.

The promises and possibilities of computer technology and the Information Age are frequent topics in the media and the news. *The Futurist* magazine routinely covers new developments and projected trends in computers, the Internet, and life changes in the Information Age. Over the last decade there have been many popular books which look at how computer technologies will change human life in the future. Nicholas Negroponte, in his book, *being digital*, presents a fascinating picture of how information and communication technologies will transform our lives in the relatively near future.[89] The information revolution is impacting education, business, politics, culture, and many other spheres of human reality. Michael Dertouzos, former head of the Laboratory of Computer Science at MIT, in his *What Will Be: How the New World of Information will Change our Lives*, describes in rich detail how computer technologies could affect healthcare, entertainment, the arts, government, the economy, and numerous aspects of everyday life. Dertouzos looks at both the potential benefits and negative effects of computer technology.[90] Ray Kurzweil in *The Age of Spiritual Machines* and *The Singularity is Near* discusses the evolutionary possibilities of computers and artificial intelligence, and how such technologies could transform the very essence of human nature.[91]

Robotic and Human/Robotic Evolution: Connected with the ongoing development of computers over the last fifty years is the evolution of robotics. Robots require computer circuitry, and as computers have become more compact and powerful, robotics has benefited from these advances. There is an ever growing robotic industry and the Japanese, in particular, are enthusiastically pursuing both robotic research and the marketing of robotic companions and toys.[92] The world wide population of robots is increasing at a faster rate than humans.

Two of the main spokesmen for the great promise of robotics in the future, Rodney Brooks and Hans Moravec, foresee robots becoming more capable and intelligent in the coming decades and working their way into many aspects of human life. Brooks, in his *Flesh and Machines: How Robots Will Change Us*, and Moravec, in *Robot: Mere Machine to Transcendent Mind*, anticipate that robots will steadily approach the full set of capacities of the human mind and body in the coming century. Robots will show purposeful, goal-directed behavior, thinking and abstract cognition, and even emotionality. In fact, robots will exceed human abilities in many respects. In essence, both Brooks and Moravec predict a robotic revolution that will transform human society and the nature of humans.[93]

Brooks and Moravec trace both the history of robotic development over the last century, including their own work in the field, and present a set of detailed predictions for the future of robotics. Moravec is somewhat more optimistic about when robots will reach human intelligence and capabilities, anticipating that this milestone will happen in this century, whereas Brooks believes it will take longer. Brooks argues that there are still some fundamental insights regarding the nature of mind, intelligence, and consciousness that are missing in our understanding. Brooks does not believe that creating a robotic intelligence equal to humans is simply a question of sufficient storage and processing capacity. Moravec presents his more optimistic predictions based on his extrapolations of when we will be able to create reasonably compact computer systems that have the speed and storage capacity of human brains. Brooks and Moravec also see themselves as approaching the challenge of creating more capable robots from different directions. Moravec has been working on creating powerful central processors that perceive, calculate, and guide the behavior of robots, whereas Brooks has designed robots with multiple parallel peripheral processors - that is, with artificial nervous systems that approximate simple mobile creatures like insects. (In fact, Brooks's robots often resemble insects.) Moravec takes a top-down approach to design, whereas Brooks takes a bottom-up approach. Yet, where they both agree is that sooner or later, robots will achieve high level mentality, consciousness, and even a sense of self.

Robots in the future will come in all sizes and shapes with different abilities and levels of intelligence, depending upon the particular tasks they are designed for. Special purpose robots are already extensively used in manufacturing and are becoming increasingly popular in the toy industry. Household cleaning robots are already available and the promise that robotic visionaries such as Brooks and Moravec make is that robots will become commonplace and ubiquitous in human society in the next few decades.

Although Brooks anticipates great debate and uneasiness over the coming rise of robots, he believes that we will eventually achieve a symbiosis with robots and be transformed both psychologically and physically in the process. As robots become more capable and intelligent, they will alter how we conceptualize and relate to machines. Brooks is presently working on robots that mimic emotional behavior and interact with humans psychologically. Further, as biotechnology, computer technology, and robotics advance, we will transform our own bodies, replacing bio-systems with robotic systems. In essence, we will become more robotic. In the future we will merge with our machines.

Moravec, though not discounting bio-robotic integration, presents a long-term, highly cosmic and visionary view of the future of robots. He foresees robots eventually exceeding human intelligence and moving into the exploration and colonization of outer space. Robots will become capable of reproducing themselves and will dramatically advance their own evolution in the centuries ahead. According to Moravec, robots are our future evolutionary children and descendents.

The Biological and Biotechnological Transformation of Humanity: Modern genetics and biology are fast approaching the time when humanity will be able to alter and manipulate its genetic make-up.[94] Not only will various diseases and obvious genetic infirmities be eliminated or diminished, but design improvements will also occur as well. New types of humans will probably emerge, not through the long-term process of natural selection, but the short-term process of purposeful genetic manipulation (germline therapy). How will such new humans transform human society? Will such new humans signify the end of our present species? Clearly a technologically engineered biological transformation in humans would be a highly dramatic event with all manner of implications and possible consequences - a paradigm instance of an evolutionary jump.[95]

Not everyone finds the prospect of enhancing humans through genetic engineering appealing or desirable. There is great controversy surrounding this growing possibility.[96] But as Gregory Stock argues in his book, *Redesigning Humans: Our Inevitable Genetic Future*, the general population, for a variety of reasons, will decide to genetically enhance their children as the technologies become available and affordable. Still, as Freeman Dyson notes, there will probably be great ethical controversy and even open conflict in the centuries ahead with the appearance and increasing number of genetically modified humans. A central issue in the debate will be over the very definition of human nature.[97]

The genetic engineering of humans is only one piece of the Biotechnological Revolution. The biotechnology industry, in general, is rapidly growing with research and production in new drugs, enhanced food products, medical treatments and technologies, artificial body parts, genetically modified life forms, and a host of other creations and innovations. Biotechnology even promises the possibility of slowing or stopping aging and opening the door to human immortality.[98] As a general future trend, biotechnology promises to significantly transform the entire ecosystem of the earth.[99] As many futurists predict, biotechnology will become one of the biggest and most profitable sectors of the economy. Many writers foresee the twenty-first century as increasingly dominated by biotechnology.

One highly stimulating overview of the area, which integrates biotechnology with ecological, evolutionary, and open systems theory, is contained in Kevin Kelly's well-known book, *Out of Control: The Rise of Neo-Biological Civilization*. As the models and metaphors of physical science dominated the Industrial Age, the coming era, according to Kelly, will be dominated by the ideas and applications of biology and ecology. We will develop various new biological and ecological technologies, e.g. life forms that eat pollution and waste, clean the house and enhance our health, and we will increasingly model our society on a biological philosophy. It is not simply that we may evolve biologically, but the whole ecological system of life on the earth may evolve with us in a new era of enhanced cooperation and interdependency. Further, the distinction between the "born" (life) and the "made" (machines, tools, artifacts) will increasingly blur, as technology and biology intertwine in numerous ways.[100]

Transhumanism: One paradigm, if not social movement, concerning the future that integrates many of the above-cited technological and scientific theories is transhumanism. One definition offered of transhumanism is philosophies "that seek the continuation and acceleration of the evolution of intelligent life beyond its currently human form and human limitations by means of science and technology, guided by life - promoting principles and values."[101] Literally meaning "to transcend humanity," transhumanism emphasizes the value and central significance of using technology and science in creating the future. Humanity - psychologically, biologically, and socially - can and should be transcended.

Extropianism - One strong version and early expression of transhumanism - lists the following seven basic principles of its philosophy: Perpetual Progress, Self-Transformation, Practical Optimism, Intelligent Technology, Open Society - Information and Democracy, Self-Direction, and Rational Thinking. The term "extropy" is defined as "The extent of a living or organizational system's intelligence, functional order, vitality, and capacity and drive for improvement."[102]

Transhumanism is clearly a pro-growth theory of the future and decidedly pro-evolutionary in its thinking. In fact, transhumanism argues for infinite or limitless evolution and growth. Humanity is a step in the grand evolutionary process and not an end point. Many of its advocates see the transcendence of humanity as involving the integration of biology and technology. Although the original transhumanist site was American-based, there is now a World Transhumanist Association and a global "transhumanist culture." Both the World Transhumanist Association and the Extropian Institute hold periodic conferences.[103]

The Technological Singularity: As introduced earlier, in a well-known article published in 1993, the science fiction writer and computer scientist, Vernor Vinge, hypothesized that, given the present growth rate of computer storage and processing power capacities, computer intelligence would probably exceed human intelligence sometime within the next fifty years or soon thereafter. Once computer intelligence reached the human level, it would quickly move beyond it and within a relatively short period of time would vastly exceed in complexity and power the capacities of the human mind. At this point, we would have passed through the "technological singularity" where our machines would become incomprehensible to us. We would be - so to speak - left in the dust. In so far as such super-intelligent computers would orchestrate and direct the future evolution of technology and all those operations and activities of society, the world around us would become increasingly impossible to understand and keep pace with. The only conceivable solution to this looming problem, short of halting further computer technology development before it's too late, is to technologically enhance ("Intelligence Augmentation") the mental capacities of the human mind (or brain). Vinge sees this option as only a short term solution, however, since to keep pace with the accelerative growth of computers after the technological singularity the human mind would have to be so drastically and continuously modified and upgraded that, in a short period of time, what constituted the human part of us would be miniscule in significance compared to the technological component.[104]

There are various other scientists and writers who agree with Vinge's general prediction of an impending technological singularity, though there is variability in estimates of the approximate date of its occurrence, how sudden and dramatic its onset will be, and what humans can (or should) do in dealing with this cataclysmic event. Both Kurzweil and Moravec agree that it is coming, but whereas Moravec believes that super-intelligent robots will thereafter become the leading edge of further evolution in intelligence, Kurzweil thinks that human minds will be able to download their personalities, intellect, and consciousness into computer systems and benefit from innumerable enhancements within such systems. There are also futurist organizations, such as the Transhumanists and the Acceleration Studies Foundation, that are very interested in exploring the different possibilities and consequences of the technological singularity. The science fiction writer Charles Stross has written two noteworthy novels, *Singularity Sky* and *Accelerando,* which explore in realistic narrative detail how human life may change as a consequence of the technological singularity. Humans may benefit from various advances in computer technology connected with approaching the singularity, such as living virtually within computer systems, or having multiple selves or lines of consciousness. As Stross speculates, humans may have to vacate the solar system if advanced artificial intelligences gains control over our world.[105]

Radical Evolution: Joel Garreau, reporter and editor for the Washington Post, in his book, *Radical Evolution: The Promise and Peril of Enhancing Our Minds, Our Bodies - And What it Means to be Human,* attempts to synthesize the various social, psychological, and even spiritual implications of the technological augmentation and modification of the human species.[106] He highlights four basic technologies - genetics, robotics, information technology, and nanotechnology - which he collectively labels "GRIN technologies" and discusses how the application of these technologies could transform the nature of humanity. In essence, Garreau pulls together many of the ideas and theorists discussed above, including Kurzweil, Vinge, Kelly, Stock, and the transhumanists. What he particularly wants to emphasize is that humanity, if not life in general, is at a critical point in history and that the potential technological enhancement of humans will represent a monumental step forward in evolution. For humans, this looming transformative event will create the biggest change in our species in the last 50,000 years. According to Garreau, the technological transformation of humans is the defining social, cultural, and political issue of our time.

Instead of presenting one possible future for humanity, Garreau identifies a variety of different conceivable scenarios for the future: The Curve, the Singularity, Heaven, Hell, Prevail, and Transcendence. These different scenarios are not absolutely distinct; the Singularity and Heaven scenarios build on the Curve scenario and the Transcendence scenario builds on the Prevail scenario. Both the Heaven and Hell scenarios assume the Curve scenario, but describe diametrically opposite consequences resulting from the Curve. Briefly, the Curve scenario predicts continued exponential technological growth; the Singularity scenario predicts (as inevitable) the emergence of super-human intelligence - an "intelligence explosion" - resulting from technological exponential growth; identifying Kurzweil as one of the strongest proponents of the following view, the Heaven scenario predicts "unimaginably" good and positive results for humanity, achieved as a consequence of exponential technological growth and the realization of the Singularity - immortality and the capacity to guide the future evolution of the entire universe are two significant developments within this scenario; the Hell scenario predicts "unimaginable" human disaster and possible extinction as a consequence of continued technological growth and the emergence of super-human technological intelligence; the Prevail scenario, contrary to both the Heaven and Hell scenarios, predicts that humans will gain control over technology and guide its development and its applications to human enhancement - this scenario assumes an uncertain future filled with surprises, as well as various mistakes committed along the way; and finally, the Transcendence scenario predicts, building upon the Prevail scenario, that humans will transform their nature through technological enhancement - Garreau cites the transhumanists as supporting this vision of the future.

Although Garreau discusses all these different scenarios, his preference, as the ideal and perhaps most realistic future, is the Prevail/Transcendence scenario, where humanity, though "muddling" through the uncertain and dangerous waters of the future, guides its own evolutionary development. For Garreau, we have a degree of control over our future. Contrary to those who would argue that humans should not attempt to enhance or significantly modify our nature, Garreau believes that human nature is not fixed but rather dynamic and transformative. Technological enhancement does, though, greatly increase our power to change ourselves. Because of the competitive advantages of human enhancement, he thinks that advances in this area will continue to emerge and that many people will enthusiastically embrace these possibilities for improvement. Within the Prevail/Transcendence scenarios, we need to stay conscious of the potential dangers of technological growth (highlighted extensively within the Hell scenario), yet it is part of human nature "to steal fire every chance we get." We are inclined, if not driven, toward self-improvement. Further, quoting Václav Havel, "transcendence [is] the only real alternative to extinction." We grow or we die.

Garreau supports the general theory of purposeful evolution. Though it is GRIN technologies that will empower us to change ourselves, hopefully humans will guide the evolutionary process through sound ethical values and thoughtful decision making.

He is particularly concerned over whether humanity, socially and psychologically, will catch up with the rate of technological change, understand it, and intelligently guide it. Can we make sense of all the new possibilities in front of us? Can we foresee which technological advancements will be of benefit and which will be too dangerous to pursue? How do we find meaning and happiness in this new world? Garreau foresees the need for a new Enlightenment, with new theories and new stories that will provide us with the wisdom to appropriately manage our impending transformation. In general, Garreau emphasizes the human element in our potential evolution. Following the ideas of Jason Lanier, Garreau argues that it is more interesting and valuable to consider how humans will change, rather than how computers will change in the future. For Lanier, the main thrust of our transcendence will be social and psychological rather than technological.

The Cosmic Adventure of Outer Space: Within this general perspective, the future of humanity is envisioned within a cosmic context. The emphasis within this perspective could be on identifying the general principles and processes of the evolution of the cosmos and how we fit into the universal scheme - perhaps finding inspiration and meaning within this cosmic context and our possible role in the future of the universe.[107] Or, more concretely, the future of humanity could be seen as involving the exploration and colonization of the cosmos - either alone or in cooperation with other forms of intelligence within the universe.

Traveling into outer space became a popular idea with the emergence of science fiction and the writings of Jules Verne and H. G. Wells. Many of the greatest science fiction novels have been about humans venturing outward to the planets and the stars.[108] In the images of science fiction, humanity spreads outward through the cosmos, colonizes other worlds, perhaps encounters and battles alien intelligences, and, if fortune favors, truly becomes citizens of the universe. Within contemporary popular culture many of the challenges of this cosmic adventure were captured and brought to the screen in the *Star Trek* TV and movie series.

The exploration and colonization of outer space has not simply been an interest of science fiction writers though. Scientists, engineers, and visionary futurist thinkers have been developing plans and designs for space craft and space settlements over the last hundred years. A good overview of past efforts and future possibilities is contained in Nikos Prantzos's *Our Cosmic Future: Humanity's Fate in the Universe.*[109] Progress in moving out into space has not been altogether successful, with various financial, political, and technological problems slowing down or even halting efforts. Yet over the last fifty years, humanity has reached the moon and sent probes and, in some cases even robots, to many of the planets and moons in our solar system.

Different writers, both fiction and non-fiction, have presented a variety of reasons, including economic, technological, cultural, psychological and even spiritual, as to why humanity should and will move into outer space.[110] According to such advocates of space travel, in the long run, space exploration will pay itself back a thousand, a million, a billion fold; the cosmic adventure will be fueled by humanity's ceaseless desire and need to explore. Perhaps it is as necessary and inevitable as the baby bird leaving the nest to find its place in the world. Dorian Sagan, in his *Biospheres: Metamorphosis of Planet Earth,* presents a biological-ecological explanation of which factors are moving us toward outer space and its colonization.[111] *The Mars Society* has actually compiled a detailed list of reasons, as well as plans and proposals, for traveling to Mars and creating a permanent human settlement. Kim Stanley Robinson, in his award-winning Mars trilogy, has described, in vivid detail, the possible challenges in terraforming Mars to make it habitable for human life. Closer to home, *The Artemis Project* has articulated reasons and plans for a settlement on the moon.[112] And for a cosmic and dramatic argument for the exploration and colonization of the entire universe, based on scientific and spiritual grounds, including technological proposals for how to do it, see Frank Tipler's *The Physics of Immortality*.[113]

Metaphorically, humanity's journey to the stars is a realization of the ancient dream of the journey into heaven.[114] For many writers who enthusiastically support the great coming adventure into outer space, it is our destiny and ultimate calling to realize this dream. Perhaps, in retrospect, the coming century will be most remembered for our transition from terrestrial, earth-bound creatures to creatures of the heavens.[115]

Theories of Ecology and Nature

The Return to Nature - The Green Movement - Environmentalism: Achieving a renewed sense of interconnectedness, balance, and appreciation of nature is central to Green and environmentalist thinking. Preserving or restoring the natural environment is a top priority. The raising of ecological consciousness across the globe among all people is a fundamental goal. The Green movement, which is the strongest and most conservative form of environmentalism, is clearly more than just a theory concerning the future; it is a paradigm with a whole set of values and guidelines for how to live; it is also a social movement, for many people attempt to live by its principles and organize to further its cause.

Anderson and Easterbrook[116] provide summaries of the Green movement and Green ideology, and Ray discusses the correlations between Green and environmentalist thinking and the basic American subcultures of Modernism, Cultural Creatives, and Heartlanders.[117] Environmentally conscious thinking and living is a strong theme within the Cultural Creatives culture.

Lovelock's Gaian hypothesis - that the earth is a single living organism - is often used as theoretical and ideological support for Green thinking. We should see ourselves as part of Gaia. We should treat the earth as alive and care for it, rather than abuse it.[118] In fact, the earth or Gaia may be worshipped as the "mother goddess" for, in effect, we are the children of Gaia. At the very least, we should see ourselves in a partnership relationship with the earth, rather than attempting to dominate or control it.

Green thinking tends to be reactionary to many modern trends in social and industrial growth - it is often highly anti-technology and anti-big business. Within Green thinking there is an emphasis on stability, harmony, and sustainability over growth in the future, and often a more local or regional vision of human life.[119] Additionally, the high production/high consumption way of life in modernized countries is frequently criticized as being the main cause of environmental and ecological problems. According to many environmentalist and Green thinkers, we should adopt a non-intrusive, light-touch approach to nature, rather than trying to control or change it.[120] The high consumption way of life is also blamed for being one of the main causes of psychological stress and life dissatisfaction in the modern world.[121] We should learn to use less and consume less. Living more in tune with nature creates a happier, more peaceful life. For radical Greens, humanity, including technology and modern civilization, is often seen as a scourge on the environment, the earth, and ecological balance. Perhaps the Industrial Era was an aberration in human history; the conquest of nature is about to fail, and we must learn to live in a way reminiscent of our pre-industrial ancestors.

Environmentalist thinking has clearly raised the ecological consciousness of humanity. In particular, there has been a focus on documenting and measuring various environmental problems and challenges, such as climate change, species extinction, pollution, depletion of resources, and the disruption or destruction of ecosystems across the globe.[122] There have been repeated publicized warnings, based on evidence collected and extrapolation on this data, that our high production/high consumption, highly intrusive modernized way of life is leading to ecological and human disaster.[123] Yet there is also considerable debate and controversy over the validity and value of such predictions and "doom and gloom" visions presented by many environmentalists.[124]

Ecology and Holism: Theoretical views within this general perspective emphasize ecological and open systems concepts and a holistic, integrative, and interdependent view of humanity, technology, and nature. Friijof Capra's vision of the future, the Integral Culture movement, and the New Age movement are illustrative examples of this perspective. From a holistic perspective, all the pieces and dimensions of life and existence fit together into a whole - everything is interdependent and interactive. Holism is usually contrasted with the analytical view of nature and human life associated with the Industrial Age and Newtonian thinking; holism highlights the properties of the whole rather than the parts. The concept of ecology is decidedly holistic since nature is described in terms of systems of interdependent living forms.[125]

The holistic and ecological themes are often combined with evolutionary thinking in theories of the future, for example, in the ideas of Barbara Marx Hubbard, Hazel Henderson, Elisabet Sahtouris, David Loye, and Erwin Laszlo. In fact, Lovelock's Gaia theory also strongly supports both a holistic and evolutionary view of life and the earth. The open systems perspective on evolution emphasizes the interactive and interconnected dimension of change. Nature evolves through interaction and according to the scientist, Harold Morowitz, all "evolution is co-evolution."[126] Sally Goerner, in fact, uses the expression "ecological evolution" to highlight the idea that the universe and nature evolve as an interactive and mutually supportive cosmic ecosystem.[127]

Newtonian science emphasized analysis, dualism, and the control of nature. Newton described the physical universe as a set of distinct units of matter. Descartes, in his philosophy and psychology, argued for a dualistic separation of mind and body. The Scientific Revolution ushered in a clear division between religion and science.[128] Eighteenth and nineteenth-century Western philosophical thinking frequently emphasized the autonomous individuality of people. The Industrial Era fostered a sense of separation and alienation from nature, and also a philosophy of controlling nature to serve the goals of humanity.

Contemporary philosophy and science are moving in an opposite direction, emphasizing the interconnectedness and holistic nature of reality. Open systems theory, chaos and complexity theory, and quantum physics all point toward a holistic view of physical reality - there are no independent parts. The physicist Lee Smolin contends that increasingly all the basic properties and entities in physics are understood and described as relational rather than intrinsic and independent.[129] Holistic medicine and neuroscience point to numerous interconnections between mind, brain, and health. New Age thinking, which embraces the philosophy of holistic medicine and health, incorporates many ecological and holistic themes, as well as challenging the separation of religion/ spirituality and science. J.J. Gibson has brought the ecological perspective into psychology, highlighting the reciprocity of the perceiver and the environment.[130] The science of ecology and the environmentalist movement emphasize the vast web of interdependencies and reciprocities among living forms and the humanity - earth relationship. In her overview of the transformation from Newtonian science to contemporary science, Goerner argues that the key new theme in scientific thinking is "connectivity" and that nature is now described as "networks."[131]

Ecological and holistic thinking about the future envisions a world in which nature and humanity co-exist in a spirit of cooperation, respect, and partnership, rather than humanity trying to dominate nature or humans trying to dominate each other.[132] The Integral Culture movement describes its main goals for the future as finding ways to again integrate and connect individuals with each other, humanity with nature, and humanity with the cosmos.[133] If, for the last three hundred years, we have lived in an era of analysis, specialization, division of labor, mechanistic metaphors, and dominance over nature, we are now moving toward a philosophy of communion, holism, cooperation with nature, and organismic metaphors.[134]

Theories of Psychology and Human Relations

Psychological Evolution: Speculative visions of humans in the future, for example in science fiction films, although often set in strange high-tech environments, usually portray humans, quite naively, as possessing the same type of psychology and mental make-up as they have today.[135] Even if humans are bio-technologically enhanced or have computer chips implanted in their brains, it is assumed that the essentials of human nature will remain relatively constant. Humans of the future are portrayed as having the same vices, the same desires, the same social relations, and the same psychological challenges as they do today. This assumption regarding the future of human psychology seems highly doubtful.

There are many reasons to think that human nature is transformative and evolving, and not some single unchanging reality. Recorded history demonstrates fundamental changes in all aspects of human existence, including society, technology, culture, belief systems, material artifacts, art, religion, and architecture.[136] All these types of change clearly have affected the human mind and the human self - we are not the same psychologically as our ancient ancestors. Based on evidence from archeology, paleontology, and anthropology, it appears that human mental capacities, behavior, and our sense of personal identity have significantly changed and evolved over time.[137] There are, in fact, many indications that human psychology is undergoing some basic changes in our contemporary world.[138]

Accelerative developments in biotechnology, the science of psychology, brain research, computer technology, education, and other areas promise to provide ways to enhance or improve human psychological capacities in the future. We can improve or transform ourselves by changing our biology through drugs or genetic engineering, by augmenting our intelligence with computer technology, by changing our environment, by enhancing or transforming our cognitive abilities, belief systems, behavioral habits, or way of life, and by modifying our culture and society, including our values and educational system. We may use technology to alter our minds and our selves, or we may transform our psychology through mental techniques and behavioral practices. We may purposefully guide psychological changes within us, or changes may be forced upon us in order to cope and adapt to a changing world.[139]

Many writers and leaders in the worlds of business, human organizations, social and political science, biology and information technology, and, most notably, psychology are talking about contemporary and potential future developments occurring in human life and human interactions. New principles of leadership, cooperation, and innovation are developing in the business world.[140] We have already noted that in the areas of biotechnology and information technology there is a great deal of attention being paid to how humans could be transformed in the future. Human diversity, global consciousness, and human empowerment are moving up the priority list of values in government, business, and social life. The New Age movement, the Human Potential and Social Potential movements, renewed spiritual and religious concerns, the Feminist movement, and many other social developments are all contributing to the idea that humanity is about to transform psychologically, socially, and even spiritually.

There are many writers and futurists in psychology and the humanities who specifically focus on the theme of a new type of self emerging in the future. Barbara Marx Hubbard and others argue for a future self that moves beyond an egocentric mindset.[141] Walter Truett Anderson, in his *The Future of the Self,* besides providing an excellent overview of the concept of the self in history and contemporary times, also presents his ideas on a "postmodern person" that possesses a pluralistic self needed for our complex and changing world.[142]

Of special note, the psychologist Mihalyi Csikszentmihalyi, in his book *The Evolving Self: A Psychology for the Third Millennium*, proposes a new vision of the self for the future. His theory of the "evolving self" emphasizes the need to transcend the egocentric constraints within us. He believes we have the inherent capacity to see beyond the limits of our present condition and transcend them. Csikszentmihalyi argues for a new type of self in the future - one that does not identify with or accept the selfish needs of genes (the body), culture, or the ego. [143]

According to Csikszentmihalyi, the key to an evolving self is the experience of flow. Flow is a state of consciousness experienced during periods of creativity. Csikszentmihalyi presents a theory of "transcenders," of people who pursue the experience of flow. Transcenders are evolving selves. Csikszentmihalyi sees the evolving self as a cosmic self that identifies and integrates with all humanity, nature, and the universe. The evolving self "flows" and "transcends" its own boundaries. It is a dynamic narrative growing and transforming with a sense of adventure and purpose. It is open to the world and to its own inner workings.

If the rate of change in our contemporary world is increasing, if things are moving faster and faster, how are we not only to survive, but, in fact, thrive in such a world? The psychologist Maureen O'Hara has developed a theory of human personality that addresses the issue of finding a balance of order and chaos in our lives.[144] According to O'Hara, there are three fundamental ways of dealing with change - defensive, psychotic, and growth responsive. The defensive response is "anxiety repressed," the psychotic response is "anxiety unleashed," and the growth response is "anxiety contained and transformed." She refers to the growth-oriented self as a "transformative self." For O'Hara, the transformative self shows flexibility, creativity, integration, balance, openness, interconnectedness, expansive consciousness, a synthesis of the rational and intuitive, a tolerance for ambiguity, a balance of cooperation and competition, empathy, and joy. The transformative self is the foundation for psychological and social evolution.

The End of Material Progress and the Growth of the Human Psyche: A view of the future that is basically the antithesis of materialist visions (technological or economic) is the theory that there will be a fundamental shift in focus toward mental development and away from physical development. C. Owen Paepke, in his *The Evolution of Progress: The End of Economic Growth and the Beginning of Human Transformation*, predicts that the recent accelerative evolution of physical technology is about to come to an end. We are about to exhaust our physical resources, which power our industrial and technological growth. Further, we are reaching the upper limits of our technological abilities. In the near future, there will be a shift of emphasis from physical improvement to biological, psychological, and social improvement. We will turn increasingly toward the development of our minds and away from both economic and technological growth.[145]

Though this view of the future may seem rather radical or extreme, it does resonate with a popular line of thought in the minds of many people who think and write about the future. Although Hazel Henderson does not predict an end of material progress, she does think that our values are going to shift away from economic growth toward themes like national welfare, the quality of life, and the satisfaction of basic human needs.[146] Peter Russell believes that our focus on external material objects as the pathway to happiness and life satisfaction is counter-productive and that we need to shift toward the inner development of our minds, if we are going to achieve happiness in life. In fact, Russell believes that it is through psychological and spiritual development that we will realize a higher, more rapid-paced level of evolution and change. The key to continual growth lies in the mind and not in the world of matter.[147] In a set of articles for the *Encyclopedia of the Future*, William Van Dusen Wishard and Graham Molitor point out that one pervasive and growing criticism of the modern world is that our values are too materialistic and economic and that we need to shift our attention more toward the human spirit and psychological issues. Human civilization should change direction from a materialistic emphasis to a humanistic and psychological emphasis.[148] Finally, in looking at the research of Ray and Anderson, the rising Cultural Creatives subculture clearly seems to be putting more of an emphasis on psycho-social and spiritual values than technological and economic values.[149]

Positive Psychology, Optimism, Happiness, and Virtue: One of the most influential new developments in the science of psychology is the "Positive Psychology" movement. Instead of focusing on psychological problems and disorders, such as stress, anxiety, psychosis, and depression, the Positive Psychology movement focuses on human strengths, such as love, wisdom, self-respect, and hope, and how to further enhance such strengths.[150] Although in some ways an evolution of Self-Actualization Psychology, popularized by Abraham Maslow and Carl Rogers in the 1960's, the Positive Psychology movement adds many important new elements and is highly relevant to futurist concerns and issues.[151]

Perhaps the most well-known spokesperson for Positive Psychology is Martin Seligman. The two topics that Seligman has studied over the last decade that are most noteworthy are optimistic versus pessimistic attitudes about the self and the future and the relationship between virtues (or character strengths) and happiness and purpose in life.[152]

Optimism and pessimism are two of the most fundamental themes underlying attitudes and theories about the future.[153] I have already reviewed a variety of theories that take an optimistic, a pessimistic, or some mixed perspective regarding present trends and potential future developments in human society. Supporting optimistic and pessimistic attitudes are the basic human emotions of hope and fear. Is the future approached with hope or with fear? What Seligman has studied are the causes and corollaries of optimism and pessimism and how to strengthen an optimistic attitude. Although pessimism can serve some value (pessimists are more realistic and accurate in their assessments of themselves and the environment and provide an important function of hyper-vigilance to potential danger and disaster), pessimists tend to see themselves as helpless and suffer from depression, which leads to inaction and loss of hope regarding the future, and they behave in ways that fulfill and confirm their negative attitudes. Optimists think about the world differently than pessimists; they see themselves as much more capable of effecting positive change, are less likely to blame themselves for failures and bounce back much quicker from failure and misfortune. Seligman discusses the social trends and causes behind pessimism and depression and, in particular, is highly critical of the heightened level of individualism in contemporary society and the waning of community and social support, which he believes has significantly contributed to the more negative attitudes and feelings about life and the future.

In an extensive study of values and virtues across different cultures and historical periods, Seligman identified six fundamental character strengths or virtues that appear universal for all people in all times. These six virtues are wisdom, courage, love and humanity, temperance, justice, and transcendence. Seligman's argument is that "authentic happiness" in life is built upon the exercise and development of these character virtues. Seligman believes that authentic happiness is a relatively enduring quality and is not necessarily associated with short term pleasure at all. Momentary pleasures tend to diminish quickly, for people adapt to the frequent experience of a repeatable pleasure. Character virtues on the other hand require effort and challenges. Hence, authentic happiness is something that must be worked at and the pathway involves an ethical growth in the individual. Psychological happiness and mental health is achieved through ethical development.

For Seligman, meaning and purpose in life involve both the development of character virtues and the identification with some reality or goal "beyond oneself." The virtues serve a 'transcendent reality" rather than just being self-serving. (Note the similarity with Csikszentmihalyi on this point.) Consequently, extreme individualism works against finding meaning and purpose. We should note that transcendence is one of the primary character virtues listed. In many ways, transcendence is anathema to our modern emphasis on the ego, self-gratification, and subjectivism - there is something beyond our private realities that needs to become our center of gravity and our standard of truth and value.

The End of Male Domination and the Rise of Femininity: For at least the last four or five thousand years, almost all of modern human civilization, East and West, has been male-dominated. In particular, religious, philosophical, and academic ideas have been, primarily, the creation of the male mind. In the West, as well as the East, we have had a male-controlled economy, government, and technology supported by a male ideology and mindset.[154]

Often presented as a challenge to the "masculine" domination of human history, human values, and society, feminist theories of the future argue for equality and justice for women in the future. One view frequently presented is that the supposed gender differences between men and women are a product of cultural indoctrination and stereotyping, and that women, given equal opportunities, will be able to perform in all spheres of life at the same level as (if not better than) men. Another view is that women possess special gifts and talents and that these qualities should be accorded equal status and value in the world. In the past, women have frequently been treated as inferior to men and whatever unique qualities women possessed, these qualities were seen as less important than those presumed unique qualities of men.[155] Hence not only should women have equal opportunity and power, but if there are certain values and ways of thinking strongly associated with a woman's point of view, these values and ways of thinking should be given equal importance in future human society.

Just as men have dominated political and intellectual history, they have also dominated thinking about the future. Up to the last few decades, most popular theories of the future were created by men. Although not always the case, such masculine views of the future have frequently highlighted technology, economic growth, expansionism, and the conquest and control of nature. Visions of the future created by women put a different slant on things. In a survey of women's thinking on the future in the *The Futurist,* the dimensions highlighted by those women questioned were intuition, partnership, humanism, equality, community, and the family.[156]

Riane Eisler's theory of human history and the future is one good example of a feminist view of the future.[157] Eisler argues that human societies over the last few thousand years have been organized in terms of dominance hierarchies with men in the top positions of power and women relegated to subservient roles and inferior status. Eisler proposes that human society as a whole needs to move beyond this dominator mentality to a partnership model where women and men have equal status and that social relationships should be created in the spirit of mutual respect and cooperation, rather than domination and submission. Eisler sees men, as well as women, as victims of the dominator mentality; a partnership mentality is better for everyone. She believes that many, if not most of today's problems in the world are a result of a dominator mentality and social order, and that a shift to a partnership society would help in correcting environmental, political, and social ills.[158]

Presently modern human society is in a state of flux and revision. We may be reaching the end of a male-dominated culture and civilization.[159] Not only are women having an increasing impact on the economy and business practices and beliefs, but women are also challenging various practices and beliefs in social, psychological, religious and philosophical areas.[160] If the male mindset is one of hierarchical dominance and control, female writers are proposing a more participatory and cooperative ideology.[161] According to feminist futurists, we are moving into an era of balance between the sexes.[162]

Wheatley, Human Organizations, and Leadership: As human society is being transformed, business organizations and the world of work are changing as well. In the contemporary business world there is a strong movement away from the traditional philosophies and practices of the past toward new ideas coming out of science, technology, psychology, social thinking, and business and leadership theory.

The traditional model of business organizations, as Toffler notes in *The Third Wave*, was built on the hierarchical, top-down, standardization concepts of the Industrial Era. Managers stood at the top and directed the activities of workers in an organization. Freedom of thought, individual initiative, and group participation were not reinforced. The paradigm business was the factory and the paradigm job was assembly work.[163] As Margaret Wheatley emphasizes, in her book *Leadership and the New Science*, traditional business organizations were highly influenced by Newton's view of reality. They were analytical, mechanistic, and excessively rationalistic.[164]

In the last few decades, some of the most noticeable changes include businesses moving toward flatter organizations with less middle management, managers being transformed into leaders who coordinate and empower individuals, the movement from more local to more global economies, and the infusion of information technology into all aspects of business and work. Further, production is diversifying from standardized to more customized goods.

Wheatley proposes a new theory of business organizations reflective of contemporary trends and especially the new ideas of Post-Newtonian science. According to Wheatley, business organizations of the future need to become much more fluid and dynamic. Intelligent workers require more opportunity for creativity. Also, future organizations need to be able to keep pace with the rapid rate of change in human society. Obsessive predictability and control are out; businesses need to appreciate the value of chaos and loosen the reins of control. Order and purpose can grow from within a group of intelligent and empowered employees; order and purpose do not need to be imposed from above. And still, a certain amount of disorder within an organization is beneficial; it keeps the system moving and growing. Business organizations of the future should be learning organizations, more open to outside influences and ready to grow and transform.[165]

Also, Wheatley's image of future business organizations connects such organizations more intimately with human society and the needs and values of people. Wheatley, in fact, advocates for a strong humanization of future business. Not only is the employee to be seen as a thinking, self-directed individual rather than simply a "cog-in-a-wheel," but the business organization of the future must adopt a whole new set of humanistic values and ideals. There should be a concern over business ethics and social responsibility. Corporations and companies can no longer see themselves as simply economic realities - they are social institutions, integral to and necessarily part of human society.

Wheatley wishes to emphasize the importance of vision and culture within business organizations. Instead of an organization having rigid and specific rules for its employees, a business corporation should organize around a vision of its future. This vision should be the mutual creation of all its employees and should motivate and inspire its employees, rather than rigidly control them. Out of a vision arises a culture, an attitude, and a way of life within the organization. The culture provides both professional and personal meaning and a psychosocial atmosphere for the organization. In essence, the business corporation of the future must find positive and non-authoritarian ways to move its employees motivationally, personally, and emotionally.

Humanism - The Compassionate Era - The Age of Light - The Solar Age: Robert Theobald's "Compassionate Era" theory of the future[166] and Hazel Henderson's vision of the "Age of Light" and the "Solar Age"[167] combine an emphasis on humanistic values with criticisms of high tech, economically focused views of the future. Theobald is critical of the Information Age view of the future because it highlights the impersonal and the technological. Rather, he sees our changing world as calling for a new culture of honesty, responsibility, humility, love, and a respect for the mysteries of life. According to Theobald, we need to live with passion and care about our world and ourselves. Theobald calls himself "a courageous realist" rather than an optimist because he thinks that the future is possibilities rather than certainties, and it will be our choices that determine tomorrow. Clearly acknowledging the great changes occurring around us, Theobald wants to stress the emotional, psychological, moral, and social dimensions of the future.

Hazel Henderson, who has been previously cited on numerous occasions, is an advocate of open systems thinking, a global perspective on humanity, and ecological - biological concepts and values. She is also critical of the Information Age theory because, as she says, "Information doesn't necessarily enlighten." For Henderson, the Information Age is a continuation of the Industrial Age mindset with its emphasis on hardware, mass production, consumption, efficiency, and competition. She believes that our future needs a philosophy and metaphorical language that underscores our connection with nature and the importance of mental wisdom and vision. Further, she advocates for a logic of complementarity - a *Yin-yang* philosophy of both/and - to replace the either-or, dualistic logic of the West.

Henderson's approach to the future could also be classified under the ecological and holistic perspective. She strongly supports the Gaian hypothesis, arguing that understanding Gaia will provide moral and intellectual guidance for creating a global and planetary civilization. She thinks holistically, viewing all the basic dimensions of human and natural reality as interdependent. Further, she advocates for justice and equality among all humans, believing that the capitalist economic growth model has intensified human inequality and overall is having a negative effect on the quality of life. Henderson wishes to emphasize cooperation among people and stresses the importance of achieving greater harmony in the world. Finally, she sees the solar system as an ecosystem with Gaia embedded in the context of the radiance of the giver of life

- the "mother Sun." Humanity needs to define itself within this broader, more cosmic context, rather than in terms of individualistic or nationalistic identities.

Shlain's Word and Image: Leonard Shlain sets his theory of the future in the context of a broad and systematic vision of human history. In *The Alphabet and the Goddess*, Shlain argues that there have been two fundamental and opposing forces in the history of human civilization. One force, which is more strongly associated with women and the female perspective, uses the image as the basic means of communication and symbolic representation. The second force, more strongly connected with males and the masculine mindset, is anchored to the use of the word and abstract language in communicating and thinking. An emphasis on the image brings with it a more concrete way of seeing the world and seems to support more egalitarian and peaceful societies, whereas, according to Shlain, the introduction and spread of linguistic literacy and alphabetic representation brings with it more violence, war, competition, and domination of men over women. Authoritarian and rule-governed societies seem to suppress the use of the image in culture and emphasize rigid memorization and adherence to sacred texts. Although the emergence of science, presumably a positive development in the history of human civilization, is connected with the rise of literacy in the West, the increasing persecution of women and children (the witch hunts) and the terrible wars among enlightened Western countries were also associated with increasing literacy.[168]

With the increasing influence of computer technology and the popular media, Shlain sees a decided shift going on in contemporary culture away from the word toward the image. Although many modern writers bemoan the apparent loss in literacy and interest in reading and writing, Shlain believes that this trend may actually have a beneficial effect. In Shlain's mind, more of a balance is needed between the influence of the word and the influence of the image. (Shlain also supports a *Yin-yang* philosophy of balance.) He connects this shift toward the image with the increasing voice and power of woman in the contemporary world, as well as with their more concrete, cooperative, emotional, and peaceful mindset. The introduction and pervasive spread and influence of movies, TV, and most recently videos and DVD's, the growing ubiquity of screens and visual displays in our world, the rapid advances in computer animation, the rising popularity of video and computer games, and our youth's increasing visual-graphic intelligence over linguistic intelligence are all significantly connected with a new mindset and way of life that is emerging in modernized countries. Given these trends and the strong possibility that such trends will only intensify in the coming years, Shlain is very hopeful about the future.

Theories of Society, Culture, and Morals

Globalization and the Global Society: One of the most popular and influential theories of the future is the view that a global society is emerging in contemporary times. According to advocates of this theory, humanity is moving from a set of relatively autonomous nation states to an integrated global organization, more ruled by international corporations and economic and ecological interdependencies than national and political ideologies and issues. The emergence of the global society is being fueled by global communication, transportation, economic exchange, and the growth and spread of a global culture.

As a starting point, it should be noted that various writers have argued that the process of globalization is not a recent phenomenon but has been building throughout human history. Wright discusses in considerable depth how economic interdependencies and exchange and communication information networks have been evolving for thousands of years, leading to the present global society - the global society has been in the making for centuries, if not millennia.[169] Howard Bloom, in *Global Brain: The Evolution of Mass Mind from the Big Bang to the 21st Century*, takes an even more expansive historical perspective than Wright and describes the emergence of communication and the networking of life on earth from its beginnings and traces the spreading of humanity across the face of the earth back to our hominid ancestors. Bloom discusses how the creation of cities and trade routes, built on conquest and economic reciprocities, laid the seeds for the development of ever expanding and interconnecting human societies and cultures.[170]

One of the most widely read early globalization theorists is Richard Naisbitt. According to Naisbit, human society is experiencing a global, economic boom. There is a steady, ongoing increase in international trade and commerce. Free enterprise and open economic exchange are increasing. National economies are disappearing because major corporations are more international; products are assembled from contributions from around the world. The world economy, in fact, overpowers the ideological differences of nations. For multinational corporations, the marketplace has become the entire globe. Telecommunication has provided a means and mechanism for complex, fast-paced global economic coordination. Through computerization and global communication systems, worldwide data on resources, production, and consumption can be continuously monitored, analyzed, and disseminated around the world. Finally, more individuals in their own private businesses and personal concerns can become involved in the global marketplace and global community through personal computers and the Internet. We are becoming global citizens.[171]

Globalization also impacts culture, values, and belief systems. Due to the global economic exchange and marketing of local and ethnic products, the mass production and distribution of goods and cultural icons from modernized countries, and the ubiquitous advertising and promulgation of images, ideas, and icons through the global media, people all across the world are being exposed to the cultures, ideas, and ways of life of all humanity. There is a great intercultural mixing going on.

According to William Wishard, the defining new reality of our time is "the human community as a single entity." We can no longer identify with a unique tribe, ethnic group, or nation; we must see ourselves as one single world community of humanity. He notes that for the first time we are incorporating planetary dimensions and considerations into our economical, political, and cultural beliefs and decisions. He also thinks that a unifying view of our origins and the universe is being provided by contemporary science that transcends individual cultures. According to Wishard, this overall shift toward a global world-view and global way of life is creating an upheaval in many aspects of human society.[172]

Writers as diverse as Hazel Henderson, Thomas Friedman, and Walter Truett Anderson see globalization as the most significant trend within our contemporary world. Henderson contends that all of the most significant changes from Industrial Age to Post-Industrial thinking will involve an enhanced global emphasis. Thomas Friedman argues that globalization is the most powerful and pervasive social, economic, and political force at work in the world today, replacing the Cold War as the defining theme of our times.[173] Globalization will change all aspects of human society. Anderson predicts that globalization will be the dominant reality of the twenty-first century.[174]

In *The Lexus and the Olive Tree*, Friedman sees the central challenge generated by the sweeping phenomenon of globalization as balancing worldwide integration with local cultures that feel threatened by globalization. He believes that a balance needs to be achieved between individuality and community. As a way to metaphorically capture these two opposing dimensions of human life, he uses the "Lexus" as the symbol for modernization and globalization and the "Olive Tree" as the symbol for local roots, family, and individuality. The "Lexus" represents the benefits of globalization, including increasing wealth and connectivity, but the "Lexus" also brings with it homogenization and standardization, and thus the regional and local traditions symbolized in the "Olive Tree" may be wiped out as modern globalization spreads across the world.

Overall, Friedman sees globalization as a positive trend. For example, he connects globalization with the spread of democracy around the world. Education and finance become democratized with globalization, and political democracy, according to Friedman, is a necessary condition for countries that want to globalize. He believes that modernized countries, and especially the United States, will benefit most from globalization and, consequently, such countries should take a disproportionate share in being responsible for the development and maintenance of a global society.[175]

In his more recent book, *The World is Flat: A Brief History of the Twenty-first Century*, Friedman further develops his ideas on globalization and its positive impact on human society.[176] In *The World is Flat*, Friedman begins by outlining a three-phase theory of the history of globalization. In "Globalization 1.0," which occurred during the period of 1492 to 1800, European nations explored, conquered, and settled many parts of the world, gaining power and control over territories, people, and resources across the globe; in "Globalization 2.0," supported by falling transportation and telecommunication costs, during the period of 1800 to 2000 the process of globalization became increasingly driven by multi-national corporations in search of new markets and cheap labor; finally, according to Friedman, we have now entered into "Globalization 3.0" where individuals, with access to the Internet, computer and communication technologies, and new computer software, have become empowered to productively enter the global arena of business, social collaboration, and informational exchange. Hence, across these three phases there has been a shift in power from nations to corporations and now to individuals. As a general point in his book, globalization is now impacting the thinking, behavior, and livelihood of individuals - individuals have gone global.[177]

Based on this historical analysis, Friedman's central thesis concerning contemporary times and the immediate future is that the world is "flattening," where individuals around the world have the opportunity to enter a level playing field to create businesses and accrue wealth. Everyone, potentially, has access to the global marketplace and the tools and capacities to take advantage of this expanded economic arena. Friedman identifies the "triple convergence" of the recent development of a Web and technology-enabled global economic playing field, new software for horizontal collaboration, and the opening of new significant players or societies, such as China, India, and Eastern Europe into the global arena, as the primary instigating causes behind the flattening of the world. Where Globalization 1.0 and 2.0 were largely a result of European and American initiatives, Globalization 3.0 brings in many non-Western players. Friedman also describes a related set of ten factors or forces contributing to the flattening of the world. He includes the fall of the Berlin Wall, the advent of PC's and Windows software, the emergence of public Web browsers, open-sourcing, outsourcing, off-shoring, supply-chaining, in-sourcing, and various "amplifying technologies." According to Friedman, the triple convergence of new players, a new playing field, and new collaborative processes are the most important forces shaping global economics and politics in our time.

Friedman acknowledges, though, that there are problems and challenges regarding the flattening of the world. For one thing, the world has not become entirely flat; there are large regions and huge populations of people that are economically and technologically impoverished and unable to participate, as of yet, in the global economy. As one critic of Friedman puts it, "much of the world is flat - flat broke."[178] (Also see Florida's analysis later in this section.) Additionally, Friedman sees terrorist and antiglobalization groups as a significant threat to the flattening of the world; terrorists, in fact, are being empowered through globalizing technologies. In essence, the battle between the Lexus (globalization and flattening) and the olive tree (nationalism, ethnicity, religion, and identity) is still ongoing.

Still, Friedman clearly supports the movement toward globalization and is highly critical of "tribalism." Within a flattened and global society, nations need to be open and not "build walls." In this regard, he is concerned about recent trends in the United States toward protectionism and heightened security. America should be a "dream factory" rather than a "fortress." For Friedman, the defining event of our recent past should be the falling of the Berlin Wall (11/9) and not the terrorist attacks (9/11). He is also worried over negative trends in education in the United States, especially in science and engineering. According to Friedman, the United States is at a crisis point in its history; will the United States flounder, or will it continue to take a leadership role in the ongoing globalization of humanity?

There are various criticisms of Friedman's ideas. Contrary to his "flattening" theory, the market is not free but dominated by a small group of countries - power is still highly localized; although Friedman blames the Middle East for not integrating into the globalization movement, it could be countered that modernized countries, under the banner of globalization, are actually attempting to impose their values on the Middle East; he puts too much faith in the power of technology to transform the world; and finally, he puts too much emphasis on economic and material factors in understanding human behavior and issues concerning the quality of human life.[179]

In an effort to be comprehensive regarding the nature of globalization, Walter Truett Anderson outlines a multi-dimensional theory in his book *All Connected Now: Life in the First Global Civilization*. Anderson not only discusses economic globalization, but also includes in his book treatments of cultural and political globalization; biological and ecological factors connected with globalization; the significance of the information network in the creation of a global society; the rise of global consciousness, increased human mobility and migration; and the huge growth in numbers of human organizations, and especially, international organizations. Anderson sees "a world of open systems" as the general theme running through all these dimensions of globalization. Although there is resistance to globalization, where some cultures and organizations wish to remain closed, according to Anderson, the overall world-wide trend is toward increasing openness and interactivity. In this respect, his view is similar to Friedman's.

Some of Anderson's main arguments and hypotheses include: There is a general ongoing trend toward multiculturalism, in spite of efforts to preserve integrated pure cultures; the twentieth-century discovery of ecology and the interdependence of the earth, life, and humanity has intensified global consciousness; there has been a significant rise in international corporations and a reciprocal rise in global governance to control and monitor these corporations; there has been a globalization of human rights and human laws; the global society is an "open society" that emphasizes individual responsibility and choice rather than dictatorial rule; a "cosmopolitan citizen" has emerged who does not identify with any particular nation; and the global society is multi-centric with many centers of power. Anderson predicts a series of global societies in the future, as the human community struggles with the challenges and inherent conflicts brought on by globalization. In the coming century, the big issue is going to be globalization itself - its pros and cons.[180]

The Metaman Theory: Gregory Stock, in his book *Metaman*, argues that life on earth has gone through three major evolutionary changes or jumps.[181] The first major change was self-replicating bacterial creatures (the origins of life); the second step was the symbiotic organization of nucleated single-cell animals and plants; and the third stage was the aggregation of single-cell life forms into multicellular organisms. In each case, the evolutionary jump involved a progressively higher order integration of individuals into greater wholes. Order and complexity evolved through the formation of larger and consequently more intricate biological systems.

Stock's Metaman theory proposes that humanity is presently going through the fourth big jump in the evolution of life. Stock believes, as do many others, that we are integrating and merging with our technology. Stock goes even further though.

Technology is integrating at a global level, forming a vast and intricate network of resource and information exchange. A global machine is forming. Collectively, humans are integrating with this global machine. Humans, together with information technology and computer network systems, are emerging as the functional nervous system of this world-wide system. Humans and the global machine are becoming so interdependent that they are synthesizing into a global super-organism.

This view of the evolutionary transcendence of humanity is presented in a very upbeat manner and style. The emergence of such a complex and intelligent global system is enhancing the individual lives of more and more people around the globe. This human-technological system brings with it increased affluence and material wealth, more freedom and opportunities, access to more information and ideas, and greater individual power. Stock's idea is that Metaman is spreading out, in a manner analogous to a growing web of living tissue, from the technologically modernized nations toward the less advanced areas. As Metaman grows, it will steadily assimilate more and more of humanity, bringing with it the benefits of the global society and economy.

In his Metaman theory, Stock combines the theme of globalization with ideas from technological and biotechnological theories of the future. The emerging global society will be like a single global organism, made up of both biological and technological components. Stock's ideas are also, in many ways, a natural extension of H. G. Wells's projection that the future will see the emergence of a Global Brain and a Global Mind. There is clearly considerable evidence that a Global Brain and Mind are emerging in our world, as human organizations, computer networks, and communications systems integrate into greater functional wholes. Such a system will vastly exceed the intelligence and complexity of individual nervous systems and minds.[182] If we were to take a Gaian perspective on the earth, assuming that the earth possesses many of the properties of a living organism, the emergence of Metaman adds an integrative layer on top of Gaia, providing a human-technological global system that monitors and coordinates Gaia.

Although such a pervasive and highly integrated global system might seem highly controlling and repressive of freedom and individuality, Stock presents the argument that the system will actually benefit human freedom and individuality. A similar argument has been made regarding the impact of a global society on individuality.[183]

Bloom's Global Brain: Howard Bloom presents the theory that life on earth has evolved as a collective global whole. He sees two fundamental processes at work in evolution. One process is integration, generating conformity and unity, and the other process is differentiation, generating diversity and individuality. Although these two processes are oppositional, integration and differentiation also work in tandem, mutually stimulating each other, and producing increasing complexity as a result. The two processes, in Bloom's mind, work toward the benefit of the whole. Integration produces coordination and order, while differentiation produces variability, which is necessary for creative experimentation in the evolutionary process. The rich and varied, yet equally interdependent network of living forms on the earth is a result of these two processes. For Bloom, a complex and intricate global brain has been evolving on the earth since the beginnings of life.

Humanity is part of this multi-species network, requiring the presence and utilization of many other living forms. According to Bloom, it is a mistake and an illusion to see humans as "isolated entities." Further, human history is filled with examples of both "conformity enforcers" and "diversity generators," and he sees our modern day philosophies of individuality and freedom versus unity and order as simply intellectual expressions of these two opposing tendencies within us and all of life.

The contemporary conflict between rigid fundamentalist groups and multicultural modernized nations is also a reflection of these two forces within us. The great cultural mixing of the last century, due to multiple waves of migration and global communication and exchange, brought with it new freedom and opportunities and a sense of hope, but it also instigated counter-reactions out of fear, for the stability and security of the past seemed threatened by the Postmodern world. For Bloom, both Muslim and Christian fundamentalism are paradigm examples of "conformity enforcers" that wish to bring order and homogeneity through authoritarian control. He calls them the "new Spartans."

Bloom believes, though, that a balance needs to be struck between integration and diversification. A police state that produces a regimented paradise would sap the inventiveness out of humanity. According to Bloom, the solution to our present problems and challenges involves a combination of self-control and social freedom. Bloom feels that the fundamentalist strategy is to control the other rather than the self and, in his mind, this approach will not work. Hence, although Bloom sees all of life and humanity as a collective whole, he believes that the further evolution of this collective whole, following the dialectic pattern of the past, is to balance conformity and diversity. He sees fundamentalism as a significant threat to this balance as well as a threat to human freedom and creativity.

Bloom describes living forms as "complex adaptive systems" and the whole global network of life as one vast "complex adaptive system" that learns and evolves. He particularly emphasizes that bacteria and microbial life, since early on in the history of the earth, integrated into a global adaptive system. With the development of human civilization and modern globalization, a new global mind, coordinated by humans and human technology, is emerging on the earth. Bloom foresees the greatest future challenge facing humanity as finding ways to more cooperatively work together with the primordial global brain of bacteria. Although he sees humans as "evolution incarnate," Bloom argues that humans should see themselves as the "neurons of an interspecies mind" that will involve the participation of all living forms, and especially the bacterial underpinnings of all earthly life.[184]

Economic Materialism: This general perspective on the future includes those theories and visions that find their inspiration in the Western secular concept of progress, as well as physical science and materialistic-economic models of social growth and quality of life. This perspective is often associated with Industrial Age thinking, specifically at it applies to human society. Toffler, Capra, and Fukuyama provide a representative sample of social, economic, and philosophical discussions of the historical rise of this view of reality and society.[185] It is the central view and theme, however, that is criticized by numerous Post-Industrial and Postmodern theories of the future; e.g. Toffler, Henderson, Slaughter, Elgin, Theobald, Eisler, and Zohar and Marshall all critique this worldview for various reasons as a starting point in the development of their alternative visions of tomorrow. It is frequently blamed for the various problems and challenges associated with modern times, including pollution, social conflict, depletion of natural resources, human inequality, Western imperialism, and dehumanization. Ray and Anderson, though, note that the largest sub-culture within the United States still most strongly identifies with the values, beliefs, and lifestyles of this perspective.[186]

As a philosophical doctrine, **materialism** states that all of reality is composed of physical matter - there is no second spiritual or mental reality, except perhaps as manifestations of the world of physical matter. In modern times, this all-encompassing theory of reality was inspired by the continued success of Newtonian physical science in explaining more and more features of reality. If everything is physical matter, then human progress can be defined as the increasing control of matter, the industrial production of more types of material goods, and the cumulative acquisition of such material goods by the general population. In general, progress is material economic growth, both in production and acquisition.

Presumably, all those important dimensions of human life, such as happiness and life satisfaction, meaning and purpose, human relationships and social well being will improve if material economic growth occurs. Further, the level of advancement of a human society can be measured in terms of its material economic output and wealth. For those future-oriented thinkers who support this theory, their vision of a bright and better future is continued materialist growth, usually facilitated through improvements in physical technology, and an improved quality of life measured in terms of material wealth and possessions.

The Big Business - Corporate - Economic Growth Model: The philosophy of secular progress promised continued growth and improvement in the material quality of life. As a reflection of this philosophical perspective, present measures of progress and the quality of life often emphasize financial, economic, and material wealth and luxuries.[187]

In the eighteenth and nineteenth centuries, the materialist-secular concept of progress became closely associated with the philosophy of capitalism. Capitalism provided a theory of how economic and material progress could be achieved through individual enterprise, competition among businesses, and a selection process involving the discriminative purchasing of competing products by consumers.[188] Though the practice of capitalism goes back much further than the eighteenth century, in the last two centuries it has become one of the most powerful driving forces in the growth of human civilization. Supported by continued advances coming out of science and technology, capitalist businesses have steadily grown in power and size. Of particular note, the spreading of modernization across the globe has critically involved a progressive expansion of free enterprise capitalist businesses and corporate organizations that produce and distribute products and services, driving an accelerative growth in the production and accumulation of wealth and material goods.[189]

In the last century, trans-global corporations have been growing in power and influence and developing into an international network that is weakening the centrality of nation states in human society.[190] This corporate-economic network not only brings with it a set of standards of growth and progress, but it also supports a particular set of values and principles of human organizations. There are futurists such as Toffler, Drucker, and Wheatley[191] who argue that the principles and values of business corporations and organizations are changing, yet at present within the corporate world there is a clear emphasis on competition, profit, consumerism, and economic growth as the central defining values and features of human life. All of human life is reduced to economic and business metaphors and measures, and the promise, hope, and expectation of this approach is for continued development along the same lines into the future.

A theory of the future that integrates economic and capitalist concepts of progress with the themes of globalization, democratization, and high technology is presented in *The Long Boom: A Vision for the Coming Age of Prosperity* by Peter Schwartz, Peter Leyden, and Joel Hyatt.[192] The authors of this book argue that we are witnessing a long-term boom in economic growth around the world that hopefully will continue for many decades into the future. According to these three, technological change, rapid economic innovation, global integration, and spreading democratization are creating a new world economy that will impact all aspects of human civilization. The key principles behind this new economic boom are globalization, openness, lack of top-down control, a pro-growth philosophy, learning and adaptation, the valuing of innovation, establishing connections and being inclusive, and an overall sense of confidence in the future. This global economic trend will transcend nationalist concerns and overcome regional political differences and conflicts.

Schwartz, Leyden, and Hyatt see the "long boom" as a continuation and evolved expression of the modern vision of secular progress. They acknowledge, though, that the two world wars, the economic depression, and nationalist conflicts in the last century significantly unsettled the optimism of the nineteenth century and created much pessimism and nihilism about the future around the world. Yet, beginning in the

1980's, there was a shift in attitude back toward economic and social optimism. Though there is still significant resistance to globalization and economic growth, especially from environmentalist, human-rights, and anti-American groups, the authors believe we are on a new upward swing in attitude and social development.

The authors contend that there are many positive features to this new global economic boom. It will produce more jobs, increased wages, and reduce the gap between the rich and the poor worldwide. In general, they think that economic growth and social justice are not in conflict with each other. Also, with the shift from an industrial to an information-knowledge economy and the introduction of cleaner, more advanced technologies, our economic system is becoming more in balance with nature. Presenting ideas similar to those espoused by Toffler and Wheatley, they see a new form of business organization emerging that replaces the centralized bureaucracies of the past with decentralized networks that empower individual workers rather than alienate and suppress them. The authors see the introduction of computer technology into the business and economic worlds as having many benefits, including better communication and the monitoring of trends. Additionally, education and learning have become more important within the new knowledge economy. In general, they believe that the only viable way to help the poor and the destitute around the world is with a pro-growth economic philosophy.

As one last point, Schwartz, Leyden, and Hyatt believe that the United States is in an ideal position within this new global economy. The United States has the largest economy and the largest market and is a leader in both research and education. According to them, the axis of innovation in the world has now moved to California, which embraces a multicultural and global mentality. In the long run, to adopt a more "selfless" attitude would be in the best interest of the United States. By helping to facilitate the spread and development of the new economy through more parts of the world, more consumers will be created.

D'Souza's Virtue of Prosperity: Dinesh D'Souza, in *The Virtue of Prosperity: Finding Values in an Age of Techno-Affluence*, presents his views on past, present, and future in the form of a dialectic or debate between what he takes to be the two fundamental opposing viewpoints in contemporary times.[193] He refers to these two opposing viewpoints as the "Party of Yeah" and the "Party of Nah." D'Souza believes that the conflict between these two views supercedes all earlier or more traditional conflicts, such as Left versus Right or conservative versus liberal. Though he acknowledges some valuable points put forth by the "Party of Nah," D'Souza basically ends up supporting the "Party of Yeah" as the best approach to the future.

According to D'Souza, the "Party of Yeah" believes in the value and beneficial power of technological capitalism. The new leaders of this party (technologists and business entrepreneurs) want to direct and liberate the world, based on a renewed faith in capitalism and the power of technology. Members of this party value the making of money. They see themselves as optimistic and visionary about the future, believing that humanity is at a pivotal point in history, ready for a significant jump forward in human civilization and the quality of life. The "Party of Yah" believes that information technology empowers people and that the new economy will both solve social ills and eliminate scarcity and poverty. They think that the Internet will build new human communities and bring humanity together. Generally, they argue that technology and economic prosperity have done much more good than harm for both humanity and the environment. Technological capitalism will do the best job of improving the environment. And of special note, the modern world of capitalism, free enterprise, and technology has been responsible for many moral gains in human society, for the "Party of Yah" people in modernized countries live in the best society ever within human history, and their belief is that the future will be even better.

Basically, the "Party of Nah" believes that modern humanity has been in a state of moral decline for several centuries and that things are only getting worse. According to D'Souza, the "Party of Nah" consists of cultural pessimists, environmentalists, conservatives, religious fundamentalists, and techno-skeptics. The "Party of Nah" points out that the distribution of wealth is not evening out, but becoming more imbalanced - financial inequality is on the rise. Also, for some critics, we live in a "toxic culture" - a culture that no longer values literacy - and in the modern world we have become increasingly demoralized in spite of all the economic progress of the last century. Further, there has been a loss of community and a sense of neighborhood; the increasing mobility of modern humans has destroyed stable local communities. D'Souza, citing Daniel Bell's classic work, *The Cultural Contradictions of Capitalism*, points out that another criticism of the "Party of Nah" is that our consumerist culture has become increasingly hedonistic. The right wing contingency of the party, in fact, contends that modern affluence has produced moral degeneracy rather than moral advance. According to this faction, technological capitalism is motivated by greed and selfishness and is inherently evil, leading to materialism, the proliferation of vice, broken families, the vulgarization of culture, and uprooted communities. Because of the consumer lifestyle in the modern world, we value things more than ethical principles or other people. Overall, our earlier societies and communities of "solidarity" have been replaced by a society of commerce, and our personal relationships, once built on affection and caring, are now built on contract. For left-wing critics, capitalism, industry, and technology have hurt the environment and we have become more isolated and alienated from nature. Although not all members of the "Party of Nah" blame all the ills of the world on technology and capitalism, they don't believe that the solution to our present social and moral problems lies in more technology or more wealth. As is frequently pointed out, after fundamental human needs are met, there does not seem to be any correlation between increasing wealth and human happiness. As a final criticism of the "Party of Nah," which D'Souza thinks is perhaps the most significant, the techno-capitalist world has destroyed the connection between one's purpose in life and the moral order of the cosmos. We no longer see or perhaps even care, how our lives and our vocations connect with the big picture of things.[194]

As noted above, after weighing the arguments of both the "Party of Yeah" and the "Party of Nah", D'Souza generally supports the vision of the future of the "Party of Yeah." He believes that modern humans can live both a good life and a life that is good and in so doing demonstrate the "virtue of prosperity." As he points out, the poor are not morally superior to the wealthy; affluence does not generate increasing evil or vice. Further, before the modern rise of technological capitalism, people were not more moral than today; our modern world has not caused some significant moral decline. In fact, D'Souza agrees with the "Party of Yeah" that there have been great moral advances in the last couple of centuries in those modernized nations that have supported capitalism and technological advancement. D'Souza, in fact, believes that affluence is morally beneficial, for it enhances one's power to help others or help the environment. (He cites that it is the rich who have done the most to improve the environment.) Further, he points out that selfishness - the psychological foundation of capitalism - is an inherent human quality and can't simply be eliminated from the human psyche. Capitalism "civilizes" greed and makes it into something that can benefit the other; the capitalist is motivated to producing a product or service that others find desirable or appealing. Although grounded in the selfishness of the producer and consumer, a capitalist economy is built on reciprocities - of transactions of mutual benefit. D'Souza also notes that although there are many critics of techno-capitalism who live in the modern world, these very critics want to enjoy the benefits of the very system they find objectionable, and that the poor and destitute who live in undeveloped countries would, much more often than not, enthusiastically embrace the opportunity to live in the more affluent countries. Finally, D'Souza argues that it is only with financial wealth that individuals have the opportunity to seriously reflect on the meaning and purpose of life. Few could ask this fundamental question before, since basic survival needs dominated their lives and, generally speaking, the answer to this question was provided to them by authoritarian religious and political systems. With affluence and a democratic society, more of us now have the freedom and the time to ask the important big questions. Many find this freedom disquieting and, perhaps, would wish to retreat to the traditional and idealized certainties of the past; they are thus critical of modern times. Although technology and capitalism may not provide the answers to the big questions, we can at least now ask them.

Jihad versus McWorld: Integration and Unity versus Pluralism and Diversity: Benjamin Barber in *Jihad versus McWorld* weaves together the issue of globalization versus regionalism and cultural pluralism with contemporary concerns over the influence of capitalism and the erosion of democracy and community in a consumerist society.[195] Barber's central argument is that the spread of global capitalism - of "McWorld" - through the growth in influence and geographical reach of international corporations is destroying democracy, homogenizing the world, turning citizens into consumers, replacing community values with selfish needs, and provoking a widespread counter-reaction - a "Jihad" - in an attempt to preserve local customs and values. Although the term "jihad" is frequently associated with Islamic terrorism, Barber uses the term in a more general sense to refer to all those regional counter-reactions, violent or not, to the spread of the global capitalist culture of "McWorld."

Barber argues that "McWorld" presents the illusion of choice and self-determination in the great array of products and services it offers the consumer, but in actuality, it diminishes human freedom, individuality, and democracy. "McWorld" creates a homogenized mono-culture that threatens cultural and ethnic diversity, thus working against individuality and freedom. Through advertisement, marketing, and the control of production there is an ongoing manipulation and fabrication of human needs and lifestyles - hence, although the marketing message is to "have it your way," - this is illusory, for advertising is a method for influencing people to believe and feel that they need or desire the products being advertised. Further, the increasing centralization and monopoly of control, in the hands of a smaller and smaller number of giant corporations and business leaders, works against a democratic and pluralistic distribution of power in the world. Barber, in fact, sees transnational corporations as totalitarian, determining all facets of human life, and the infotainment sector as the center or hub of this ubiquitous system of influence and control over people.

Within "McWorld" there is a subjugation of all values under profit and economic growth, and a trivialization and commodification of higher values. As all values are reduced to monetary worth, all people are reduced to customers and consumers. Barber is especially worried that the consumer has been elevated over the citizen. Instead of keeping abreast of the problems and challenges facing humanity and considering how to constructively participate in the improvement of human society, people in modernized countries spend their time attentive to new gadgets, products, and services and the purchasing and use of such commodities. Furthermore, trans-global corporations undermine the power of national states and government institutions which at least attempt to deal with social problems and challenges. Our most powerful human organizations are more concerned with making money, expanding and creating new markets, and generating economic growth than creating a better world.

Global capitalism is not producing equality, but rather is increasing the schism between the wealthy and the poor, both within individual countries and between modernized and undeveloped countries. "McWorld" leads to the exploitation and oppression of workers in poorer countries to support the lifestyles of those in wealthier countries. Fueled by a philosophy of continued economic growth, we are in a race between escalating needs and limited resources. In Barber's mind, it is quite understandable why many people, especially those in undeveloped poorer regions, often react violently to the insidious spread and increasing power of "McWorld."

What Barber argues for to counteract the growing power of "McWorld" is neither terrorism nor regional authoritarian cultural movements. He believes that people in modernized countries need to become less self-centered and more socially conscious. Unless we re-energize our sense of citizenship and see ourselves more as members of a social community, then we really will end up in a world in some ways very similar to Huxley's *Brave New World*, addicted to pleasures defined by economic corporations, oblivious to our loss of freedom, and ruled by an elite few.

Affluenza and Conscious Consumerism: One of the most thorough-going critiques of our modern capitalist and consumerist society is contained in the book *Affluenza* by John DeGraaf, David Wann, and Thomas Naylor. The authors not only provide an analysis of the causes and main features of the "epidemic" of affluenza, but they also provide a "treatment plan" for the future - of how to change our modern lifestyle and values in a way that cures us of this disease and produces healthier and happier lives.[196]

DeGraaf, Wann, and Naylor believe that modern humans are literally afflicted with a psycho-social disease, which they label "affluenza", and define as "A painful, contagious, socially transmitted condition of overload, debt, anxiety, and waste resulting from the dogged pursuit of more." There are many causes behind the spreading of affluenza including human greed; the expansion of production and the creation of new needs; choosing work, money and consumption over leisure, liberty, and time; the planned obsolescence of products; the growth of credit and the advertising industry; the commercialization of everything; and the popular belief that consumption somehow equals freedom, self-actualization, and self-reward.

The symptoms of this disease include "shopping fever" and "mall mania," shopping as a form of therapy, increasing debt and bankruptcy, unrealistic and swollen expectations, clutter and chronic congestion, stress and time urgency, a loss of community, resource exhaustion, obesity, and an overall loss of meaning in life - a "poverty of the soul."

What the authors propose as a treatment plan - as a new way to live in the future - includes the following: Recognizing and assessing the severity of the problem; learning to live a life of moderation and balance instead of excess; shifting from a materialistic philosophy of life to a life that emphasizes spiritual, aesthetic, and cultural values; becoming more environmentally conscious and getting back out into nature; interacting with people more and things and gadgets less; working less, slowing down, and spending more time in leisure; and, in general, reflecting on the question of "what is life for" and pursuing quality in life. The authors see the need to redefine the nature of progress, including more humanistic values in addition to the economic values that presently dominate measures of progress. Our lives and central defining values need to change in the future.

Cultural Pluralism and Human Diversity: The emphasis within this perspective on the future is the variability and uniqueness of human belief systems and ways of life. As noted above, there are strong regional and ethnic counter-reactions to the perceived homogenizing effects of globalization and global capitalism. Especially in non-Western societies, many people are attempting to preserve the richness and diversity of their unique cultures. In particular, cultural and anti-globalization, nationalistic trends have emerged in Islamic, African, Asian and South American countries.

Non-Western advocates of the philosophy of cultural pluralism argue that Western visions have dominated thinking on the future (for example, the Western philosophy of secular progress), and that we need to open our view of tomorrow to the richness and variety of systems of thought and value within non-Western cultures. Cultural pluralism in this sense is not one theory, but many views - united in their opposition to the perceived myopia of Western futurists and social thinkers. The future should be pluralistic rather than dominated by one culture.[197]

There is evidence that there are certain fundamental differences among the cultures and people of the world. In his well-known study of world cultures, Harry Triandis came to the general conclusion that cultures tend to vary on the collectivist versus individualist dimension; some cultures put more of an emphasis on conformity and maintaining sameness whereas others emphasize individual freedom and uniqueness. Interestingly, it is Western cultures (those often accused of attempting to homogenize the world) that are more individualistic and it is Eastern cultures that are more collective.[198] In their ongoing survey of world values, Inglehart and Baker find another systematic difference among cultures. Modernized and economically developed countries value secularization, rationalism, and self-expression, whereas economically undeveloped countries put more value on tradition, religion, and survival needs.[199] A third important study, conducted by the psychologist Richard Nisbett, seems to indicate that Western people think differently in certain important respects from those in the East. For example, Westerners are more linear, analytic, and individualistic, whereas Eastern people tend to think more in circular logic, and are more holistic and communal.[200]

The importance of human diversity in the future is a theme that also gets highlighted in descriptions of modernized countries. In *The Third Wave*, Toffler argued that the mass media and mass society are disappearing, being replaced by a diversity of styles, tastes, philosophies, and ways of life within the modern world. Cultural and psychological human diversity are being more readily accepted, if not encouraged, for example, regarding race, religion, age, and sexual preferences.[201] As mentioned above, it is Western and modernized countries that value individuality and self-expression, and seem to value human diversity within their populations the most.

The argument is frequently made that modern globalization is actually a force that supports human diversity and individuality, and that cultures which resist globalization in the name of preserving their distinctive ways of life actually value conformity and homogeneity. Global theorists, such as Friedman, Anderson, and Henderson, see globalization as actually leading to a more democratic and liberal world - the exact opposite of what the anti-globalization forces say they fear. Hence, globalization theorists could argue that they are the ones who most strongly support human diversity and cultural pluralism.

Victoria Razak and Sam Cole, in their article "Culture and Society," delineate three possible future scenarios for global human diversity. Cultural polarization would be a world of separatism and antagonism between different human cultures - a world of diversity with no unity; cultural assimilation would be a world of homogeneity and a single world culture - a world of unity with no diversity; and cultural pluralism would be a world in which a single global society found a way to preserve and value the unique qualities of each human culture - a world of unity with diversity.[202]

Conflictual Pluralism: Samuel Huntington's *The Clash of Civilizations and the Remaking of World Order* is one of the most controversial recent books written on the themes of cultural pluralism and globalization. The central thesis of his book is that culture and cultural identity will be the primary determinant in the political dynamics of the near future. In particular, Huntington foresees significant political conflict among the main world civilizations, such as the West, Islam, and China. He does not believe that a universal or global civilization is emerging in our times - instead he sees tension, conflict, and division.[203]

Huntington presents a history of humanity as a foundation for his theory of contemporary and future society. According to him, in pre-modern times there was not a great deal of contact among the major world civilizations, but beginning around 1500, Western European countries, while continually fighting among themselves, began to conquer and colonize most parts of the rest of the world. After World War II a major global division emerged between Soviet Communism and Western countries, and the Cold War dominated world politics for the next fifty years. During the Cold War, the major "fault line" in the world was ideological - between collective communism and democratic capitalism. Although many nations won their political independence from Western dominance during this period, modernized Western countries simultaneously began to push for increasing economic and cultural globalization - as a way to maintain control over non-Western countries. According to Huntington, the philosophy of globalization is basically a Western creation with Western cultural ideals at its core. With the end of the Cold War and the collapse of the Soviet Union, a new central conflict has emerged in the world between Western modernized civilization and multiple non-Western civilizations. Global politics has reconfigured along cultural lines.

For Huntington, there are presently nine major world civilizations, including Western, Islamic, Latin American, Sinic (Chinese), Buddhist, Hindu, Orthodox, Japanese, and African. Each of these civilizations has at its core a relatively distinct culture and each of these cultures is most strongly determined by the religious traditions of that culture. Although nations are significant political entities in the world, nations will tend to align together along cultural lines and oppose other nations that fall within different cultures and civilizations. Culture both unites people and nations within a civilization and sets up oppositions between civilizations. The revitalization of religions across the globe, in particular, has reinforced cultural differences and created increasing world tension. The rise of terrorism, especially coming out of Islamic civilization in opposition to the West, appears to be an implication of Huntington's theory.

Although the West has been the dominant civilization in the world over the last few centuries, according to Huntington the power of the West is declining. Beginning in the twentieth century, a revolt against Western domination began, and especially over the last couple of decades the Western ideal of a universal global civilization is being actively resisted by other major cultures and civilizations. Huntington acknowledges the spread of modernization throughout the world, but he does not believe that modernization inevitably leads to Westernization or globalization. Even with the introduction of new technologies, capitalist economic principles, consumer products, cultural icons, and global media coming out of modern Western civilization into non-Western countries, he thinks that the core culture, beliefs, and values in non-Western countries have not changed. Western products do not transform non-Western mindsets. Increasing interaction among world civilizations is not bringing more peace; there are still significant levels of conflict around the world. For Huntington, non-Western countries are adopting the benefits of modernization while still maintaining their regional cultures. And in fact, modernization may make non-Western civilizations more powerful and able to increasingly resist and challenge the supremacy of the West. In the final analysis, Huntington argues that the world is becoming more modernized but less Western.

Postmodernism: The philosophy of Postmodernism grew out of numerous critical reactions in the nineteenth and twentieth centuries against secular modernism. Steven Best and Douglas Kellner, in their multi-volume study of the rise of Postmodernism, argue that what all these reactions have in common is a rejection of reason as the authoritarian source of truth and value.[204] There have been romantic, conservative, mystical, and naturalist anti-modern movements, up to and including, the counter-culture 1960s movement. The dream of the Enlightenment - of a rationally ordered world - has been attacked and rejected by all these perspectives. In the second half of the twentieth century, Postmodernism, embracing many features of these critical movements, and especially Nietzschean and existentialist philosophy, mounted a multifaceted attack on rationalism and Enlightenment philosophy.

In general, Postmodernism involves a critique on the authority and power of modernist thinking. Aside from the anti-rationalism theme that runs through many reactions against modernism, the related theme of oppression and excessive control is another dimension of modernism that has been attacked. The rationalism of modernism has been seen as repressive by romanticists and mystics alike.[205] Further, according to Best and Kellner, many Postmodern thinkers see capitalism, industrialism, male chauvinism, commercialism, Western imperialism, and even objectivism as power-centered, oppressive features of the modern West. Robert Nisbet lists totalitarianism, racism, military conquest, and destruction of the environment, all forms of excessive control and domination, as additional negative consequences of progress that have been strongly criticized over the last century.[206] From a philosophical point of view, Postmodernism has attacked the ideas of "grand narratives," universal theories, absolute truths, and even the progressive view of history as monolithic, constraining, and controlling.

Postmodernism, which by its very name means "after" or "against" modernism, has as one of its central values individualism. But it inherited this value from modernism. In fact, the Postmodern critics of the oppressiveness of modernism are in fact creations of the heightened individualism that modernism bequeathed to the contemporary world. Yet Postmodernists seem to think that the espoused individualism of modernism has actually been highly selective and restrictive. Modernism has been accused of being repressive and domineering toward women and all non-Western cultures. It is only Western males, and usually more wealthy ones at that, who have reaped the full benefits of individual determination and expression. Postmodernism also points out that in spite of the professed value of individualism and freedom, there are also other ideals in modernism that are more aligned with order and control. For example, reason and science presumably lead to singular truths, and the idea of secular progress defines a particular and limiting mindset regarding the direction of the future. In its development secular modernism actually supported opposite ideals - both toward freedom and individuality and toward unity, order, and a single truth.[207] Hence, in spite of the fact that modernism instigated increased individualism in the West, it also pushed toward the constraint and control of humanity in other ways. Postmodernism argues that the constraints have to be further loosened up.

In general, according to Best and Kellner, Postmodern writers see a new society emerging in contemporary times. Although there are continuities with the past, the dominant philosophy and way of life of secular modernism is falling apart. The new paradigm of life reflects many recent changes and developments in the world, including computers and high technology, the rise of the media culture, decolonization, cultural fragmentation, and the decline of nations as centers of power. The new paradigm also reflects certain fundamental changes in thinking.

Best and Kellner identify four dominant themes in Postmodernism. The first theme is a rejection of unifying and universal schemes of thought. Postmodernism emphasizes pluralism, differences, complexity, multiplicity, and the end of any philosophies of life that apply to everyone. This first theme encompasses the Postmodern emphasis on individuality and the Postmodern rejection of reason and secular progress as universal and absolute guides for life. The rejection of "grand narratives" also falls under this theme. The second major theme involves an abandonment of closed, rigid, and fixed structures of meaning in favor of indeterminism, uncertainty, ambiguity, chaos, and contingency. The meaning of life is not certain - in fact, nothing is certain - and there are no literal or straightforward readings of any text or philosophical principle. Everything is ambiguous. The third theme involves a rejection of naïve realism in favor of perspectivism. All truths and values are relative. There is no absolute truth or objectivity – all beliefs are expressed from a perspective - historical, personal, and cultural. Finally, the fourth theme is a breaking down of discipline boundaries and absolutist distinctions. Fact and value, theory and practice, and science and society can not be separated from each other.

Many find the philosophy of Postmodernism liberating, exciting, and exhilarating, in its breaking free of the constraints of the past; others find Postmodernism threatening, confusing, nihilistic, irrational, and anxiety-producing. Interestingly, many of these criticisms of Postmodernism would actually be seen as strengths from within the Postmodern camp. For Postmodernists, it is a mistake to think that life is stable, secure, entirely rational, and perfectly orderly. As Best and Kellner note, even within the Postmodern movement, there are multiple interpretations of the meaning and significance of the movement. Because of its emphasis on individualism, multiplicity, and perspectivism, Postmodern thinking, by its very nature, does not yield unanimity of opinion. Is Postmodernism a relatively coherent philosophical position or a plethora of different views? Is Postmodernism an uplifting and liberating system of thought or a nihilistic and depressing point of view? Even among its advocates there is debate on these questions.

Chaos, Creative Disorder, and Cyberpunk: As noted above, not everyone believes that we are moving toward some great global society, global super-organism, or universal paradigm of thinking and living. Quite the opposite – according to many, we are moving into an era of increasing diversity, freedom, anarchy, and invention. An idea that has gained increasing importance in contemporary times is chaos. Rather than chaos being seen as a "bad thing," for many writers humanity is finally abandoning the delusion of some grand unified world system - a dream that has obsessed numerous conquerors, dictators and would-be world leaders since before Alexander the Great. The creative, unconstrained, and complex features of chaos are seen as positive features of the contemporary world; chaos should be embraced and reinforced.

To recall from the discussion on contemporary evolutionary theory, chaos has become an object of recent scientific study and appears to be an integral dimension in the evolution of complex systems.[208] Chaos is perhaps a fundamental feature of all natural systems; it is connected with indeterminism, unpredictability, and creativity within nature. Although the modern study of chaos arose in the natural sciences, the idea has been picked up in social and organizational thinking. For example, in her discussion of business organizations, Wheatley identifies chaos as a key principle in understanding how to facilitate growth and creativity in human groups.[209] In opposition to the Newtonian idea of an orderly, deterministic, and harmonious universe, the Industrial Age social philosophy of an orderly society, and the rationalist philosophy of Enlightenment, social chaos theorists, which include many Postmodern philosophers, view chaos as an essential and beneficial feature of human reality.

One particularly interesting collection of readings on the theme of creative disorder is *Mondo 2000: A User's Guide to the New Edge - Cyberpunk, Virtual Reality, Wetware, Designer Aphrodisiacs, Artificial Life, Techno-Erotic Paganism, and More.*[210] *Mondo 2000* was a San Francisco-based magazine that covered the rising computer culture. The magazine had a definite anti-establishment flavor and, in particular, emphasized the growing cyberpunk culture emerging in the world of computers and electronic reality. Although the magazine stopped publication in 1998, the magazine and the anthology still capture much of the inventiveness, color, and electricity of cyberpunk culture.

The reality presented in *Mondo 2000* is a variegated stew of chaos and open systems theory, virtual reality and virtual sex, electronic music, the growing wonders of psychopharmacology ("smart drugs," etc.), robots, artificial life, and brain implants. The book is intentionally a smorgasbord of ideas and trajectories, rather than an integrated treatise. It models in style the fast-paced, glitzy nature of our times. *Mondo 2000* presents an image of future reality, where what is "real" and what is "electronic and virtual" become intermixed. Further, it presents a "dynamist" vision (to use Postrel's term) of human life that emphasizes the value of creativity and chaos over order and control.

As a vision of the future, cyberpunk, to recall, first emerged in the science fiction writings of William Gibson. In the 1980s and 1990s numerous other writers, including Bruce Sterling and Neal Stephenson, contributed to this new science fiction vision of the future. Cyberpunk paints a picture of the future where information technology permeates most or all aspects of life and often the sense of social order and reality is totally disrupted. The cyberpunk vision is closely associated with features of Postmodernism; objectivity and truth are often difficult if not impossible to ascertain, the world is mysterious and confusing, and logic and linearity are replaced with free associative streams of consciousness.

Wired magazine is another good example of a creative, techno-oriented perspective on the future that clearly reinforces the cyberpunk culture. In fact, *Wired* has become, in the last few years, the premier popular publication of computer culture. Again, as in *Mondo 2000*, there are strong elements of Postmodernism, contemporary scientific theory, the integration of information technology into all areas of life, and a philosophy of freedom and change, all set within a visually stunning, almost helter-skelter format of electronic colors and computer graphics. *Wired* addresses all aspects of human life from politics and business to psychology and literature in the Information Age. Yet reflecting the contemporary Western world, there is also a strong commercialized feel to the magazine. Kevin Kelly (the author of *Out of Control*) is the former Executive Editor of *Wired* and Rudy Rucker and R.U. Sirius (of *Mondo 2000*) have been contributing writers.[211] Bruce Sterling and William Gibson (both cyberpunk science fiction writers) have also regularly written for the magazine.

The Triumph of Democracy: Francis Fukuyama, in his controversial book *The End of History and the Last Man*, presents a rich and comprehensive social-political history of humanity and a clear vision of future human society.[212] For Fukuyama, both political and economic democracies, as ideal human conditions, have been progressively spreading worldwide over the last few centuries. This progressive social evolution, though, is dialectical, with ups and downs and defeats and victories. New democracies emerge across the world and then often collapse back into authoritarian social systems. But according to Fukuyama, there is a definite progressive trend; though there have been significant setbacks throughout history, in modern times more and more countries are acknowledging and supporting the rights of individuals to determine their government and to determine their economic livelihood in an open market economy.[213]

For Fukuyama, the "end of history" will be when democracy finally triumphs worldwide. He contends that the global realization of liberal democracy will constitute the end point of mankind's ideological evolution (what we have been struggling toward throughout human history) and be the final form of government. Consequently, there will be no further basic social or political changes and this event will constitute the "end of history".

Fukuyama believes that human history can be seen as a single coherent evolutionary process - that there has been an overall direction to the course of human events. In the spirit of early Christian thinking and, more recently, Hegel and Marx[214], Fukuyama believes that there is a "universal history" to humankind - a basic pattern of change and destination toward which all humanity has been heading. Fukuyama's candidate for defining a universal direction in human history is freedom, democracy, and the acknowledgement of individual value. He believes that this is a discernable trend in history, and that this trend has a positive value.

Commenting on the increasing failure of centralized states, Fukuyama remarks that centralized economies cannot create or maintain post-industrial economies in which information and technological innovation are essential. But he also says that the failure of so many dictatorships in recent times is due to a lack of perceived legitimacy. Totalitarianism is an attempt to control the total fabric of society and culture - a central command under which everything is coordinated and directed and a unifying idea that determines both psychosocial values and economic practices. The old, strong states of the past have failed on both economic and ideological grounds - not only did they flounder in the new economic system, but they also failed as convincing and satisfying ideologies. A government must support both the mental and material welfare of its people, or it will collapse. Centralization failed as an economic practice and as a political ideal and value.

According to Fukuyama, the world-wide liberal revolution emphasizes economic privatization and free trade. Economic liberalism is the recognition of individual rights of free economic activity and exchange based on private property and markets. Liberal trade theory says that anyone who participates in such a system will benefit. Supporting this theory, Fukuyama observes that Asian countries that came into the free enterprise system late in this century have grown rapidly. Fukuyama believes that modern economic society will work for all countries; it will create an integrated economy and a socially prosperous system.

Fukuyama's concept of democracy rests on the idea of reciprocal recognition. We agree to value the opinions and input of everyone. No one is treated as a slave to the wishes of others. Everyone has self-worth and basic rights as a human being. A liberal society involves the reciprocal agreement among its citizens of their mutual self-worth. For Fukuyama, governments must address people's needs for individual recognition. People require a sense of value in their lives; people require a sense of mutual respect.

The idea of reciprocal recognition plays a central role in Fukuyama's approach to the future of human society and the future of government. Fukuyama's hope is that in the future we will see an end of imperialism - of lordship and bondage among nations. Fukuyama believes that the way past our present international paranoia is through the creation of a world government based upon the principle that its members be democratic nations who believe in the principles of democracy. Nations need to adopt a position of the universal recognition of the value of all other nations and relinquish any sense of dominant pride or superiority. Further, the law of nations must be democracy itself. Decisions should be reached consensually and adhered to by all of those nations within the world government. The key to future world peace and cooperation is very simple - we respect each other and recognize the value of our differences.

Fukuyama's theory highlights an important thesis about the future. He contends that the driving force of change in history is not technology or economics, but psychological and ethical values, an idea he shares with many other futurists. Also, it is interesting that Fukuyama believes human history shows a fundamental moral advance that will continue into the future. In fact, the central goal for the future evolution of human society, in Fukuyama's opinion, is not technological or economic advance, but moral progress.

The Triumph of the Individual: Naisbitt and Aburdene, in *Megatrends 2000*, argue that the underlying and most basic trend in contemporary times, supporting all other changes, is "the triumph of the individual." Instead of a dystopian world of mass conformity ruled by an authoritarian world government, the future is materializing as a world of diverse individuality where people have more individual power and more freedom. It is the tide of growing human freedom, above everything else, that is carrying us into the twenty-first century.[215]
Naisbitt and Aburdene point out the following factors contributing to the individualist trend:

- The computer and global communication systems are empowering individuals to create and market their products and ideas on a global scale.

- Information, as the fundamental resource supporting individual power in our world, can be possessed by many, whereas the Industrial Era central resource of physical wealth could only be possessed by a few.

- A global marketplace affords a much larger arena of opportunity than a local or national marketplace.

- A philosophy of individual responsibility has replaced the "follow-the-leader" mentality of the Industrial Age.

- Individual entrepreneurs play an increasingly more powerful role in the world economy.

- The global communication and monitoring system makes it increasingly difficult for tyrants and dictators to control large masses of people.

- The global communication system and the global distribution of products, ideas, and images offer people a much wider range of choices.

Naisbitt and Aburdene's theory of the triumph of the individual aligns with various themes discussed in previous theories, including human diversity, democracy, Postmodernism, and even globalization. Naisbitt, along with other contemporary writers, believes that globalization actually contributes to individual empowerment.[216]

Putnam and *Bowling Alone*: Individualism and separatism, if carried to an extreme, can have a variety of negative effects, both individually and collectively. There may be a loss of social trust and sense of community and a weakening of social connectedness and networking, or as some writers identify it, "social capital." The main thesis of Robert Putnam, in his highly popular book, *Bowling Alone: The Collapse and Revival* *of American Community*, is that social capital has dramatically declined in the United States over the past three decades and that this trend is a cause for great concern and needs to be counter-acted and reversed. Putnam documents, through an extensive array of empirical studies and surveys, the decline in social engagement, group participation, and social networking in the United States. He believes this trend toward social isolationism is damaging to democracy, our communities, our social institutions, and the overall quality of life.[217]

Putnam defines social capital as the value associated with social networks and associations. Social networks support the development of trust among people, facilitate cooperation, communication, and the exchange of information, establish and maintain "norms of reciprocity" or principles of mutual aid, and stimulate the growth of a broader sense of personal identity - people within social networks have a strong sense of "we" and not simply "I" or "me." All of these effects of social networks are highly beneficial. Further, Putnam argues, again supporting his view with empirical studies, that social and public institutions, especially those within democratic societies, are much more effective in dealing with problems and challenges when these organizations are supported by a high level of civic engagement on the part of the general population. Democracies work best when the citizens are actively involved. Public and institutional efforts to reduce crime and drug abuse, improve education, and constructively address poverty and unemployment benefit from an actively involved, socially networked community population. Putnam also brings up the point that economic vitality, innovation, and productivity are directly related to high levels of social networking - a similar point is made by Richard Florida in his *Rise of the Creative Class*; when people gather together both at work, or informally after work to socialize and share ideas, this positively impacts economic development. Finally, the level of social capital and networking in a society has a positive impact on physical and mental health - people reduce by half their chance of dying in the next year if they join and participate in one new social group.

In *Bowling Alone*, Putnam presents the results of a huge variety of studies that would indicate that social networking and consequently social capital in the United States have seriously deteriorated in the last few decades. Overall, membership in all types of associations is down twenty-five percent; family ties, with the increasing divorce rate and the weakening of both the nuclear and extended families, have significantly weakened; volunteer work has also substantially dropped; neighbors do not associate with each other anywhere near as much as they did in the recent past; people entertaining and inviting their friends to their homes is down fifty percent; Americans have become psychologically disengaged from politics and political activities and report being less trusting of both public institutions as well as other individuals. Even church-related groups, which are the most common form of social associations, declined in membership in the 1960s and have not recovered since - in fact, religion has become less tied to institutions and more self-defined (for example, New Age spirituality). As the inspiration for the title of his book, Putnam reports that although bowling, an extremely popular American recreational activity, is up ten per cent in the last few decades, bowling in leagues is down forty per cent - it is more likely that we "bowl alone." Putnam's general assessment of American public life resonates with Benjamin Barber's argument that people in modernized consumerist nations have lost their sense of civic engagement and citizenship.

Putnam identifies a variety of possible explanations for the decline in social networking and social capital. Increasing mobility and transience in the population, suburban sprawl, and the return of women to the workforce are listed as potential causes (the bulk of membership in many social associations has been women, hence if women spend more time at work they have less time to participate in social associations), but Putnam highlights the technological privatization of leisure as the most important factor eroding social capital. Over the last several decades the biggest change in how Americans spend their time is a dramatic increase in watching TV. Watching TV is basically a passive and solitary activity. Putnam also points out that the newest wave of technological innovations and products with increasing customization and individualization further exacerbates the problem. The eventual result could be a population that does not venture out into the public world, where each of us is immersed in our own private virtual reality entertainment center. One could add, in support of this point, following the arguments of Barber and the authors of *Affluenza*, that a world of consumers, lost in their own individual and customized pleasures, undermines and destroys a civic-minded society.

Hence, Putnam presents a diagnosis of a critical problem in American society and a warning for the future: The vitality and quality of our democratic way of life, our economy, and our personal well-being are all in danger of severe deterioration. For Putnam, something should be done to change this present course of events. Since writing *Bowling Alone*, Putnam has been traveling around the United States further assessing the level of social engagement in the country and discussing with people how to revitalize the American community. He has authored, together with Lewis Feldstein, a new book, *Better Together*, in which he reports on a variety of local efforts that bring people back together again to solve social problems. It is noteworthy that the new social networks Putnam describes emerged from the bottom-up, collectively through the interaction of many individuals, rather than being dictated or directed from some higher institution or government policy. In *Better Together*, Putnam and Feldstein distinguish between two kinds of social capital: "Bonding" or inward social capital, which involves creating and developing connections among similar types of people with similar interests, and "Bridging" or outward social capital, which involves forming connections between people of diverse types or interests. They argue that the development of bridging social capital is more difficult, but more important, in a society of increasing diversity. In an article titled "Bowling

Together," Putnam discusses the effects of the terrorist attack on the World Trade Center and notes that this event seemed to trigger a significant increase in political consciousness and engagement. Social trust in both institutions and other individuals is higher now in the United States than in the past three decades - even trust across ethnic groups in the United States is up. As Putnam argues, a dramatic enough crisis, such as the terrorist attack, will bring people together. Yet, TV viewing (perhaps as a way to closely follow ongoing events) also increased following the terrorist attacks.[218] Overall, Putnam expresses a guarded optimism about the future of America, but on the condition that people further increase their efforts to network and associate together more than in the recent past.

Florida and the Creative Class: Richard Florida, in his influential analysis of the contemporary economic world, *The Rise of the Creative Class,* ties together the themes of social networking and social separatism. In some essential ways he presents a view of the present and the future that is the direct antithesis of Friedman's theory of the "flattening of the world." The economic playing field is not leveling but becoming increasingly imbalanced, where power and productivity are concentrated, both nationally in the United States and globally in a limited set of big urban centers.[219] Specifically critiquing Friedman's theory in a more recent article, Florida states that the world is not flat - "The World is Spiky."[220]

In *The Rise of the Creative Class,* Florida presents a theory of which social and cultural factors support economic productivity in the contemporary world. He argues that the fast growing "creative class" is primarily responsible for the economic boom of the last few decades. The creative class, which accounts for approximately thirty per cent of the total workforce in the United States, includes those individuals who are involved in the creation and design of new ideas and products, are paid to think innovatively and independently, or are responsible for managing and directing companies that produce new services and goods. The creative class tends to be more educated than the general population and includes entrepreneurs who start up new businesses and novel enterprises, and artists and writers who invent new concepts, visions, and experiences defining the cutting edge of culture. The creative class is at the core of the growing "knowledge economy," doubling in numbers in the last twenty years and increasing tenfold since the beginning of the twentieth century. In Florida's mind what is powering our present economy is innovation and the creative class is the source of this generativity. He refers to this explanation of economic productivity as the "creative capital theory."

Florida contends that the creative class is progressively concentrating in certain selective cities that support their values and lifestyle. The creative class values openness, meritocracy, diversity, uniqueness, and authenticity. He also states that the most successful places, both in terms of attracting members of the creative class and generating economic productivity and growth, are those urban centers that combine the three T's: -- tolerance, talent, and technology. Cities that have vibrant and indigenous street-level cultural centers and active participatory recreational areas, with larger percentages of "gays" (which signifies tolerance and openness) tend to draw the creative class. Florida particularly highlights that the creative class values social interaction with people of diverse points of view and ways of life. Overall, there is a desire for mental stimulation and the exchange of different ideas among members of the creative class and they will move to those cities that provide the greatest opportunity for this occurring.

According to Florida, a cultural and social divide is emerging in the United States, and even globally, between big urban centers that possess these qualities and much of the rest of the world - often rural and less modernized areas - that is floundering and remaining relatively unproductive. In the United States, a cultural divide is growing between the "Blue" states and regions which value diversity, innovation and what is new, and "Red" states and regions which are more traditionalist and rural. The same pattern is evident in other countries, such as China and India, where their centers of economic productivity and concentrations of the creative class are localized in a relatively small set of metropolitan areas, such as Shanghai, Beijing, and Mumbai.

In order to assess the relative creative strength of a city, Florida has developed what he refers to as a "Creativity Index." The Creativity Index is a combination of four sub-indices: The Creative Class Index, the High-Tech Index, the Innovation Index, and the Diversity Index. Using these indices, Florida ranks San Francisco, Austin, San Diego, Boston, and Seattle as highest on the Creativity Index, seeing these metro-areas as leading the economic growth and level of innovation within the United States.

Based on his analysis of the relative levels of creativity, innovation, and economic productivity around the world, Florida attempts to demonstrate that the world is spiky and becoming even more so. The bulk of economic productivity is concentrated in the United States and East Asia, notably in Japan, and similarly the level of innovation is localized in the United States and East Asia. Further, although the members of the creative class are highly mobile, they tend to move among the main centers of creativity and productivity, rather than spreading outward into the rest of the world. So the creative class tends to cluster together, valuing social exchange, networking, and the sharing of diverse ideas, while at the same time tending to separate and segregate off from the bulk of the rest of the world population.

Aside from this trend toward increasing division and the concentration of talent and innovation continuing in the near future, Florida foresees an escalating competitive battle among regions of the world to recruit and draw creative people to their areas. Since he believes that recent governmental and social changes are making the United States, overall, more restrictive, protective, and conservative, he fears that the United States could lose its economic advantage and leadership edge in the world. Members of the creative class will neither choose to stay in the United States nor move here to begin with from other countries - in essence, unless the United States becomes more receptive and open it will suffer a talent and brain drain in the foreseeable future.[221]

Quinn's *Beyond Civilization*: In his book, *Beyond Civilization: Humanity's Next Great Adventure*, Daniel Quinn proposes that humanity should move toward a tribal social system of organization and abandon the present social system of "civilizations." Quinn defines a "civilization" as a hierarchical organization where those at the bottom of the hierarchy work like "pack animals" primarily for the benefit of a relatively small ruling elite at the top of the organization. Although tribalism preceded the emergence of human civilization and was the dominant form of social organization for most of human history, since the beginnings of the Agricultural Age and the rise of city states and empires, civilizations have controlled the lives of most people around the world. For Quinn, civilization, as a social order, has many harmful and negative effects and should be replaced with tribalism.[222]

Quinn describes a tribe as a group of individuals where everyone works toward the benefit of the whole, and although there are leaders in a tribe, everyone is equal in terms of the benefits received within the tribe. Civilizations, on the other hand, all involve a centralized ruling caste that subjects the bulk of the population to serving and benefiting its needs and desires. Quinn uses the example of pyramid building in the Egyptian empire as a paradigm case of life in a civilization; most people are involved in the task of moving stones and boulders to erect monuments to the pharaohs. Quinn argues that we must take back the world from the pharaohs.

Although advocates of the value of human civilizations would argue that this form of social order is inevitable, the supreme creation of humanity, and the "right way" for people to live, for Quinn, civilizations destroy the environment, generate poverty and homelessness, and produce in people more and more "drug like" diversions to cover up the stress and life dissatisfaction of moving boulders for the pharaohs. In the last ten thousand years, since its creation, civilization has brought us face to face with our extinction. Civilizations raise self-denial to an ethical ideal, frequently promising, through their religious institutions, reward in an afterlife for being slaves in this life. Civilizations make people dependent on the system rather than on each other and produce relatively isolated individuals who live and work behind closed, locked doors. The fact that people often have trouble with living in such a world is seen as a flaw in humans, rather than the social system. Humans are entrapped in an economic prison within civilizations, for there doesn't seem to be any other way to make a living than to participate in the slavery and drudgery of the system. For Quinn, we can not fix the system with more social programs, which simply create more problems - we must jettison the system.

Quinn does not advocate that people attempt to overthrow the present social order. Rather, he believes we should walk away from it. He distinguishes between a commune - where people live together and hold common beliefs - and an occupational tribe, such as a circus - where people may not hold similar beliefs but collaborate to support each other and make a living. According to Quinn, the counter-culture of the 1960s failed because its advocates missed the important point that they needed to discover a way to make a living, with the result that they all ended up "working in the pyramids." As a basic principle in his philosophy, in contrast to the idea that there is a "right way" to live for humans, the Tribal Revolution is based on the idea that "There is no one right way for people to live." In essence, Quinn is an advocate for human freedom and self-determination against the forces of institutional slavery, but he believes that this goal will be realized through economic collaboration.

Spiritual, Religious, and Mystical Theories

The earliest recorded theories of the future are mythological and religious.[223] Beginning in the ancient cradles of civilization, in Egypt, Mesopotamia, China, and India, mythological tales and philosophical meditations on the origin, saga, and destiny of the cosmos and humankind were created that gave people a sense of meaning and purpose for their lives and all existence.[224] Of particular note, the individual lives of people were connected with the great workings of the cosmos. Religion and myth provided conceptual and dramatic schemes for understanding reality and giving ethical direction to life.[225] Although there are numerous and varied ancient religions and myths that dealt with past and future, and the values and goals that should guide and direct human action, there are two fundamental perspectives on time and the future coming out of these ancient traditions that dominate mainstream religious and spiritual thinking today. These two views - one Eastern and one Western - are perhaps best captured in the simple contrast of whether reality and time is seen as a circle or seen as a line.[226]

Zoroastrian-Abrahamic Visions - The Theory of Armageddon - The Triumph of Good over Evil - Paradise, Heaven, and Hell: Within the West and the Middle East, beginning and probably inspired by the apocalyptic vision of Zoroaster, Judaic, Christian, and Islamic theories of the future have all involved the idea that time is linear and progressive and is basically characterized by an ongoing struggle between the forces of good and evil. The founding writings and prophecies of Judaism involved the idea that in the future a Messiah would come who would lead the chosen people of God to victory over the evil forces in the world. Christianity and Islam, both believing that the ultimate prophet and "Messiah" has come (Christ and Mohammed respectively), anticipate a great final world battle (Armageddon) where God and the forces of good will triumph over Satan and the forces of evil. Further, both Christians and Muslims believe that those people who believe in and follow the directives of God will be rewarded with eternal life and ultimate happiness in Heaven, while disbelievers will suffer eternal damnation in Hell. Within Christianity, many believe that Christ will return - a "Second Coming" - to lead the forces of good against a prophesized Anti-Christ and the forces of evil, and that Christ will reign over a thousand-year-long earthly paradise (the "Millennium") that will follow after the final battle before deserving souls ascend into Heaven.

This general theory of the future has provided the moral and metaphysical core of a general paradigm for life and the future. There are clear directives within this perspective for how to live and how to prepare for the future. Further, this theory and paradigm has generated two of the most powerful social movements within human history. Christianity and Islam, as world religions, are embraced and practiced by billions of people worldwide; in fact, the two most popular religions in the world are Christianity and Islam. In the name of the beliefs, prophecies, and values of their respective doctrines and creeds, followers of Christianity and Islam have conquered, colonized, and ruled over countless nations and millions of other people worldwide in an attempt to convert non-believers to their religion. In many ways very similar in their fundamental belief systems and visions of the future, Christianity and Islam probably offer the most influential theories of the future over the last two thousand years of human history.

Though progressive in their view of the future - believing that life in the future will be better than today (for believers of course) - the road to this better reality will involve great conflict, turmoil, war, and chaos. There is a passage through the darkness as a prelude to reaching the light. Yet the general result of this upheaval and dark journey is destined or predetermined. God and the forces of good will triumph. It is the individual destinies that are (at least in some sense) uncertain and contingent upon the choices and free will of humans. Still, in the final analysis, there are really only two possible future scenarios for each person, happiness and bliss in Heaven or misery and torture in Hell. This vision of the future is very black and white.

There is a highly bi-polar quality to the Christian-Islamic view of the future. First there is the dark side: A vision of war and great destruction, as well as punishment and eternal damnation for those individuals who reject God. But then, there is a positive side. In Christianity, for example, there is an emphasis on love, forgiveness, and kindness as critical moral qualities in living one's life. It is through these positive qualities that a better future will be realized. Also, there is a clear sense of cosmic justice in Islam and Christianity, as well as Judaism. Although life may be unfair, God will enact justice, rewarding those who live a good life and punishing those who lead an evil life. Finally, in spite of all our imperfections, failings, and times of unhappiness in our present mode of existence, there is a promise of eventual perfection and complete happiness once we ascend to Heaven.

As one final point, there is a decidedly other-worldly and metaphysical dimension to this theory of the future. The forces that will ultimately determine the future are supernatural and at a higher plane of existence than the natural world; these same forces were also responsible for the creation of the world, as well as the general plot, drama, and direction of events in the world. Humans existing on the material plane and within time cannot affect the overall outcome of things; we can only chose in which of the two predetermined metaphysical realms we spend eternity. Secondly, the ultimate future for humanity is therefore not something within the physical world of time, but in a metaphysical sphere. We may attempt to improve the reality around us, but this reality will eventually pass away. Our destiny lies elsewhere.

Eastern Visions – Taoism – The Dance of Shiva - Reincarnation – Nirvana: In contrast to Western linear thinking, both Taoist and Hindu views of reality involve a cyclic theory of time. Within Chinese Taoism, reality is described as an oscillation of the complementary forces of *Yin* and *Yang*. Reality involves balance, of darkness and light, of male and female principles, in reciprocal harmony over time. As the psychologist Richard Nisbett documents in his cross-cultural study of patterns of thinking across the world, people from

China and the Far East tend to see time as circular; they anticipate downward turns if things have been going well, and anticipate upward turns if recent events have been negative. (This is in stark contrast with people in the West, who tend to expect further good times if life recently has been going well.) Similarly, within Hindu cosmology, a common belief is that the God *Shiva* will bring worldly existence to a catastrophic and fiery ending, but then the God *Vishnu* will dream the world anew and a new cycle of creation and destruction will begin. Temporal existence is ultimately an endless repetition of birth and death and this cycle of creation and destruction is often referred to as "The Dance of *Shiva*."

There is, though, a progressive dimension within Hindu thinking, as well as in its philosophical offspring, Buddhism. The Hindu doctrine of reincarnation involves the notion that the souls of individuals who have died will return to an earthly and temporal existence incarnated in a new body. This part of the theory is cyclic. Yet, the new body and temporal existence of the soul will depend upon how ethical a life the soul led in the previous life. This is the theory of *karma*. A good life brings good *karma* and a better existence in the next life. An ethically bad life produces negative *karma* and leads to a lower existence in the next life. If, in its new incarnation, a soul can eliminate through ethical actions whatever bad *karma* it accrued in its past lives, the soul can ascend into Nirvana and break free of the cycle of life, death, and rebirth. Nirvana is a higher existential state - an eternal and spiritual state involving unity with the absolute oneness of *Brahman*, the ultimate source of all being. This is the progressive dimension of the theory.

As can be seen in the Hindu ideas of reincarnation, *karma*, and Nirvana, the direction of the future is determined by ethics; positive moral behavior leads to a better future - negative moral behavior leads to a worse future. Further, cosmic justice determines the direction of the future. Good deeds are ultimately rewarded and bad deeds are punished. Finally, the positive future that awaits morally good individuals lies outside of the temporal and material realm, in the higher spiritual reality of Nirvana. In this regard, Eastern and Western religious views of the future are the same. Both promise a spiritual and eternal existence after physical death. The ultimate future is beyond the physical world.

One significant difference, though, between Eastern and Western religious visions of the spiritual future can best be illustrated in the teachings of Buddhism. Within Buddhist philosophy, it is desire and an attachment to the individual ego that produce misery and unhappiness in the world. In fact, desire is a consequence of conscious attachment to an ego. The goal of enlightenment is to see beyond the ego. Once we achieve this enlightened state - to see reality as a one without a separation of the ego and the world - we achieve Nirvana. Hence, while Western religions like Christianity and Islam promise individual immortality, Buddhism views Nirvana as precisely the lost of individuality. In Hinduism also, the emphasis is on achieving oneness with Brahman, rather than the preservation of an individual ego. There is clearly more of an emphasis on individuality in the West than in the East and this difference shows up in their different views of the ultimate future. As another basic cultural difference revealed in Nisbett's studies, Westerners highlight their individuality, whereas Easterners emphasize their togetherness and collectivity.

As one final contrast between East and West, none of the major Eastern religions envisions the final chapter of humanity as involving some great global conflict between the forces of good and evil. Although *Shiva* may destroy the world in the end, there is no great war that is fought between different spirits or factions of humanity. In Taoism there is ultimate balance throughout time; one does not find some fundamental component of reality being defeated or eliminated. In the West, there is an evil or corrupt side of reality that must be eradicated through destruction and violence.

New Age Spirituality: New Age thinking is both all-encompassing and highly individualistic. First it combines, connects, and attempts to synthesize the mystical, magical, mythological, scientific, and naturalistic.[227] It takes ideas from contemporary popular science, holistic medicine, human potential and self-actualization psychology, ancient and non-Western religions, occult and magical practices, witchcraft, esoteric cosmologies, ecology, the folklore of ghosts, spirits, demons, and angels, and even ideas on UFO's and space aliens, and attempts to pull it all together into a spiritual approach to life. New Age is intercultural, global, and universalist in its understanding of reality. It does have a strong Eastern flavor - meditation, chanting, and yoga are common practices - Oriental deities are frequently worshipped - and the idea of reincarnation is exceedingly popular.[228] In general though, it is a smorgasbord - a "god of a million faces."[229]

New Age thinking is often criticized as having no strong intellectual standards for it embraces almost any idea, no matter how crazy or unrealistic. Yet, in its enthusiastic openness to multiple points of view, it does aspire to overcome the various divisions and conflicts between science and religion and one religion versus the next.

Second, from this amazing array of ideas and traditions, individuals can literally pick and chose which ideas to accept and practice. New Age is highly liberal and lacking some central doctrine that all its followers must believe in. In some ways it is the antithesis of fundamentalism, if not all organized religions. It stresses that each individual must find his or her unique path to enlightenment. It is an expression of the extreme individuality and philosophy of freedom of modern times.[230]

New Age practitioners and writers are enthusiastic about predicting the future. Many clearly believe that the future can be foretold. Predictions are often highly detailed and there is a common belief that there is a strong component of destiny to the future. Contemporary New Age literature is filled with spiritual prophecies and visions of the near and far future. The foretelling of the future through dreams, crystal balls, tarot cards, astrology, and spiritual and cosmic epiphanies, all ancient mystical practices, is an extremely popular activity associated with New Age thinking. This conviction among many New Age thinkers that the future can be foretold, as well as influenced or directed through magical practices, gives the movement a rather dogmatic flavor.

New Age futurist predictions tend to be highly optimistic and upbeat. According to many visions, paradise, either on earth or within some higher spiritual plane, will be realized in the future. The expression "New Age" literally means that a new era in human existence is coming - an age that will be much better than the present one. Following on the common 1960s belief that we are entering the Age of Aquarius, the future should be a time of peace, brotherhood, enlightenment, and transcendence. Frequently, highly advanced aliens or spiritual beings will be involved in leading us to this higher level of existence. Such visions are often seen by skeptics and critics of the New Age as Pollyannaish and unrealistic. Yet, New Age philosophy is hopeful, and a good antidote to the nihilism and pessimism of our time, and New Age followers often do try to create a better world around them, less frenzied, more humane, and filled with beauty.

There is also a strong metaphysical element to the New Age. Higher or multiple dimensions of reality, often populated with spiritual and other-worldly beings, are frequently believed in. Just as in mainstream religion, followers of the New Age think that supernatural and metaphysical forces or entities influence the future. There is also a related counter-culture dimension to the New Age. As an outgrowth of trends that developed in the Hippie Culture, the New Age is highly critical of the corporate business world, of crass materialism, and the limited worldviews of mainstream society. The New Age sees above and beyond the everyday physical world in which the common person lives. Coupled with its metaphysical dimension, New Age thinking also embraces naturalism, which further reinforces its opposition to mainstream high-tech culture and urbanized human society.

All in all, the New Age is both a paradigm and a social movement - it is a way of life. Connected with its beliefs and practices, there is New Age music, New Age paraphernalia and icons, New Age jewelry and fashion, New Age nutrition, New Age centers, New Age bookstores and shops, and New Age gatherings and fairs. In spite of its anti-establishment and anti-materialist philosophy, it is a highly commercialized approach to life and the future.

The Evolutionary Omega Point Theory of God: Most major world religions and spiritual traditions assume some type of higher reality above the physical world, but such views tend to be dualistic - dividing existence into two distinct realms - a spiritual and a physical. This dualistic split supports and reinforces the spiritual-secular division in everyday life and human values. Perhaps, though, it is possible to synthesize the secular and materialistic with the spiritual. Such an integrative vision of reality has been a pursuit of many writers and thinkers during the modern era and there are contemporary philosophers, futurists, and scientists who aspire to such a synthetic and non-dualistic system of thought.[231]

Further, if there is some general type of progressive movement across time, as for example envisioned in the theory of evolution, then perhaps the "higher realm" described in religious and spiritual thinking is simply the result or culmination of the progressive direction of time.[232] Writers such as Barbara Marx Hubbard, Ken Wilber, and Frank Tipler clearly imagine an evolutionary direction to humanity and nature that will lead us to not only a physically more advanced state, but also to a higher moral, spiritual, and psychological state as well.

Could we envision a cosmic reality in the future that includes a universe of diverse intelligent forms, making use of intelligent machines, space technology, and the biotechnological tools of purposeful evolution? Could such a cosmic evolution be driven by spiritual, social, and moral values where technology and material wealth are more of a means to an end rather than an end in itself? The dualism and antagonism of the spiritual and the secular could be transcended in the future.

Most religions postulate some type of Supreme Being or God - presumably existing on a higher metaphysical level of reality. But if we follow an evolutionary logic concerning the nature of reality, this higher level of existence and Supreme Being could be seen as the culmination of cosmic evolution. God could be the direction of evolution. The future is the evolutionary ascension to God.

The French biologist, Jesuit scholar, and philosopher, Teilhard de Chardin, in his well-known book *The Phenomenon of Man*, argued that humanity would evolve into an integrated global mind and achieve a spiritual culmination in an **Omega Point** at the end of time.[233] In this theory, Chardin attempted to synthesize evolutionary science with religious and spiritual ideas - his Omega Point theory entailed a teleological interpretation of evolution. Evolution had a purposeful direction. Chardin proposed the concept of a **noosphere** - an emerging collective human consciousness enveloping the earth - that would be the medium in which the Omega Point was realized. Within this emerging noosphere, he believed that humanity would psychologically and morally come together in a collective concern for its own destiny and that of its home, the earth, and through compassion and care ascend to a higher level of consciousness. Chardin was an extreme optimist, for he saw humanity escaping eventual death and extinction through this spiritual ascension. Chardin saw this ultimate spiritual destiny as a consequence of natural evolution.[234]

Continuing on this theme of living immortality, the influential scientist Freeman Dyson, in his book *Infinite in All Directions*, argues that mind and intelligence will progressively spread through the universe and eventually be able to overcome the hypothesized heat death of the cosmos and persist forever. Within this scenario, Dyson sees the universe continuing forever, but this may or may not occur, depending on whether the universe eventually reverses its present expansion and collapses back in a Big Crunch.[235]

The physicist and cosmologist, Frank Tipler, finds his inspiration in both Chardin and Dyson.[236] Tipler proposes that intelligent life can gain control over the future evolution of the universe and generate an infinite amount of energy and information processing in directing the dynamics of the universe, thus creating an eternal and infinite consciousness. To accomplish this feat, intelligent life will need to spread throughout the entire universe and, through technological and scientific means, coordinate itself as a single collective force. For Tipler, this guided process of cosmic evolution will culminate in an Omega Point at the end of time, which, in essence, will be God - an infinite mind possessing unlimited compassion and love and capable of raising everyone who has ever lived back to consciousness from the dead. Thus Tipler attempts to unite science, space technology, and the principles of evolution with the common religious beliefs in an infinite all-powerful God and personal immortality for humanity.

This Omega Point theory of God would, though, seem to conflict with the idea that God is eternal or co-present with reality. God is manifested in time rather than existing outside of time. Also traditional religious thinking tends to identify God as the Creator of the universe at the beginning of time. The evolutionary theory of God proposes that God exists in the future - perhaps even at the end of time -- and that God is the evolutionary creation of the universe, rather than the other way around.

Tipler does offer an explanation, however, for how to resolve the contradiction between these two different views of God. In an earlier book, *The Anthropic Cosmological Principle*, co-authored with John Barrow, the argument is presented that the initial conditions of the universe and the values of physical constants in it appear finely tuned to support the evolution of life and intelligence.[237] It seems as if the universe was "designed" to produce life and intelligence. This argument is usually referred to as the "strong anthropic principle." In his later book, *The Physics of Immortality*, Tipler, argues for the validity of the strong anthropic principle. He contends that God is at the end of time; God, who is the natural culmination of the evolution of intelligence and life in the universe, sets the conditions at the beginning of time for its own realization at the end of time. God, once realized, is able to transcend time.

A similar kind of argument can be made, though, without having to postulate a God at the end of time. In his book, *Biocosm: The New Scientific Theory of Evolution: Intelligent Life is the Architect of the Universe,* James Gardner proposes that the conditions of the present universe were purposefully set to allow for the evolution of life and intelligence, but it was intelligent beings from our "mother universe" that created these conditions. Instead of seeing our universe as a self-contained reality, Gardner believes that our universe was born or created from a progenitor universe, in which highly evolved intelligent beings determined the initial conditions and physical constants of this universe. In turn, in our own future evolution, our descendents will create the conditions for the birth of a new universe that will support life and consciousness as well. [238]

What all these views have in common is the belief that a higher form of mind and intelligence, usually possessing equally evolved moral and ethical qualities, will be the result of evolution in the cosmos, sometime in the far future. The key concept in all these views is evolution, clearly a scientific idea. Religious beliefs in a higher level of reality, a higher form of mind and intelligence, and even the possibility of immortality are integrated with evolution and the mind-boggling possibilities of the far distant future.

Integrative Theories

I would define an integrative theory of the future as an abstract description and explanation of the future that strongly attempts to synthesize all the main dimensions of the future into a synthetic and comprehensive scheme of thought. An integrative theory would cover both the scientific and artistic-humanistic, the technological and the religious-spiritual, the economic and political, and the social and the psychological. It would cover both humanity and nature, including the cosmos as a whole. It would not only be descriptive, identifying trends and making (at least) tentative or general predictions, but it would also be evaluative and prescriptive, providing a normative or ethical assessment of past and present and values for the future. Following from its normative dimension, an integrative theory would include guidelines and practices for how to live to realize the vision and values of the theory. An integrative theory would have an associated paradigm and provide the foundations for the creation of a social movement.

At least some of the theories described thus far could be viewed as integrative, at least satisfying most, if not all, the criteria listed above. Some examples would be Barbara Marx Hubbard's conscious evolution theory, Tipler's Omega Point theory, Henderson's Solar Age theory, and the general theory of cosmic evolution. Perhaps one could also include religious belief systems which attempt to explain all of reality, predict all the important trends and events of the future, and provide moral rules or guidelines for how to live. Religious theories, though, often tend to slight or reject the scientific and the technological, and do not have much to say about the economic and business dimensions of life. I have included all these theories in one of the previous categories because these theories highlight some particular theme or dimension of human reality that fits with one of the previous categories. In any event, as I mentioned earlier, the placement of theories into particular categories is not absolute and unambiguous; even some of the theories presented below as integrative could perhaps be included in one of the earlier categories.

Integral Culture: The expression "integral culture" refers to a philosophy and way of life that emphasizes the interconnectedness of all reality. In this regard it is inspired by holistic science and thinking and clearly rejects the modern divisions between science and religion, the spiritual and the secular, mind and matter, and fact and value. Integral philosophy attempts to pull it all together into a comprehensive vision for the future.[239] Sally Goerner, in *After the Clockwork Universe:*

The Emerging Science and Culture of Integral Society, provides a theoretical and historical overview of the development of integral thinking, how it differs from earlier philosophies in the West, and what its main implications are regarding the future of human society.[240]

Integral Culture philosophy sees human society at a crisis point due to the excessive individualism and lop-sided materialism of the West. In essence, the central problem of our times is that human existence is disintegrated and fragmented - there is neither balance nor holism. Contemporary human society is alienated and separatist. The Integral Culture movement advocates for a new sense of connectedness among individuals and different cultures, and between humanity and nature, humanity and the cosmos, and male and female, and a new set of values associated with this philosophy of connectedness.[241] Jon Spayde sees the Integral Culture movement as part of a "New Renaissance,"[242] connecting various themes in this "New Renaissance" philosophy with the writings of Duane Elgin, Riane Eisler, Hazel Henderson, and Barbara Marx Hubbard.[243] What all these writers and thinkers have in common is a holistic and evolutionary perspective on tomorrow, and a rejection of the industrial vision of progress, including its metaphysical dualism, excessive materialism, competitive individualism, and hierarchical system of social control. Through connectedness and love, we need to see ourselves less egocentrically and become part of the greater whole. We need to engage in purposeful evolution, creating a new, more balanced perspective on life. According to many of its advocates, we need to adopt the theory of cosmic evolution as a new common creation story to give all of humanity a sense of cosmic connectedness.

Integral Culture philosophy is not opposed to either spirituality or technology. Spirituality, though, needs to be re-conceptualized in the context of modern evolutionary and ecological science and needs to move beyond the divisiveness that has characterized much of the history of world religions. Further, technological, as well as economic growth, are not necessarily bad, but these dimensions of human reality must be evaluated and guided by higher more humanistic values. It is an issue of balance and integration.

The Integral Culture movement is a network of people and common ideas. It is overlaps with other approaches to life and visions of the future. Eisler, Hubbard, and Henderson, all identified as leading figures in this movement, all have their own distinctive theories of the future, but they also share many common beliefs. The Cultural Creatives described by Ray and Anderson also show many resonant qualities with the Integral Culture movement and, undoubtedly, many individuals would identify with both social groups. _

Wilber's Integral Philosophy: Ken Wilber, a highly influential and prolific contemporary philosopher and writer, is often associated with the Integral Culture movement. Wilber, in fact, uses the expression "integral philosophy" to describe his approach to life, human understanding, and the future. His central philosophical goal is to articulate a comprehensive scheme of thought that pulls together the insights of East and West, as well as science and the great spiritual and religious traditions of the world.[244] Although Wilber can be viewed as a New Age philosopher and writer and, in fact, has contributed articles, commentary, and interviews to New Age publications, he has also significantly influenced more mainstream thinkers, such as the futurist Richard Slaughter.[245]

Wilber's philosophy emphasizes the theme of evolution and development. Reality, for Wilber, is dynamic and progressive. As some important examples of this fundamental theme in his thinking, Wilber describes human history as having developed through a series of different levels of thought, social organization, and patterns of living. He identifies five levels or stages in human history: The archaic (foraging tribes), the magical (horticultural villages), the mythic (agrarian states and empires), the rational (industrial nations), and the existential (planetary information society). More broadly, he describes five levels of evolution within the "Kosmos" (a term

Wilber uses to describe the totality of existence). Each of these levels has an inner and outer dimension. The levels are: Mineral/matter, plant/vitality, animal/emotion, human/mind, and mystic/intuition. Wilber also identifies three basic levels of development as humans grow and mature: The pre-personal, which is anchored to the body, instinct, and nature; the personal, which is anchored to the mind, intellect, and culture; and the trans-personal, which is anchored to the soul, intuition, and the Kosmos. Additionally, he believes that there are developmental stages in politics and political thought, beginning from the conservative and liberal, but moving up to higher levels which integrate as well as transcend the earlier levels. Finally, it is interesting that Wilber even describes his own thinking and philosophy within a developmental scheme; he identifies various stages or phases of thinking he has gone through in his philosophical and spiritual quest. He sees his philosophy as a dynamic and evolutionary reality.

A key concept in Wilber's theory of evolution is the idea of the "holon" - a concept he took from the writer, Arthur Koestler, and his book, *Janus*.[246] According to Wilber, all of reality is composed of holons. A holon is defined as an entity which is both a whole that consists of parts, but in turn is a part of a greater whole. Hence, everything in reality is both a whole and a part. Holons possess a dual drive toward agency (autonomy) and communion; that is, reality involves the complimentary processes of individuation and integration. Holons have a drive toward both preservation and self-transcendence. Reality evolves through the co-evolution of holons which cluster or come together to form more encompassing holons; wholes integrate into bigger wholes. Emergent holons both envelop and preserve their constituent parts, yet creatively transcend the parts. Although this general pattern of evolution is universal across the entire history of time, since each new level of holons is creative and transcendent relative to the previous level, the future can only be predicted in general outline. The unique qualities of emergent holons in the future can not be foreseen from our present vantage point in the present. Evolution is creative.

Based on this general framework for understanding the nature of reality, Wilber contends that evolution indeed does have a direction. He lists increasing complexity, integration and differentiation, organization, relative autonomy, and telos/purpose as key features to the direction of evolution. It is interesting to note that telos or purpose is not, in Wilber's scheme, a causative factor behind evolution, but rather an emerging property of reality created through evolution. It is also interesting to note that evolution moves in complementary directions simultaneously - toward greater integration and greater individuation. Wilber also integrates God into this general theory of the direction of evolution. Evolution is "God-in-the-making" and "spirit-in-action."

The complementary aspects that Wilber identifies in the process of evolution lead us to a second key theme in his philosophy, a theme which captures the critical sense in which Wilber's philosophy is integral. Wilber rejects the various forms of philosophical dualism, such as mind and matter, or the whole and the parts, that have been pervasive throughout human history. Further he rejects schemes of thought which emphasize one aspect of reality over the other, such as materialism versus idealism or objectivism versus subjectivism. Wilber has attempted to develop a philosophical and scientific description of reality that encompasses and integrates all the fundamental polarities of existence. For example, he believes that it is equally important to acknowledge and understand the "outer" (physical) and "inner" (consciousness) dimensions of reality. He argues that secular modernism has overly emphasized the outer, physical dimension of reality and lost sight of the inner reality of consciousness and self within all of us. Modernism is incomplete. Also, he points out that Enlightenment philosophy and modern science have attempted to uncover an objective description of truth and reality, but miss the point that there is a subjective component in all human knowledge as well. Hence, an absolutist, totally objective vision of reality is impossible. On the other hand, he rejects the extreme subjectivism, relativism, and pluralism of Postmodernism, which goes to the opposite extreme in thinking. He also rejects what he sees as the lop-sided philosophy of extreme rationalism; knowledge can also be acquired through intuition. As an important example of his integral philosophy, Wilber presents a "four-quadrant" theory (discussed in an earlier chapter) that attempts to provide a balanced and comprehensive scheme for understanding human reality. He identifies the "inner individual" revealed through introspection, the "outer individual" revealed through brain and cognitive science, the "inner collective" revealed through the study of culture, and the "outer collective" revealed through sociology. In the ongoing evolution of human reality, all four of these quadrants co-evolve together.

Wilber is a visionary. He sees humanity continuing to evolve in the future. Further, he believes that this evolution will involve a strong spiritual dimension. In fact, he predicts that the next collective stage beyond the ego-centered, rationalistic mindset of modern human society will be a mystical stage and involve an envelopment and transcendence of the individual ego. He thinks that through looking at those more enlightened individuals of today and the past, we can get a sense of the future evolution for all of humanity. He believes we will go beyond Postmodernism and find patterns and connections in our cultural and individual diversity; we are not at some final stage in our philosophical enlightenment - by far. In fact, Wilber believes that there is an infinite number of potential or possible "worlds" or "levels" ahead of us in the future.

Spiral Dynamics: Based on a comprehensive view of human development first put forth by the psychologist Clare Graves, writers Don Beck and Christopher Cowan, in their book, *Spiral Dynamics: Mastering Values, Leadership, and Change*, present a theory of individual and social growth that according to some advocates "explains everything." Basically, the theory proposes that humanity (or human nature) has developed through a series of equilibrium states of increasingly more complex bio-psycho-social stages in coping with the challenges of existence. Each new stage envelops and yet transcends the previous stage; also, the developmental process is open-ended, with no final stage of perfection, since each new stage may solve old problems but inevitably has to confront new problems, questions, and challenges. Also, through successive stages or levels there is a fundamental oscillation between focusing on the external world, the group, and self-sacrifice (interdependence mode) and focusing on the inner world, the individual, and self-expression (independence mode).[247]

Beck and Cowan have developed a color-coded system to designate the different stages of human development. Each stage is characterized by a dominant "vMeme" (a core value system) - a broad, deep, and encompassing pattern of thinking and values that structures and directs human behavior and the human mind at that level of existence. (The idea of a vMeme is similar in meaning to the concept of a paradigm.) According to the authors, the stages that have emerged or begun to emerge in human history, so far, are:

- Beige: Loose clan-based groups dominated by nature, instinct, and basic survival needs - "based on biological urges/drives; physical senses dictate the state of being."
- Purple: Tribal groups that are animistic, magical, superstitious, and ritualistic with strong ancestral and blood bonds - "threatening and full of mysterious powers, spirit beings which must be placated and appeased."
- Red: Exploitative and authoritarian groups with strong "Big Boss" leaders, heroic figures, slavery and repression, and rigid social hierarchies that are power and action-driven and egocentric - "like a jungle where the tough and strong prevail while the weak serve; nature is an adversary."
- Blue: Strong group norms and group discipline, social control through guilt and obedience to authority,

absolutist views of truth and value, high discipline, and an emphasis on self-sacrifice - "controlled by a Higher Power that punishes evil and eventually rewards good works and Right living."
- Orange: Entrepreneurial, calculating, individualistic, success-driven, materialistic, competitive, and a drive to control the environment - "full of resources to develop and opportunities to make things better and bring prosperity."
- Green: Communitarian, egalitarian, the need for social approval and contact, facilitative leaders, the importance of social harmony, and an attention to the environment - "the habitat wherein humanity can find love and purposes through affiliation and sharing."
- Yellow: Systems thinking, an emphasis on mutuality, intrinsic love of learning, ecological, and the importance and reinforcement of unique talents in individuals - "a chaotic organism where change is the norm and uncertainty a usual state of being."
- Turquoise: Holism and spiritual harmony, the integration of thought and feeling, the capacity to understand multiple points of view, and integral philosophy - "a delicately balanced system of interlocking forces in jeopardy in human hands."[248]

Some important points to highlight about this theory of human development are: Past, present, and future are integrated in this theory, as well as individual and group psychologies; each level, depending on the situation and problems being confronted, can be either constructive and positive or pathological and negative: the most dominant and powerful value systems in our contemporary world are Blue and Orange - the Blue and the Orange generally correspond with fundamentalist - traditionalist and modernist ways of thinking within Ray and Anderson's scheme; the vMemes or value systems are situationally driven - an individual person may react to one type of situation (for example, the professional sphere) with one value system and react to a different kind of situation (for example, the personal or private sphere) with another value system; people and groups can regress as well as move forward to higher levels of complexity; change occurs through passing through the different stages - one can not push or force someone to jump up several stages suddenly; growth or change to a new level occurs through envelopment rather than elimination of less complex stages - the stages are cumulative and build on each other; the levels are not totally discrete - there are transition states between the levels; the final two stages listed, the Yellow and the Turquoise, represent a significant jump over the previous six stages - a movement to a "second Tier" from "subsistence levels" to "being levels"; at this point in time only a very small percentage of the human population functions at the Yellow and Turquoise levels, but this would seem to be the direction in which humanity is heading in the near future - there is even the suggestion of Coral and Teal levels beyond the Turquoise[249]; and the overall pattern of growth embodied in the oscillation of outward/collective and inward/individual levels corresponds with similar ideas about human nature contained in the writings of Bloom, Csikszentmihalyi, Nisbet, and Triandis.

Spiral dynamics has had a growing influence on business and government. It provides various principles and concepts for understanding human organizations, the nature of leadership, and how to constructively facilitate change. Yet also, a strong resonance has recently developed between Beck's ideas on spiral dynamics and Ken Wilber's integral philosophy and psychology. Out of this sympathetic connection has grown the idea of Spiral Dynamics Integral.[250]

Zey's Expansionary Vision: The futurist, Michael Zey, in his book *The Future Factor,* presents a general theory of the future based on five fundamental forces that he believes capture the essence of how humanity will change, why these changes will occur, and where it is all heading.[251] Drawing his inspiration from a variety of scientific and philosophical thinkers, including Kurzweil, Dyson, Teilhard de Chardin, and the Russian cosmist, Vladimir Vernadsky, Zey describes his general perspective on the future as the "expansionary theory of human development."[252] The overarching or primary force identified within his expansionary theory is vitalization, which is supported by the other four forces: dominionization, species coalescence, biogenesis, and cybergenesis. Zey is a strong supporter of the importance of continued growth and the spread and expansion of the human species outward into the cosmos. He believes that humanity has a destiny - a cosmic destiny - which will involve gaining control over the fundamental forces of nature and directing the ultimate fate of the universe as a whole.

Zey defines the primary force of vitalization as the drive to transform and improve the universe and to spread life, human intelligence, and consciousness throughout the cosmos. Dominionization is the drive toward controlling the forces and elements of nature. Species coalescence is the process toward the unification of humanity, both physically and socially. Biogenesis is the enhancement of the human body through biotechnology, nanotechnology, and genetic engineering. Finally, cybergenesis involves the ongoing integration of humans and machines, especially information technologies, which will facilitate the future evolution of our species.

There is a clear sense of teleologism in Zey's expansionary theory. He supports the anthropic principle, arguing that the initial conditions and fundamental physical constants of the universe appear finely tuned to support the evolutionary emergence of life, intelligence, and perhaps even human consciousness. He believes that the universe, in some way, built into itself the capacity to create a being or mechanism that would find a way to overcome the slow death of entropy and maintain the existence of the universe. Zey thinks that potentially humans are the means by which the universe will save itself. Further, Zey believes that recent scientific evidence indicates that the universe has built into it the drive or force (perhaps intention) to self-organize into increasingly more complex entities. Evolution is not based entirely on random variations and natural selection. Humans are the result of this drive toward order and complexity within nature. Further, he sees a fundamental theoretical and ideological conflict in contemporary times between those people who support balance and sustainability and those who support growth and development. (His view on this conflict is similar to Postrel.) Zey argues that the balance theorists are going down the wrong path; our survival as well as the survival of the universe depends on humanity's growing, expanding, and reaching out into the cosmos. In Zey's mind, humans have a special role in nature and we need, through an act of will and imagination, to gain control and dominion over the cosmos. The expansionary view must triumph or humanity will become extinct. We must gain control over ourselves as well. He clearly believes that humans, through species coalescence, biogenesis, and cybergenesis, must engage in "purposive self-development."

Zey presents what he sees as an "optimistic" vision of the future. He rejects Postmodernism, nihilism, and pessimism. Humanity needs a positive cosmic image of its future. Humanity needs to feel self-empowered and important in the grand scheme of things. We have an important mission to fulfill and we need to rise to the occasion.

Kelly's *Powerful Times* and Dynamic Tensions: Eamon Kelly, the CEO of the Global Business Network, in his book, *Powerful Times: Rising to the Challenge of Our Uncertain World*, presents a theory of the dynamics of contemporary times and a set of possible scenarios for the future based on the central idea of "dynamic tensions."[253] Kelly lists seven fundamental dynamic tensions, which cover economic, political, social, ecological, and religious/spiritual issues within our present world. These contemporary tensions are the main motive forces driving us into the future.

- Clarity and Craziness - The world is becoming more transparent, there is increasing surveillance and huge amounts of information on everyone and everything; yet there is also information overload, multiple alternative perspectives on all major issues, and consequently increasing confusion, uncertainty, and ambiguity over the validity and reliability of information sources resulting in growing distrust and incredulity.
- Secular and Sacred - There is escalating conflict and tension between secular thinking and practices (dominated by science, quantification, economic materialism, and a logic of either-or) and religious/sacred views of reality, which are steadily growing in influence and number of followers. Secular thinking, inspired by the European Enlightenment, embraces freedom, liberty, and tolerance, and those countries that adopt its principles have become more wealthy, innovative, and democratic. Yet, from the religious perspective, secular thinking is incomplete as a comprehensive approach to life and has created significant problems in the world. Religious fundamentalism is particularly threatened by secularism.
- Power and Vulnerability - United States military power continues to grow and security systems have been significantly strengthened, especially in the last few years following the terrorist attacks of 9/11; military spending in the United States far exceeds (by eight fold) any other country in the world. But there is also a clear sense of increasing vulnerability to terrorist attacks and other potential threats to peace and stability, such as organized crime and global epidemics.
- Technology Acceleration and Pushback - New technologies continue to proliferate at an accelerative rate and promise to transform both

the world and the nature of humanity. Kelly cites four technological-scientific areas - information technology, nanotechnology, cognitive and brain science, and biotechnology - that are rapidly advancing. These technologies will be increasingly applied to the improvement and modification of the human body and the human mind. As Kelly notes, there is a "limitless demand for enhancement" and Americans now spend more on beauty than education. Yet there are moral and pragmatic concerns over the application of technology to human enhancement, and these concerns probably will intensify in the years ahead.

- Intangible and Physical Economics - Although the world economy is shifting toward "intangible" products, such as human services, knowledge and information, experiences, and beauty/art/aesthetics, the physical infrastructure of cities and nations is deteriorating and in need of significant overhaul and improvement.
- Prosperity and Decline - The global economic system has created new prosperity and opportunity for many more people in the world, but a significant portion of the world population is suffering from poverty and overall decline in the quality of life. The gap between the rich and the poor is growing. While countries like China, India, and Brazil have created much more prosperous economies in the last couple of decades, benefiting large numbers of their people, other countries, especially in African and Latin America, suffer from disease, corruption, crime, and internal conflict and show little if any promise of improving their situations.
- People and Planet - The final creative tension is between humanity and the earth. The more economically powerful and wealthy countries in the world are using natural resources and huge amounts

> of energy at increasing rates, but in so doing are threatening to unsettle the global climate. As more people are becoming aware of ecological issues and the connection between the economy and the environment, there is a growing concern over how to realize a sustainable economic system that meets the needs of all people.

As the central theme running through his book, Kelly supports a both/and logic in understanding our contemporary world and creating a positive future. Although he acknowledges the value of either/or logic, he believes that our times require a shift in thinking toward both/and logic. For example, each of the above tensions involves two seemingly contradictory trends occurring simultaneously in our world today. Both trends within each pair need to be acknowledged as very real and powerful and, in all probability, continuing into the immediate future. The world is complex, rather than either-or. This complexity produces ambiguity and uncertainty; it would be simple if one could legitimately state that the world is moving in one direction rather than another, but there are conflicting trends, interacting with each other and vying for dominance. Are things getting better or are things getting worse? Kelly presents convincing descriptions that both views have a degree of truth. When Kelly discusses how we are to create a better world in the future, he presents a set of complementary pairs of principles to guide our actions. He states that we need a balance of market wisdom (how to do the profitable thing) and moral wisdom (how to do the right thing); we need to be concerned about both winning and collaborating, about both growth and survival, and about both technology and people. We need both focus and an openness to ambiguity. Also, our organizations require both networking and hierarchical control. Moreover, he foresees trends toward increasing localization and globalization occurring simultaneously. Finally, Kelly advocates for an increasing emphasis on "divergent thinking," where instead of locking into one point of view, we continually work at seeing multiple points of view - often of opposing perspectives. In general, in his depiction of the present and his recommendations for the future, Kelly adopts a Taoist philosophy - for Kelly, the world is a *Yin-yang*.

Kelly proposes three different scenarios for the future: the New American Century, Patchwork Powers, and Emergence. Aside from his emphasis on both/and logic, a second major theme in his book is uncertainty about the future; in fact, he states that we should embrace this uncertainty. In this regard, he outlines three relatively distinct possibilities for our future world. In the New American Century, the Western model of government and economy will dominate, and the United States will continue, if not enhance, its strong leadership role in the world. In Patchwork Powers, the United States loses its central leadership role as, for example, China and India become increasingly powerful players in the world; political and economic power becomes more equally shared among nations, peoples, and international organizations. In Emergence, traditional, centralized sources of power fail and new sources of power develop in the world; this transformation does not bring peace and stability though - crime and lawlessness are ongoing problems. Kelly does not present these three scenarios as mutually exclusive (another example of his both/and logic), but contends that all three scenarios will probably correspond to some features of the future. Kelly's question is which reality will dominate.

In his conclusion, Kelly presents one final illustration of both/and logic as a capstone principle for creating a positive future. Either/or logic, applied at a social level, leads to the dichotomy of "us versus them." Yet, Kelly argues, in resonance with many other contemporary writers, that humanity must move beyond this dichotomous thinking and cultivate a more expanded sense of "we." We must become more inclusive and less divisive. If we are to avoid the "tragedy of the commons" - of selfishly using up all the earth's resources to the detriment of others sharing our common environment - then humanity must develop a greater sense of collective community. For Kelly, now that we have globalized our economy and our culture, we need to globalize ourselves - each of us must identify with our global humanity.

Anderson on Culture, Evolution, Technology, Globalization, and Enlightenment: Walter Truett Anderson, in a series of books over the last fifteen years, has developed a multi-faceted and relatively comprehensive theoretical analysis of contemporary trends and potential future directions for humanity. In his books, he has examined and synthesized such diverse topics as human belief systems, values, and culture, biotechnology, information technology, evolution and ecology, human psychology, society and globalization, Eastern and Western philosophy, and the past and potential future evolution of enlightenment. His most noteworthy books over this period include *Reality Isn't What it Used to Be*, *Evolution Isn't What it Used to Be*, *The Future of the Self*, *All Connected Now*, and *The Next Enlightenment*.[254]

In *Reality Isn't What it Used to Be*, Anderson presents the argument that in contemporary times there is a fundamental conflict and disagreement between absolutists and relativists regarding the nature of human knowledge and human values. The former believe that human knowledge and values are grounded in absolute and objective principles and facts; the relativists believe that knowledge and values are historically, culturally, and psychologically relative (or subjective). In essence, this is the conflict between fundamentalists (the absolutists) and Postmodernists (the relativists), though we could also include as absolutists those who believe that science provides objective and certain knowledge about reality. Anderson, in this book and later writings, clearly seems to side with the Postmodernists, at least to a degree.[255] He does believe though, contrary to many Postmodernists, that there can be progress in the growth of knowledge.

In *Evolution Isn't What it Used to Be*, Anderson argues that evolution is evolving and becoming purposeful with the introduction of biotechnology and information technology into the "augmentation" and enhancement of our species. In fact, he sees technology as permeating out into all aspects and dimensions of nature, including the monitoring and control of our environment. There can be no return to a pure or unspoiled nature. Biotechnology and information technology are increasingly intertwined and, following a similar line of thinking to Kevin Kelly, Rodney Brooks, and Andy Clark, Anderson sees a general blurring of the separation of life and technology. The whole wide world is becoming the "whole wired world."

In *The Future of the Self*, Anderson picks up the Postmodern theme again, and presents the argument that the human self is a social construction, situational specific, and pluralistic, rather than singular and absolute. Anderson argues that given the complexity and rush of change in our contemporary world, a new type of self is emerging - one that is pluralistic and much more fluid. Anderson ties together human psychology, advances in the sciences, trends in culture and society, and the impact of technology on human life and the human mind, in creating a Postmodern vision of the self.

I have already discussed some of Anderson's main ideas on globalization as presented in his book, *All Connected Now*. To recall, Anderson sees globalization as a multifaceted phenomenon, involving a strong technological dimension - a theme he carries forth from his earlier book on evolution. We are being wired together - the environment is being wired together as well. Anderson also reinforces and further develops his emphasis on the pluralistic and multicultural quality of our times - a theme he introduced in his earlier books on *Reality* and the *Self*.

One theme that runs through Anderson's books is evolution; nature and human society is dynamic, changing, and developmental. Globalization has had a history which he traces in *All Connected Now*, and in *The Future of the Self*, Anderson looks at the history and evolutionary development of the self. In fact, for Anderson, evolution is itself changing as conscious purpose and technology become increasingly important in human growth and change. In *The Next Enlightenment: Integrating East and West in a New Vision of Human Evolution,* Anderson recounts the historical development of enlightenment in both Western and Eastern cultures. Evolution is perhaps the central theme in this last book, for Anderson believes that enlightenment is "an evolutionary project" - an expression of the dynamic and growth-oriented dimension of reality. And a key element in the state of enlightenment is seeing that all is flow - that all being is becoming.

For Anderson, enlightenment involves a liberation from the egocentric constraints of viewing ourselves as a singular and absolute, unchanging self. In the *Future of the Self,* Anderson critiques this limiting idea of the self. In *The Next Enlightenment,* Anderson goes further in arguing that the most important problem of our times is overcoming this constraining view of self-identity. War, conflict, indifference, and cruelty, all arise out of conceptualizing our identity, both individually and culturally, as bounded and singular entities. Within this mindset, we fail to see the "oneness" of all humanity and the "oneness" of ourselves and the universe - instead cultures and individuals segregate and oppose each other and humanity separates itself from nature. Enlightenment involves as a central insight, this understanding and experience of oneness. In *All Connected Now,* Anderson highlights the importance of a growing sense of global consciousness and the theory of open systems (the interconnectivity of all things); in *The Next Enlightenment* Anderson discusses the idea of "cosmic consciousness" as an essential feature of enlightenment. It is important to see that enlightenment means freedom for Anderson. In his history of enlightenment, provided in the first part of his book, he reviews efforts through the ages to free the human mind from the cultural and psychological forces and assumptions that enslave and suppress us.

Anderson synthesizes a variety of ideas in his theory of enlightenment. He pulls together ideas from both the East and West. He sees a thematic connection between the Buddhist ideas of oneness and flow and the Western ideas of interdependency, interconnectivity, and evolution. He discusses "flow" and "transcendence" in the context of both the western psychology of Csikszentmihalyi and Eastern meditative practices. He sees the value of both rationality and intuition as paths to enlightenment. He supports the openness of New Age spirituality, but critiques the lack of epistemological standards in this movement. He rejects the professed certainty of fundamentalism and argues instead that a key feature of enlightenment is the courageous embrace of mystery and uncertainty in human existence. Identifying a series of "liberation movements" within human history, which include the European Enlightenment, Darwinian evolutionary theory, Freudian psychology, and the Human Potential movement of the 1960s, Anderson believes that enlightenment is a higher level of consciousness, enveloping and transcending earlier stages in the growth of the human mind, that was achieved by some people in the past. He anticipates increasingly more people achieving this state of consciousness and mentality in the future as an expression of the evolutionary development of humanity.

A New Enlightenment: Anderson is not the only futurist who foresees a new enlightenment spreading through humanity in the years ahead. I have already described the ideas of Barbara Marx Hubbard, Ken Wilber, Spiral Dynamics theorists, and Mihalyi Csikszentmihalyi, all of whom believe that a new level of consciousness and human mentality is emerging, or at the very least, given the challenges and complexity of our times, should be emerging in our species. One futurist who talks about a Second Enlightenment - the first being the European Enlightenment of the eighteenth century - is Rick Smyre, the director of *Communities of the Future*. Smyre argues that just as new social, scientific, and technological developments occurring during the rise of modernity instigated the need for a first Enlightenment, recent events and changes in our world require a new mode of thinking for handling the new challenges of our times.[256]

Smyre identifies a set of megatrends in contemporary times and a set of principles of thinking and behavior for successfully dealing with these megatrends and the overall accelerative change and complexity in our world. First, the megatrends Smyre describes are:

- Oil production will peak within ten to thirty years, requiring a shift to completely new energy methods.
- The limits to traditional representative democracy are appearing because of the increasing complexity of society, the overwhelming amount of information in real time media, and the need for big money to be reelected.
- The increasing symptoms of global warming are becoming obvious in many different ways.
- Unknown viruses and resistant bacteria are emerging that are untouched by existing medicines.
- Conflict between the trends of aging in the developed world and the increasing numbers of youth in the developing world creates an economic and social time bomb.
- There is an ongoing severe reduction in biodiversity throughout the world.
- We are approaching a moment of "technological singularity" when runaway advances outstrip human comprehension and all our knowledge and experience becomes useless as a guidepost to the future.
- The future vitality and sustainability of the economy and society will be dependent on the ability of leadership to develop cognitive complexity and continuous innovation in the capacities of the citizens.
- There will be an increasing clash of civilizations and rise of terrorism.
- There are three economies in churn at the same time: a) Last stages of the Industrial Economy; b)

Twenty-year transition stage called the Creative Knowledge Economy; and c) The early stages of a Web/Networked Economy.

- National borders increasingly will be seamless.
- Artificial intelligence is emerging which, when combined with biotechnology and nanotechnology, may very well transform the concept of what it means to be human.

Although different contemporary writers and futurists may put together different lists of major trends in our times (for example, compare Smyre's list with Cornish's super-trends or the Millennium Project's fifteen global challenges), Smyre does make an attempt to be comprehensive in his list, identifying social, economic, political, technological, and ecological variables in his analysis. Further, in his list Smyre pulls together the ideas of various other futurists and modern writers, such as Huntington, Kurzweil, Freidman, and Rifkin.

When Smyre approaches the question of which principles of thinking and behavior are needed to successfully cope with these trends, he develops a triadic model. First he lists those principles that characterized the First Enlightenment; then he identifies a second set of principles that capture the essence of his vision of a Second Enlightenment, and in general, these new principles are oppositional to the first list; he then adds a third list which attempts to synthesize the polarities of the first two lists. He identifies these principles with an "Integral Society." His triadic list of principles is:

First Enlightenment	Second Enlightenment	Integral Society
Independent (either/or)	Interdependent (and/both)	Systemic
Self-interest	Help Each Other Succeed	Concomitant Good
Linear Thinking	Connective Thinking	Synthesis and Generation
Static Structures	Modules, Webs and Networks	Dynamic Adaptability
Reductionism	Holism	Connective Analysis
Standard Education and Accountability	Unlearning, Uplearning, and Non-Linear	Transformative Learning
Meaning from Materialism	Meaning from Creativity and Spiritualism	Balance of Values
Competition	Collaboration	Generative Development
Prediction and Certainty	Anticipation and Ambiguity	Parallel Strategic and Ecological Planning

Culture Dumbed Down	Culture Constantly Upgraded	Elegance in Complexity
Mix of Goodness and Skepticism	Integration of Reason and Mystery	Truth and Discovery Coexistent
Debate	Dialogue	Futures Generative Dialogue
One Best Answer	Choices	Concept of Applied Appropriateness
Representative Democracy	Electronic Republic	Knowledge Democracy

There are certain central themes in Smyre's principles. At least seven of the principles under the Second Enlightenment revolve around the theme of connectedness and interdependence, contrasted with the First Enlightenment emphasis on independence and autonomy. Also, the themes of mystery, uncertainty, change, and transformation show up in several of the principles under the Second Enlightenment. The contrast between linearity and independence and interdependence and holism also corresponds, interestingly, with Nisbett's major distinction between Western and Eastern thinking, which would seem to imply that the Second Enlightenment involves a shift toward a more Eastern way of thinking.

One of Smyre's main sources of inspiration is Sally Goerner's *After the Clockwork Universe*. To recall, Goerner argues that a pervasive shift in both scientific and social thinking is occurring in modern times, away from the Newtonian model of reality as a machine and toward the idea of reality as a web of interdependencies. Goerner's central idea is connectedness which, in her mind, is pivotal to the emergence of a new "integral culture."[257] What Smyre attempts to accomplish with his third column of principles, which he labels as "Integral" (inspired by Goerner's use of the term), is a synthesis of the first two columns, acknowledging that there is value and truth in the First Enlightenment principles - it is simply that they need to be "integrated" and balanced with Second Enlightenment principles.

There are also various parallels between the ideas of Smyre and Goerner and Anderson's view of the "Next Enlightenment." Anderson highlights the themes of connectedness and oneness and critiques the psychological mindset of dualism and ego-separateness with the world. Further, Anderson supports an evolutionary perspective on reality and the future development of the human mind. And he believes that the challenges and problems facing us today call for a new way of perceiving and understanding the world.

Macdonald - Wisdom, Deep Understanding, and *Matters of Consequence*: Creator and editor of *The Wisdom Page* on the World Wide Web, Copthorne Macdonald presents an integrative vision of important ideas of the past, contemporary issues and trends of today, and a philosophy and action plan for future in his book, *Matters of Consequence: Creating a Meaningful Life and a World That Works*. He attempts to pull together the subjective and objective, science and spirituality, theory and practice, fact and ethical value, and philosophy with politics and economics into a comprehensive proposal for how to transform our present world into a "wisdom culture" for tomorrow. Many of his ideas resonate with theories of the future already discussed, including those of Wilber, Ray, Hubbard, Eastern philosophy, holism, and evolution. Macdonald is especially concerned with facilitating both a global social movement and a transformation of individual lives.[258]

The central theme in Macdonald's writings is wisdom. In his earlier book, *Toward Wisdom*, he states that the development of wisdom is critical to the creation of a better world in the future and that the lack of wisdom and its application is the main cause of contemporary world problems. According to Macdonald, wisdom is a complex, multifaceted set of capacities. It is a holistic quality reflecting the total psychological make-up of a person. He highlights certain key features of wisdom, such as a reality-seeking attitude (a desire to know), a non-reactive mode of behavior and acceptance of reality, a holistic perspective that goes beyond the immediate here and now and the personal, a realization of the oneness of reality, and a disposition to act for the benefit of others and the whole. The wise person sees things clearly, is prudent in action, possesses a deep understanding of the human/cosmic relationship, and, through the application of wisdom to life, achieves happiness and peace of mind. Macdonald also argues that the self-actualizing personalities described by the psychologist Abraham Maslow are paradigm examples of the wise person. Maslow describes self-actualizing individuals as creative, inner-directed, detached, more concerned with things outside themselves, able to give love, and living by a unique set of values, including truth, playfulness, beauty, and self-sufficiency.[259]

In *Matters of Consequence* Macdonald focuses on "deep understanding," which is a particular kind of wisdom. Deep understanding has three fundamental components: A comprehensive intellectual understanding of reality, an inner directed and intuitive self-awareness and understanding, and a capacity for significant "doing" or action. The first two components roughly correspond to scientific and spiritual-psychological perspectives on reality - the objective and the subjective - and, according to Macdonald, both are necessary for deep understanding and, consequently, wisdom. He believes that the insights acquired through these two perspectives can be aligned and integrated into a coherent whole - science and spirituality are not incompatible. For Macdonald, science in general provides answers to factual questions, whereas spirituality provides answers to questions of meaning and ethics. But he also repeatedly states that a deep understanding of the subjective and the objective realms will lead individuals to insight regarding what is ethically correct and what needs to be done to improve the world - a deep understanding of "fact" generates an understanding of value and a call to action. Macdonald connects this theory of deep understanding with the future in arguing that the pervasive development across all humanity of deep understanding (and consequently wisdom) is the next step in human evolution.

In laying the groundwork for his theory of deep understanding and the creation of a positive future, Macdonald provides a comprehensive overview of reality, of both the objective-scientific and the subjective psycho-spiritual. He discusses philosophical ontology (the nature of Being), evolution, complexity, the nature of mind and consciousness, cosmology, culture and society, economics, and biology and ecology. He not only includes reviews of contemporary and classical theories for each of these areas, he also critiques, along the way, various dominant belief systems and practices and proposes fundamental values and principles to guide humanity in the transformation of existing social institutions and modes of behavior. In so doing, he provides a critical analysis of contemporary human problems and creates a set of fundamental principles for transforming the world.

Macdonald believes that it is exceedingly difficult to predict the future of human society. Hence, he suggests that we should focus on envisioning what we believe is a preferable or ideal future and then work toward realizing this vision. In this regard he provides a detailed "Year 2050 Vision," which includes an extensive list of transformative principles and actions for achieving this vision. First, he envisions humanity developing a much greater cosmic and global understanding and appreciation of reality (following from his concepts of wisdom and deep understanding). He foresees an upswing in the quality of life for humanity and much less of a concern with material possessions and consumption. Sustainability will continue to increase in importance as a fundamental value of human life. There will be an increase in creativity, learning, and psycho-spiritual development. In general, we will do more with less and increasingly pursue a "full rich life of mind." Also, Macdonald foresees the universal provisioning for all people and a more equitable distribution of wealth. In general, he argues that the pursuit of wealth is one of our most problematic contemporary values and needs to be significantly diminished in importance. He questions the need for continued material growth –has it bought human happiness? – and argues that our overall sense of the quality of life needs to incorporate more psychological, social, and spiritual values with less of an exclusive emphasis on economic and technological development. There will be a new balance of work and leisure, with ample time for the pursuit of individual interests and self-development. Our economy will move more toward a focus on the inner life and inner experience, and eventually human transformation. For Macdonald, the ideal world of 2050 will be much more politically stable, with the emergence of world governance, "pooled sovereignty" among nations, and a dominant spirit of cooperation and collaboration over war and competition. He envisions the continued replacement of dictatorships with democracies and increasing equality and general human rights for all people. Macdonald even suggests that intellectuals, philosophers, cultural leaders, and writers should pursue political careers, for our future needs wise leaders who possess deep understanding, highly developed ethical characters, and a capacity for seeing the long view of things. As another basic point, Macdonald supports Rabbi Michael Lerner's contention that the main world problem is the "globalization of selfishness" and that the ideal future should involve a "globalization of spiritual consciousness." For Macdonald, we live in a "spirit-denying society" and this attitude needs to change. Finally, Macdonald wishes to emphasize that his envisioned world transformation needs to occur at both the individual level and the collective global level – in essence, a combination of top-down and bottom-up approaches.

Following from his ideas on wisdom and deep understanding, Macdonald clearly supports certain fundamental values in both his critique of contemporary times and his vision of an ideal future. He values cooperation and concern with the whole over selfishness and self-centeredness. For Macdonald, we must identify with the "ONE" and the "ALL." He values a balance of spirituality and the inner life with science and materialism, though he is highly critical of the excessive materialism of the modern world. He values expansive and deep understanding as an absolute necessity in guiding our actions and creating our future. He speaks of a "mind-directed" evolution for humanity. For Macdonald, the human mind needs to evolve. Although he argues for the value and importance of science, ultimately he sees humanity as "spirit" - a manifestation of the cosmic spirit in evolution. Finally, he strongly argues for "massive change" in our present world if we are to realize a positive future.

Summary and Conclusion: Future Consciousness, Evolution, Reciprocity, Spirituality, Virtue, and Wisdom

In the final section of this chapter I highlight some of the main themes contained in the preceding review of theories and paradigms. Also, I highlight some of the main issues and disagreements among the different viewpoints. As I proceed through the summary I evaluate the theories and identify what I believe are the most valid and valuable ideas among the different viewpoints.

From the first section on theories of time and change, one central disagreement concerns whether human reality should be viewed as fundamentally transformative or static. This disagreement is what Postrel refers to as the clash of "stasis" and "dynamism." Although there may be features of the universe that are stable (for example the laws and physical constants of nature, but even this is open to scientific debate and further investigation[260]), it seems that the overwhelming evidence supports a transformative view of human reality and the universe as a whole. Both nature and humanity have a dynamic and transformative history.

Static visions often wish to uphold and defend values, belief systems, and ways of life that are grounded in traditional or ancient cultures and doctrines. Clearly there is considerable wisdom to be found in our diverse cultural heritages, and although we shouldn't, as many of their followers would argue, accept such traditional ideas and practices as authoritative and absolute, we should thoughtfully consider their potential validity and value. The future should not reject the past - at least to some degree the future needs to envelop and learn from the past.

Throughout recorded human history, there have been repeated conflicts between those who wish to preserve the status quo, if not even return to past practices and beliefs, and those who argue and push for change and revision. Often it is those in power who wish to preserve continuity and those who wish to gain power that advocate for something new and different.[261] The future seems to unfold as a result of the ongoing clash between the old and the new - each stance, in fact, seems to trigger and intensify its antithesis. In a world of rapid change, it is understandable that many people wish to defend and preserve tradition and traditional sources of power. It is highly probable that the future will continue to exhibit this "dynamic tension" (to borrow a term from Eamon Kelly) between the philosophies of stability and change. It is a basic dynamic of the human mind.[262]

Based on evidence accumulated from a great variety of different sources and scientific disciplines, it seems that evolution is unequivocally the best theory we presently have regarding the transformative nature of reality. Although we may not understand all the details, or even all the basic principles, involved in the evolutionary process, both our history and our future are probably best understood in evolutionary terms. Further, evolution seems to apply not only to biology, but to the broad features of the cosmos as a whole, as well as human history and the ongoing transformation of our species. As Wright and others have argued, aside from biological evolution, there has also been cultural evolution. Although there is ongoing debate over whether evolution is progressive, there appears to be a variety of indicators, such as increasing complexity, individuation and integration, and even intelligence, that demonstrate a progressive direction to evolution. Also, evolution seems to be evolving and accelerating, and humans will probably play an increasing role, in a purposeful and conscious way, in guiding evolution. There is, of course, a debate over whether we should attempt to technologically alter ourselves or nature, but the argument against purposeful evolution seems to rest upon a static view of humanity and nature, which I believe is factually wrong. Another reason for thinking we will engage in purposeful evolution in the future is that we have been engaged in it throughout our entire history, along a variety of psychological, social, biological, technological, and even spiritual lines. It is not something new; we simply are developing more powerful means for doing it. Yet, given the conflict and dynamic tension over stability and change within human society, it is probable that many individuals will resist transformation, creating an amalgamation of the old and the new in future humanity.

The future evolution of humanity is not a given, though; there is an element of uncertainty and possibility in the future. Although there are general laws of nature that structure and determine, to a degree, the course of events in nature, there are also elements of indeterminism, emergent novelty, and choice in the unfolding of time. Based on both contemporary thinking regarding the limits of human knowledge, as well as modern scientific theory regarding the probabilistic and open-ended nature of reality, it seems to me that no one can have absolute or certain knowledge of the future. Future consciousness is tentative and contingent. As Macdonald "strongly suspects...no one has ever been completely out of the dark." Contrary to teleological views of the future, I think that the future is yet to be determined and no one can possess certain knowledge about it. It is yet to be. Teleological views of the future provides mental security, but the future is inherently risky and an adventure.

Even if it makes sense to support change and further growth and evolution in the future, growth and change need to be sustainable in the long run. Thinking about long term sustainability is an expression of advanced future consciousness. Although our modern world, with all of its positive qualities, is a result of the progressive theory of the future developed during the European Enlightenment, it seems clear that there were elements of short-sightedness in this vision of the future that need to be rectified (for example long term environmental impacts of industry were not sufficiently thought through). Also, the philosophy of growth and change should to be tempered and guided by considerations concerning the overall quality of human existence. If change and growth is frenzied, thoughtless, or motivated by questionable short term values it can damage both human society and the world at large. Our future evolution needs to be guided by our scientific, social, and ethical understanding of human nature and the environment in which we live. Critics of our fast paced world and our modern obsession with continued growth and development make some valid points.

Developments in science and technology are of critical importance in understanding the future. I believe that built into us is a drive toward increasing our knowledge of the cosmos and ourselves, and that knowledge empowers and benefits the human species. Macdonald lists cosmic understanding through science as one of the key features of wisdom. Science is open-ended, though, and there is probably no end to its future development. Still, in the last 300 years, science has revealed more about the nature of reality than was discovered throughout all of previous human history. Science may not be the only valid way of knowing, but it certainly has demonstrated its power and value. Further, technology seems intrinsic to human nature; we make things and use our inventions to enhance and enrich our ways of life. As Andy Clark argues, we are "natural-born cyborgs" and we have been since the origins of our species. Both science and technology have had a powerful impact on human society, and all indications point to an even greater influence on us in the future. We are in the midst of a Second Scientific Revolution and this revolution will change the way we think about ourselves and reality and will stimulate, in the coming years, a whole host of new technological developments. In the relatively near future, technology will in all probability be used to transform human nature and perhaps provide the means for the creation of other intelligent and conscious beings. As Garreau and others have argued, computer technology, nanotechnology, biotechnology, robotics, and cognitive/brain science will all contribute to the "radical evolution" of humanity. I think that Kurzweil and others are correct - technological development and its growing interface with biology is an integral and newly emerging feature of the evolutionary process.

We can expect that various waves of technological innovation - running the full gamut from computers and robots to genetic and ecological engineering - will repeatedly transform the world around us. As Wells stated, we may be only at "the beginning of the beginning." We are developing a global nervous system - the Internet - which will support the monitoring and coordination of both human and ecological activities around the world. Will this lead to some type of global mind or Omega Point, as Chardin envisioned? Further, it seems as if robots will become ubiquitous throughout all spheres of human life in the near future. And if nanotechnology fulfills its promise, the world in the future will be transformed into a highly fluid, intelligent, and flexible medium for evolution and creation. And there will be other technologies, yet to be perfected, or even imagined, such as quantum computing, that will emerge in the decades and centuries ahead.

I would propose, though, that humans and technology form a reciprocity - each influences the other and quite literally depends upon the other for its existence and continued development. Humans and technology co-evolve. Hence, the future is not just advancements in technology - the human mind, human values, and the human spirit must also be brought into the equation in predicting the future. In creating a realistic and ethical vision for the future, as well as a holistic (or integral) image of human existence, we need to strike a balance between humanistic and technological concerns. Advances in technology should inform our ethical decision-making, but our psychology, our social concerns, and our ethics needs to guide our technological development. One of the great issues and challenges of our time is to put more of an emphasis on the humanitarian uses of technology.

Ecological and naturalistic theories of the future highlight the importance of taking a holistic, interdependent, and interactive view of humanity and nature. The principle of reciprocity again seems to be a key idea in understanding the nature of the world in which we live. We must think about the totality of nature and the interdependency of all things in envisioning the future. Yet in so doing, I think it is a mistake to view nature as static and intrinsically harmonious - as something that can be preserved. As I argued above, nature is dynamic and evolutionary and, I should add, filled with competition and conflict. Perhaps we can guide nature, but we can not contain it. Further, to believe that we can retreat to a pre-technological state and live in harmony with nature is regressive and naïve. We are technological beings and we are inextricably in the business of influencing nature; (but then all of nature is in the business of influencing nature - there are no innocent bystanders); we may as well influence nature in a thoughtful and ethical fashion.

We need to journey into outer space to realize our place in the cosmos and achieve a cosmic understanding of reality. We will never grasp the big picture if we stay situated in one tiny spot in the universe. Knowledge requires exploration. To break out of our narrow and egocentric view of reality, humans need to explore and settle the solar system, the stars, and even the galaxies and the universe. This mind-boggling journey is an essential expression of our drive to know; it is perhaps, as many futurists argue, how we will actively participate in the evolution of the universe. It is a spiritual quest as well. Perhaps it will be robots, or genetically re-engineered and technologically augmented humans who will travel into the cosmos, but in one manner or form, we need to spread our wings and set sail into the ether.

Finally, on the science and technology theme, a real question for the future is whether humanity will be transcended by intelligent or "spiritual machines" – the thesis of Kurzweil. Clearly humanity will be significantly transformed through technology in the coming centuries, but perhaps as Vinge argues, there will come a point where our descendents will be so technologically enhanced and different from us, that indeed, from our present point of view, we will no longer be human and will have become "machines." But then I think that what a machine is will dramatically change in the future. I think that the "technological singularity" will sooner or later come, at least in the sense that minds and collectives of minds of the future will, in some important ways, be incomprehensible to us from our vantage point today. This event will mark a watershed in evolution, as great in significance as any other event in the history of the earth, if not the cosmos.

One thing seems certain - human nature is not a constant, and with the ongoing changes in technology and human society, it is critical that we psychologically evolve. Theories of how humans can or should evolve in the future need to be an essential feature in any comprehensive theory of the future. Psychological theories, such as those proposed by Seligman, Csikszentmihalyi, Wilber, Anderson, and O'Hara, need to be integral to the ongoing dialogue on the future. Seligman, I believe, makes a provocative suggestion in connecting psychological change with ethical development. Human evolution will involve technological augmentation, biological enhancements, psychological and mental transformations, and ethical and spiritual developments. Individual growth will occur in reciprocal interaction with social and cultural changes. Human organizations will be transformed as well, as human psychology evolves.

One key area of debate within social and cultural theories of the future revolves around the relative importance of unity, globalization, community, and conformity on one hand, and individuality, diversity, cultural pluralism, and freedom on the other hand. This debate goes back at least as far as the European Enlightenment. There is the ongoing worry that unification undermines individuality and freedom. This concern over assimilation into the whole is described in Friedman's *The Lexus and the Olive Tree* and Barber's *Jihad versus McWorld*. But there is the reverse concern, perhaps best expressed in Putnam's *Bowling Alone* that Americans, at least, are becoming too disconnected and our sense of community is deteriorating. While many fear that we are becoming too interconnected (without any space to breathe), others critique the excessive, self-absorbed individualism of the modern world. Following Bloom and Graves, as well as many earlier writers, we can interpret these opposing concerns and values as a fundamental and ongoing dialectic in the history of humanity. Using Kelly's terminology, we could describe individualism versus communalism as a fundamental "dynamic tension" in contemporary times.

There are some writers, such as Henderson, Naisbitt, and Stock, though, who see this as a false dichotomy, and believe that greater social order and connectedness will strengthen diversity and freedom. A common expression in our times is "unity through diversity." For Friedman and Naisbitt, globalization empowers the individual. The dialectic or tension of individualism and communalism perhaps serves both ends of the equation; through the continual conflict of individuating and integrating forces in human reality, both dimensions are further developed and enhanced. Paradoxically, we become increasingly integrated and diverse over time. It seems to me that this tension in human affairs will continue indefinitely into the future, and even if it is oppositional in nature, it is a basic *Yin-yang* of human existence where each side fuels the evolution of the other. Individuality and social organization reciprocally evolve.

There is uncertainty concerning the degree to which humanity can or even should further integrate in the foreseeable future. There is the argument presented by Postmodernism and Creative Chaos that a multiplicity of points of view and ways of life is unavoidable and, in fact, beneficial to the freedom and flexibility of the species. There is the argument that war and violent conflict is either intrinsic to human nature or a deeply entrenched feature of human nature.[263] Finally, there is the strong possibility that with the potential dispersion of humanity into outer space, as well as the increasing use of genetic engineering in the human species, humanity will differentiate and separate even further than today.

Connected to the theme of integration and differentiation is the general question of how human society in the future will achieve a better balance of power and affluence among all the diverse people and cultures of the world. The West has been criticized for creating a lopsided distribution of power, wealth, and social status that centralizes control in aggressive and highly competitive white males. Feminist thinkers like Eisler argue for an equality of status and influence between females and males and a balance of partnership and competitive modes of thought. Many critics of globalization argue that it is a Western creation and it is the wealthy in the West who most benefit from it. Economic globalization is a form of Western imperialism and colonialism. In spite of Friedman's claim that the world is flattening, I think Florida is more on target in arguing that the world is exceedingly spiky. Most writers on this issue argue that future humanity needs to be much more egalitarian and democratic than the elitist, power-centered, and supremacist social organizations of the past. Poverty versus prosperity is one of Kelly's basic tensions and one of the key challenges identified in the Millennium Project.

Another ongoing issue in social-cultural theories of the future, which in fact goes back at least to the nineteenth century, is the power and value of capitalism. Critics point out how capitalism generates increasing inequality, turns everything into a commodity, reinforces greed, and produces a society of consumers. The writers of *Affluenza* strongly argue for all these points, but so do many other social critics associated with New Age, Cultural Creative, Integral Culture, and environmentalist perspectives. Supporters of capitalism, for example, the writers of *The Long Boom* and Thomas Friedman, argue that this economic system has an overall beneficial effect on the quality of human life. D'Souza summarizes the varied arguments of both sides of this controversy. In general, capitalism seems to be a mixed blessing. I think, though, that the most valid criticism of contemporary capitalism and consumerism is that we live in a world where the economic sector has a disproportionate amount of control over human life - it dictates too much how we live and what we ultimately value. Many visionaries of the future, past and present, wrestle with how to rectify this failing of the system, how to counter-balance its incredible power, and what to put in its place. The making of money and the buying of things should not be the central defining answer to the meaning of life.

Religious and mythological visions of the future are the oldest recorded theories and paradigms of the future. Such visions, in spite of all the changes that have occurred in human life over the last five thousand years, are still the most influential views of the future among humanity as a whole. Modernists and secular thinkers frequently see traditional religions as archaic and authoritarian systems of thought control. Yet, there clearly is something about religious belief systems and practices that connect with the human soul and psyche. As I argued earlier, religious myths provide personalization, narrative meaning, ethical direction, and they resonate with the heart and human emotion. Further, many of the themes in religion and myth provided the conceptual and theoretical foundations for more modern ideas about reality and the future.[264] Religion, myth, and spirituality, I believe, will not disappear in the future.

I think, though, that religion, myth, and spirituality need to evolve.[265] In our contemporary world, as Kelly, Anderson, and others have pointed out, there is the ongoing, if not intensifying conflict between fundamentalist religions and secular - humanistic - scientific ways of thinking. Additionally there is the conflict between fundamentalism and mainline religion, which emphasize conformity and obedience to a singular doctrine and creed, and New Age spirituality, which supports individuality and the incorporation of multiple and varied religious views. It seems to me that fundamentalist religions have the strengths of bringing unity, focus, and clear purpose to their followers, but at the expense of remaining closed and authoritarian, oblivious or antagonistic to science, and opposed to individual thought and expression. The clash of creationism and theologically motivated intelligent design thinking with evolutionary science is one paradigm example of how fundamentalism doggedly resists coming to terms with science and opening itself to growth and transformation. Fundamentalist religion has lost the sense of mystery and adventure, critical to our attitude toward the future. Although there is a great value in the diversity of religious and spiritual points of view in the world, the absolutist quality of fundamentalist religions works against any progress in finding common ground among religions or realizing a global spirituality. Fundamentalism opposes diversity of points of view and creates divisiveness. New Age thinking, in its liberalism, at least works toward a sense of mutual appreciation of points of view.

Hence, it seems to me that the keys to growth and constructive evolution in religion, myth, and spirituality involve the incorporation of scientific ideas about reality, the relinquishment of absolutist thinking, and the willingness to open up to the potentially equal value and validity of different religious points of view. As I suggested earlier, science fiction provides a way to integrate futurist imagination and scientific and contemporary thinking with the "power of myth." I think that efforts to integrate evolutionary and scientific thinking with concepts of spiritual growth and the nature of God, as for example presented by Chardin, Tipler, Macdonald, and Wilber, are of great value in this regard.

Following from the ideas of balance and reciprocity, valid and constructive theories of the future should take into account and integrate all the major dimensions of human reality, and further, should place the ongoing saga of human history within a cosmic context. As holistic philosophy and open systems theory emphasizes, everything in nature is interconnected and, hence, we should understand how all the different aspects of human reality impact each other, and how the whole human condition connects with the cosmos. Recall that one of Macdonald's key defining features of wisdom was a comprehensive and cosmic understanding of humanity and reality. A wise approach to the future will be comprehensive and cosmic in scope. This is not to imply that individuality, disharmony, pluralism, and chaos are not important features of human reality and nature. (This issue has already come up in the previous discussion of human society.) These features of existence must also be incorporated into an "integral theory" - part of the whole is the dimension of fragmentation. Still, the general point is that placing too much emphasis on one dimension of reality, be it technology, economy, science, politics, or religion, is to miss essential features of the total reality of the future and the overall interdependency of things. Efforts to create an integral culture or philosophy for the future have value. Efforts to place the future of humanity in a cosmic context, such as that created by Zey, Tipler, and Macdonald, are essential - our vision of the future must not be earthbound.

From the writings of Nisbett, Wilber, Shlain, and Macdonald, among others, it is clear that there is a certain fundamental balance in thinking that needs to be cultivated now and in the future. Following from Nisbett, a balance of Eastern circular-holistic and Western linear-analytic thinking is critical for our global future. Wilber and Macdonald highlight the importance of both subjective and objective perspectives on reality. Shlain focuses on the importance of both visual-image and linear-linguistic thinking. From my arguments earlier in this book, the mythic and the scientific should be balanced and integrated, as is the case in science fiction. Kelly makes the general point that the future requires both/and thinking, as well as either/ or thinking - which in essence, is Nisbett's position. But then all of these arguments for complementarity and balance are examples of both/and thinking - captured in the Eastern idea of the *Yin-yang*. (One could include the masculine and the feminine in this set as well.)

Kelly's ideas on dynamic tensions and both/and thinking underscore another key feature to keep in mind when we theorize and think about the future. Throughout human history there have been innumerable dynamic tensions - conflicts of opposites as Hegel would describe it. Taoist philosophers conceptualized the fundamental polarities of human existence in terms of the *Yin-yang* - opposites that are united in an ebb and flow of dominance and submission. As in the past and the present, the future will be a both/and - an ongoing tension and opposition of conflicting forces and belief systems. In the broadest, deepest sense there will never be total peace and harmony.

As noted earlier, one ongoing conflict in human history is between stability and the past, and change and the future. Yet in order to understand the future, the past and future must be brought together; we can not see the future very well from the narrow confines of the present. Understanding history and placing the contemporary transformation in a historical, developmental, and evolutionary context is an essential dimension of an integral philosophy for the future. Writers and futurists such as Barbara Marx Hubbard, J.T. Fraser, Robert Wright, Ken Wilber, and Howard Bloom all attempt to ground their theories of the future within historical and developmental frameworks. Part of future consciousness is historical consciousness.

The future should be conscious of the past, envelope it, and incorporate its wisdom and teachings, but the future needs to transcend the past. A common argument among futurists, social commentators, and other visionaries is that we need a new way of thinking to live and thrive in the future. In spite of numerous economic, technological, and social advances in our contemporary world, many people argue that we are beset with a host of problems and challenges that require a significant change in our mindset for their solution. Writers such as Anderson and Smyre believe we need a New Enlightenment. Both writers derive at least some of their inspiration from the past, but they also both incorporate new ideas, for example, open systems thinking and evolution, into their descriptions of the basic principles of this New Enlightenment. I think that it is particularly valuable that Anderson tries to pull together Eastern and Western views of enlightenment and that he sees enlightenment as an ongoing quest and evolutionary direction within human history.

What would be the basic principles and features of a New Enlightenment? Based on what Anderson and Smyre argue, and incorporating many of the key themes highlighted in this summary of theories of the future, I would suggest that a deep sense of the dynamic, fluid, open-ended, mysterious, self-transcendent, and evolutionary nature of reality is essential to the New Enlightenment. I think that this first dimension of the New Enlightenment provides the key to synthesizing the valuable insights of science, religion, and spirituality. We are on a journey, an odyssey, with a sense of beyond-ness toward the transcendent. Macdonald lists this dynamical and open-to-mystery perspective on reality as a key feature of wisdom. Second, a clear sense of the reciprocity of all things, of the balance between the complementarities of nature, such as individuality and integration, and order and chaos, and between the complementarities of the human mind, such as reason and intuition and analysis and synthesis, would also be a critical feature of the New Enlightenment. (Although the expressions "connectedness" and "oneness" are often suggested to capture the holistic nature of reality, part of a truly holistic sense of reality is to see the individuation and disconnectedness within reality as well.) Third, there should be an expansion of consciousness - a moving away from an egocentric, immediate here and now perspective - toward a global and cosmic consciousness and an enhanced temporal consciousness of both past and future. Fourth, and this point connects back with the idea of reciprocity, the New Enlightenment, on one hand, will undoubtedly be bound up with a technologically enhanced human mind and human society, and on the other hand, an evolved ethical dimension within humanity. It will be cyborgs - highly virtuous cyborgs - within a technologically interconnected environment that will realize the New Enlightenment. Following Seligman, I would propose that the pursuit and development of virtues is the key to finding happiness, meaning, and purpose in life, and creating a positive future. The evolution of virtue will provide the necessary counter-balance to technology and materialism in generating a more comprehensive perspective on the meaning and quality of human life. Part of the New Enlightenment will be an ethical evolution – a development in human virtues.

I particularly think that a new and re-energized conception of the virtue of wisdom, incorporating many of the basic features of the above description of the New Enlightenment, will be critically important to the future of our species. In support of Macdonald, I think that the cultivation of wisdom is an essential ingredient to creating a positive future. Wisdom integrates intellect, emotion, and action. Wisdom is grounded in an expansive awareness of the whole that acknowledges and values other people and their points of view, and involves the recognition of human fallibility and the need for courage, faith, and tempered optimism in the face of the uncertainty of the future.[266] Wisdom is the highest expression of human development and future consciousness. If our minds are evolving and we are moving toward a New Enlightenment, then I would suggest that the essence of the New Enlightenment will be the individual and collective development of wisdom.

References

[1] Kuhn, Thomas *The Structure of Scientific Revolutions.* Chicago: University of Chicago Press, 1962.

[2] Anderson, Walter Truett *Reality Isn't What It Used To Be.* New York: Harper, 1990; Anderson, Walter Truett "Four Different Ways to be Absolutely Right" in Anderson, Walter Truett (Ed.) *The Truth About the Truth: De-Confusing and Re-Constructing the Postmodern World.* New York: G.P. Putnam's Sons, 1995.

[3] Ray, Paul "What Might Be the Next Stage in Cultural Evolution?" in Loye, David (Ed.) *The Evolutionary Outrider: The Impact of the Human Agent on Evolution.* Westport, Connecticut: Praeger, 1998; Ray, Paul and Anderson, Sherry *The Cultural Creatives: How 50 Million People are Changing the World.* New York: Three Rivers Press, 2000.

[4] Postrel, Virginia *The Future and Its Enemies: The Growing Conflict Over Creativity, Enterprise, and Progress.* New York: Touchstone, 1999, Introduction and Chapter One.

[5] Friedman, Thomas *The Lexus and the Olive Tree: Understanding Globalization.* New York: Farrar, Straus, and Giroux, 1999; Barber, Benjamin *Jihad vs. McWorld.* New York: Ballantine Books, 1995, 2001.

[6]Marty, Martin "Fundamentalism" in Kurian, George Thomas, and Molitor, Graham T.T. (Ed.) *Encyclopedia of the Future.* New York: Simon and Schuster Macmillan, 1996.

[7] Anderson, Walter Truett, 1990; Armstrong, Karen *The Battle for God.* New York: Ballantine Books, 2000.

[8] Kelly, Eamon *Powerful Times: Rising to the Challenge of our Uncertain World.* Upper Saddle River, New Jersey: Wharton School Publishing, 2006, Pages 55 - 62.

[9] Lombardo, Thomas "Ancient Myth, Religion, and Philosophy" in Lombardo, Thomas *The Evolution of Future Consciousness.* Bloomington, Indiana: Author House, 2006.

[10] Miller, Kenneth *Finding Darwin's God: A Scientist's Search for Common Ground between God and Evolution.* New York: Perennial, 1999; Creation Science - http://emporium.turnpike.net/C/cs/index.htm; Gardner, James *Biocosm: The New Scientific Theory of Evolution: Intelligent Life is the Architect of the Universe.* Makawao, Maui, Hawaii: Inner Ocean, 2003.

[11] Postrel, Virginia, 1999.

[12] Prigogine, Ilya and Stengers, Isabelle *Order out of Chaos: Man's New Dialogue with Nature.* New York: Bantam, 1984.

[13] Toffler, Alvin *Power Shift: Knowledge, Wealth, and Violence at the Edge of the Twenty-First Century.* New York: Bantam, 1990.

[14] Wright, Robert *Nonzero: The Logic of Human Destiny.* New York: Pantheon Books, 2000; Christian, David *Maps of Time: An Introduction to*

Big History. Berkeley, CA: University of California Press, 2004, Chapters Eleven to Thirteen.

[15] Toffler, Alvin *Future Shock*. New York: Bantam, 1971.

[16] Bishop, Peter "Change" in Kurian, George Thomas, and Molitor, Graham T.T. (Ed.) *Encyclopedia of the Future*. Simon and Schuster Macmillan, 1996.

[17] Toffler, Alvin, 1990.

[18] Stock, Gregory *Metaman: The Merging of Humans and Machines into a Global Superorganism*. New York: Simon and Schuster, 1993.

[19] Bertman, Stephen *Hyperculture: The Human Cost of Speed*. Westport, Connecticut: Praeger, 1998; Gleick, James *Faster: The Acceleration of Just About Everything*. New York: Pantheon Books, 1999,

[20] Bertman, Stephen "Cultural Amnesia: A Threat to Our Future", *The Futurist*, January-February, 2001.

[21] Didsbury, Howard F. "The Death of the Future in a Hedonistic Society" in Didbury, Howard F. (Ed.) *Frontiers of the 21st Century: Prelude to the New Millennium*. Bethesda, Maryland: World Future Society, 1999.

[22] Kurzweil, Ray *The Age of Spiritual Machines: When Computers Exceed Human Intelligence*. New York: Penguin Books, 1999; Moravec, Hans *Robot: Mere Machine to Transcendent Mind*. Oxford: Oxford University Press, 1999; KurzweilAI.Net - http://www.kurzweilai.net/; When will computer hardware match the human brain? - http://www.transhumanist.com/volume1/moravec.htm.

[23] Stross, Charles *Accelerando*. New York: Ace Books, 2005.

[24] Christian, Peter, 2004, Chapters Fourteen and Fifteen; Moore, Stephen and Simon, Julian *It's Getting Better All the Time: 100 Greatest Trends of the Last 100 Years*. Washington, D.C.: Cato Institute, 2000; Gleick, James, 1999.

[25] Acceleration Studies Foundation - http://accelerating.org; http://accelerating.org/articles/whatibelieve.html.

[26] Russell, Peter *The White Hole in Time: Our Future Evolution and the Meaning of Now*. New York: HarperCollins, 1992.

[27] A good overview of the history of emergent novel entities and processes in the universe since the beginning of time can be found in Morowitz, Harold *The Emergence of Everything: How the World Became Complex*. Oxford: Oxford University Press, 2002.

[28] Lombardo, Thomas "Ancient Myth, Religion, and Philosophy" in Lombardo, Thomas *The Evolution of Future Consciousness*. Bloomington, Indiana: Author House, 2006.

[29] Nisbet, Robert *History of the Idea of Progress*. New Brunswick: Transaction Publishers, 1994; Lombardo, Thomas "Enlightenment and the Theory of Secular Progress" in Lombardo, Thomas *The Evolution of Future Consciousness*. Bloomington, Indiana: Author House, 2006.

[30] Watson, Peter *The Modern Mind: An Intellectual History of the 20th Century*. New York: HarperCollins Perennial, 2001, Chapter Fourteen;

Best, Steven and Kellner, Douglas *The Postmodern Turn*. New York: The Guilford Press, 1997; Easterbrook, Gregg *The Progress Paradox: How Life Gets Better While People Feel Worse*. New York: Random House, 2003; Heinberg, Richard "Toward a New Definition of Progress" *The Futurist*, July-August, 1997.

[31] Dawkins, Richard *The Blind Watchmaker*. New York: W. W. Norton and Company, 1986; Dennett, Daniel C. *Darwin's Dangerous Idea*. New York: Simon and Schuster, 1995; Miller, Kenneth, 1999; Laszlo, Erwin *Evolution: The Grand Synthesis*. Shambhala, 1987; Laszlo, Erwin "One Man's Quest: A Brief History" in Loye, David (Ed.) *The Evolutionary Outrider: The Impact of the Human Agent on Evolution*. Westport, Connecticut: Praeger, 1998.; Capra, Fritjof "Evolution: The Old View and the New View" in Loye, David (Ed.) *The Evolutionary Outrider: The Impact of the Human Agent on Evolution*. Westport, Connecticut: Praeger, 1998; The Darwin Project - http://www.thedarwinproject.com/temp.html.

[32] Gould, Stephen Jay *Wonderful Life: The Burgess Shale and the Nature of History*. New York: W. W. Norton, 1989; Gould, Stephen Jay "The Pattern of Life's History" in Brockman, John *The Third Culture*. Touchstone, 1995; Stewart, John *Evolution's Arrow: The Direction of Evolution and the Future of Humanity*. Canberra, Australia: The Chapman Press, 2000.

[33] Watson, Peter, 2001, Chapters Fourteen and Thirty-Nine; Brockman, John (Ed.) *The Third Culture*. New York: Touchstone, 1995, Part One; Morris, Richard *The Evolutionists: The Struggle for Darwin's Soul*. New York: W.H. Freeman and Company, 2001.

[34] Davies, Paul *The Cosmic Blueprint: New Discoveries in Nature's Creative Ability to Order the Universe*. New York: Simon and Schuster, 1988; Goerner, Sally *Chaos and the Evolving Ecological Universe*. Luxembourg: Gordon and Breach, 1994; Smolin, Lee *The Life of the Cosmos*. Oxford: Oxford University Press, 1997; Hubbard, Barbara Marx *Conscious Evolution: Awakening the Power of Our Social Potential*. Novato, CA: New World Library, 1998; Adams, Fred and Laughlin, Greg *The Five Ages of the Universe: Inside the Physics of Eternity*. New York: The Free Press, 1999; Morowitz, Harold, 2002.

[35] Eldredge, Niles and Gould, Stephen "Punctuated Equilibria: An Alternative to Phyletic Gradualism" in Schopf, T. J. M. (Ed.) *Models in Paleobiology*. Freeman Cooper, 1972.

[36] Prigogine, Ilya and Stengers, Isabelle, 1984; Goerner, Sally, 1994; Kauffman, Stuart *At Home in the Universe*. Oxford: Oxford University Press, 1995a; Kauffman, Stuart "Order for Free" in Brockman, John *The Third Culture*. New York: Touchstone, 1995b.

[37] Margulis, Lynn *Symbiosis in Cell Evolution*. 2nd Ed. New York: W. H. Freeman, 1993; Wright, Robert, 2000; Stewart, John, 2000; Bloom, Howard *Global Brain: The Evolution of Mass Mind from the Big Bang to the 21st Century*. New York: John Wiley and Sons, Inc., 2000; Sahtouris,

Elisabet *EarthDance: Living Systems in Evolution*. Lincoln, Nebraska: IUniverse Press, 2000.

[38] Dawkins, Richard "A Survival Machine" in Brockman, John *The Third Culture*. New York: Touchstone, 1995.

[39] Fraser, J.T. *Of Time, Passion, and Knowledge: Reflections on the Strategy of Existence.* Princeton, N. J.: Princeton University Press, 1975; Fraser, J.T. *Time as Conflict*. Basel and Stuttgart: Birkhauser Verlag, 1978; Fraser, J.T. *The Genesis and Evolution of Time: A Critique of Interpretation in Physics*. Amherst, Massachusetts: University of Massachusetts Press, 1982; Fraser, J. T. *Time, the Familiar Stranger*. Redmond, Washington: Tempus, 1987; The International Society for the Study of Time - http://www.studyoftime.org/.

[40] Fraser, J. T., 1987, Pages 188 - 221, 310 - 342.

[41] Wright, Robert, 2000.

[42] Henderson, Hazel *Paradigms in Progress: Life Beyond Economics*. San Francisco: Berrett-Koehler Publishers, 1991; Eisler, Riane *The Chalice and the Blade: Our History, Our Future*. San Francisco: Harper and Row, 1987; Eisler, Riane *Sacred Pleasure: Sex, Myth, and the Politics of the Body*. San Francisco: HarperCollins, 1995.

[43] Lombardo, Thomas "The Origins of Future Consciousness" in Lombardo, Thomas *The Evolution of Future Consciousness*. Bloomington, Indiana: Author House, 2006.

[44] Stewart, John, 2000.

[45] Hubbard, Barbara Marx "Conscious Evolution" in Kurian, George Thomas, and Molitor, Graham T.T. (Ed.) *Encyclopedia of the Future*. New York: Simon and Schuster Macmillan, 1996; Hubbard, Barbara Marx, 1998; Csikszentmihalyi, Mihalyi *The Evolving Self: A Psychology for the Third Millennium*. New York: Harper Collins, 1993. Csikszentmihalyi, Mihalyi *Creativity: Flow and the Psychology of Discovery and Invention*. New York: HarperCollins, 1996.

[46] Evolve - A Global Community Center for Conscious Evolution - http://www.evolve.org/pub/doc/index2.html.

[47] Tipler, Frank *The Physics of Immortality: Modern Cosmology, God, and the Resurrection of the Dead*. New York: Doubleday, 1994.

[48] Laszlo, Erwin, 1987; Loye, David, 1998; The Darwin Project - http://www.thedarwinproject.com/temp.html; Club of Budapest - http://www.clubofbudapest.org/.

[49] Loye, David (Ed.) *The Great Adventure: Toward a Fully Human Theory of Evolution*. Albany, New York: State University of New York Press, 2004.

[50] Meadows, Dennis "Global Environmental Problems" in Kurian, George Thomas, and Molitor, Graham T.T. (Ed.) *Encyclopedia of the Future*. New York: Simon and Schuster Macmillan, 1996; Meadows, Dennis, Meadows, Donella, et al. *The Limits to Growth*. New York: Universe Books, 1972;

Meadows, Dennis, Meadows, Donella, and Randers, Jorgen *Beyond the Limits*. Toronto: McClelland & Stewart, 1992.

[51] World Watch Institute - http://www.worldwatch.org/. World Watch Institute *Vital Signs: 2001*. New York: W.W. Norton and Company, 2001; World Watch Institute *State of the World 2003*. New York: W.W. Norton and Company, 2003.

[52] Fukuyama, Francis *The End of History and the Last Man*. New York: The Free Press, 1992; Henderson, Hazel, 1991; Theobald, Robert *Turning the Century*. Indianapolis: Knowledge Systems, Inc., 1992.

[53] Slaughter, Richard *The Foresight Principle: Cultural Recovery in the 21st Century*. Westport, Connecticut: Praeger, Adamantine, 1995; Slaughter, Richard (Ed.) *The Knowledge Base of Future Studies*. Volumes I, II, and III. Hawthorn, Victoria, Australia: DDM Media Group, 1996.

[54] Kahn, Hermann *On Thermonuclear War*. Princeton, New Jersey: Princeton University Press, 1960.

[55] Ellison, Harlan "A Boy and His Dog" (1969) in Ellison, Harlan *The Beast that Shouted Love at the Heart of the World*. Signet, 1969; Miller, Walter *A Canticle for Leibowitz*. New York: Bantam Books, 1959.

[56] Wagar, W. Warren *The Next Three Futures: Paradigms of Things to Come*. New York: Praeger, 1991.

[57] Costanza, Robert "Four Visions of the Century Ahead: Will It Be Star Trek, Ecotopia, Big Government, or Mad Max", *The Futurist*, February, 1999.

[58] Moran, Richard *Doomsday: End-of-the-World Scenarios*. Indianapolis, Indiana: Alpha Books, 2003.

[59] Rees, Martin *Our Final Hour*. New York: Basic Books, 2003.

[60] Adams, Fred and Laughlin, Greg, 1999.

[61] Sartre, Jean Paul *Nausea*. New York: New Directions. 1938, 1964; Sartre, Jean Paul *Being and Nothingness*. New York: Washington Square Press, 1953; Barrett, William *Irrational Man: A Study in Existential Philosophy*. Garden City, New York: Doubleday and Company, 1958; May, Rollo, Angel, Ernest, and Ellenberger, Henri *Existence: A New Dimension in Psychiatry and Psychology*. New York: Simon and Schuster, 1967; Hergenhahn, B.R. and Olson, Matthew *An Introduction to Theories of Personality*. 6th Edition. Upper Saddle River, NJ: Prentice Hall, 2003, Chapter Sixteen.

[62] Gell-Mann, Murray *The Quark and the Jaguar: Adventures in the Simple and the Complex*. New York: W.H. Freeman and Company, 1994; Miller, Kenneth, 1999.

[63] Cornish, Edward *Futuring:The Exploration of the Future*. Bethesda, Maryland: World Future Society, 2004; Bell, Wendell *Foundations of Future Studies: Human Science for a New Era*. New Brunswick: Transactions Publishers, 1997, Vol. I.

[64] Rucker, Rudy, Sirius, R.U., and Queen Mu, *Mondo 2000: A User's Guide to the New Edge - Cyberpunk, Virtual Reality, Wetware, Designer*

Aphrodisiacs, Artificial Life, Techno-Erotic Paganism, and More. New York: Harper Collins, 1992.

[65]Prigogine, Ilya *From Being to Becoming: Time and Complexity in the Physical Sciences*. San Francisco: W. H. Freeman and Company, 1980; Prigogine, Ilya and Stengers, Isabelle, 1984; Pagels, Heinz *The Cosmic Code: Quantum Physics as the Language of Nature*. New York: Bantam, 1982; Pagels, Heinz *The Dreams of Reason: The Computer and the Rise of the Sciences of Complexity*. New York: Simon and Schuster, 1988; Goerner, Sally, 1994; Goerner, Sally *After the Clockwork Universe: The Emerging Science and Culture of Integral Society*. Norwich, Great Britain: Floris Books, 1999.

[66] Zohar, Danah *The Quantum Self: Human Nature and Consciousness Defined by the New Physics*. New York: William Morrow, 1991; Zohar, Danah and Marshall, Ian *The Quantum Society: Mind, Physics, and a New Social Vision*, New York: William Morrow and Co., Inc., 1994; Capra, Fritjof *The Turning Point*. New York: Bantam, 1983; Henderson, Hazel, 1991.

[67] Goerner, Sally, 1999.

[68] Principia Cybernetica - http://pespmc1.vub.ac.be/.

[69] Wilson, E.O. *Consilience: The Unity of Knowledge*. New York: Alfred A. Knopf, 1998; Wilson, E.O. "Back from Chaos" *The Atlantic Monthly*, March, 1998b.

[70] Wilson, E.O. "The Biological Basis of Morality" *The Atlantic Monthly*, April, 1998c.

[71] Wilson, E. O. *Sociobiology: The New Synthesis*. Cambridge, MA: Harvard University Press, 1975; Hergenhahn, B.R. and Olson, Matthew, 2003, Chapter Twelve.

[72] Kaku, Michio *Visions: How Science will Revolutionize the 21st Century*. New York: Anchor Books, 1997.

[73] Zey, Michael G. *Seizing the Future: How the Coming Revolution in Science, Technology, and Industry Will Expand the Frontiers of Human Potential and Reshape the Planet*. New York: Simon and Schuster, 1994.

[74] Brooks, Rodney A., and Flynn, Anita "Fast, Cheap and Out of Control: A Robot Invasion of the Solar System," *Journal of the British Interplanetary System*, 42, 1989; Brooks, Rodney *Flesh and Machines: How Robots Will Change Us*. New York: Pantheon Books, 2002; Bass, Thomas "Robot, build thyself" *Discover*, October, 1995; Conway, McKinley "Super Projects: Rebuilding and Improving Our Planet" *The Futurist*, March-April, 1996; Conway, McKinley "The Super Century Arrives", *The Futurist*, March, 1998; Drexler, K. Eric, Peterson, Chris, and Pergamit, Gayle *Unbounding the Future: The Nanotechnology Revolution*. New York: William Morrow, 1991; Halal, William "The Top 10 Emerging Technologies" *The Futurist*, July-August, 2000; Halal, William, Kull, Michael, and Leffmann, Ann "Emerging Technologies: What's Ahead for 2001-2030" *The Futurist*,

November-December, 1997; Riley, Robert Q. "Specialty Cars for the 21st Century: Downsized Cars with Upscale Appeal" *The Futurist*, November-December, 1995; Hiemstra, Glen "Driving in 2020: Commuting Meets Computing" *The Futurist*, September-October, 2000; Pearson, Ian (Ed.) *The Macmillan Atlas of the Future*. New York: Macmillan, 1998; Pearson, Ian "The Next 20 Years in Technology: Timeline and Commentary" *The Futurist*, January-February, 2000.

[75] Kaku, Michio, 1997; Anderson, Walter Truett *Evolution Isn't What It Used To Be: The Augmented Animal and the Whole Wired World*. New York: W. H. Freeman and Company, 1996; Naam, Ramez *More Than Human: Embracing the Promise of Biological Enhancement*. New York: Broadway Books, 2005.

[76] Joy, Bill "Why the Future Doesn't Need Us" *Wired*, April, 2000.

[77] Kurzweil, Ray, 1999; Kurzweil, Ray *The Singularity is Near: When Humans Transcend Biology*. New York: Viking Press, 2005; Brooks, Rodney, 2002.

[78] Gray, Chris Hables "Our Future as Postmodern Cyborgs" in Didsbury, Howard (Ed.) *Frontiers of the 21st Century: Prelude to the New Millennium*. Bethesda, Maryland: World Future Society, 1999; Clark, Andy *Natural-Born Cyborgs: Minds, Technologies, and the Future of Human Intelligence*. Oxford: Oxford University Press, 2003; Naam, Ramez, 2005.

[79] Kelly, Kevin *Out of Control: The Rise of Neo-Biological Civilization*. Reading, MA: Addison - Wesley, 1994; Anderson, Walter, Truett, 1996; Naisbitt, John *High Tech - High Touch: Technology and our Accelerated Search for Meaning*. London: Nicholas Brealey Publishing, 2001.

[80] Postman, Neil *Technopoly: The Surrender of Culture to Technology*. New York: Vintage Books, 1992.

[81] Csikszentmihalyi, Mihalyi, 1993; Brooks, Rodney, 2002.

[82] Dertouzos, Michael *What Will Be: How the New World of Information will Change our Lives*. New York: HarperEdge, 1997; Kurzweil, Ray, 1999.

[83] Tipler, Frank, 1994; Negroponte, Nicholas *being digital*. New York: Vintage Books, 1995.

[84] Kelly, Kevin, 1994.

[85] Toffler, Alvin *The Third Wave*. New York: Bantam, 1980; Toffler, Alvin, 1990.

[86] Naisbitt, John *Megatrends: Ten New Directions Transforming Our Lives*. New York: Warner, 1982; Naisbitt, John and Aburdene, Patricia *Megatrends 2000*. New York: Avon Books, 1990.

[87] Castells, Manuel *The Information Age: Economy, Society, and Culture. Vol. I. The Rise of the Network Society.* Second Edition. Oxford, UK: Blackwell Publishers, 2000; Castells, Manuel *The Information Age: Economy, Society, and Culture. Vol. II. The Power of Identify*. Second Edition. Oxford, UK: Blackwell Publishers, 2004; Castells, Manuel

The Information Age: Economy, Society, and Culture. Vol. III. End of Millennium. Second Edition. Oxford, UK: Blackwell Publishers, 2000.
[88] Drucker, Peter F. *Post-Capitalist Society*. New York: Harper Business, 1993; Drucker, Peter F. "The Age of Social Transformation" *Atlantic Monthly*, November, 1994; Bell, Daniel *The Coming of Post-Industrial Society*. New York: Basic Books, 1973; Bell, Daniel "Introduction: Reflections at the End of an Age" in Kurian, George Thomas, and Molitor, Graham T.T. (Ed.) *Encyclopedia of the Future*. New York: Simon and Schuster Macmillan, 1996.
[89] Negroponte, Nicholas, 1995.
[90] Cornish, Edward "The Cyber Future: 92 Ways Our Lives Will Change by the Year 2025" *The Futurist*, January-February, 1996; Dertouzos, Michael, 1997.
[91] Kurzweil, Ray, 1999; Kurzweil, Ray, 2005.
[92] Robots.Net - http://robots.net/ ; Robotics Society of America - http://www.robotics-society.org/rules.shtml; San Francisco Robotics Society of America - http://www.robots.org/.
[93] Brooks, Rodney, 2002; Rodney Brooks Page - http://www.ai.mit.edu/people/brooks/index.shtml - http://www.edge.org/3rd_culture/bios/brooks.html; Moravec, Hans, 1999; Hans Moravec Home Page - Robotics - http://www.frc.ri.cmu.edu/~hpm/.
[94] Elmer-Dewitt, Philip "Cloning: Where Do We Draw the Line" *Time*, November 8, 1993; Elmer-Dewitt, Philip "The Genetic Revolution" *Time*, January 17, 1994; Kaku, Michio, 1997; Naisbitt, John, 2001; Stock, Gregory *Redesigning Humans: Our Inevitable Genetic Future*. Boston: Houghton Mifflin Company, 2002; Naam, Ramez, 2005.
[95] Anderson, Walter Truett, 1996; Naisbitt, John and Aburdene, Patricia, 1990.
[96] Naisbitt, John, 2001; D'Souza Dinesh *The Virtue of Prosperity: Finding Values in an Age of Techno-Affluence*. New York: The Free Press, 2000; Fukuyama, Francis *Our Posthuman Future: Consequences of the Biotechnology Revolution*. New York: Picador, 2002.
[97] Stock, Gregory, 2002; Dyson, Freeman *Imagined Worlds*. Cambridge, MS: Harvard University Press, 1997.
[98] Kurzweil, Ray and Grossman, Terry *Fantastic Voyage: Live Long Enough to Live Forever*. U.S.A: Rodale, 2004.
[99] Easterbrook, Gregg *A Moment on the Earth: The Coming Age of Environmental Optimism*. New York: Viking, 1995.
[100] Kelly, Kevin, 1994; Stock, Gregory, 1993.
[101] Transhumanist Resources and Alliance - http://www.aleph.se/Trans/index-2.html.
[102] The Extropy Institute - http://www.extropy.org/; Natasha Vita More Universe - http://www.natasha.cc/.

[103] The World Transhumanist Association - http://transhumanism.org/index.php/WTA/;Transhumanist Arts and Culture Center - http://www.transhumanist.biz/.
[104] Vinge, Vernor "The Coming Technological Singularity: How to Survive in the Post-Human Era" *Vision-21: Interdisciplinary Science and Engineering in the Era of Cyberspace NASA-CP-10129*, 1993 - http://www-rohan.sdsu.edu/faculty/vinge/misc/singularity.html.
[105] Stross, Charles *Singularity Sky*. New York: Ace Books, 2003; Stross, Charles, 2005.
[106] Garreau, Joel *Radical Evolution: The Promise and Peril of Enhancing Our Minds, Our Bodies - And What it Means to be Human*. New York: Doubleday, 2005.
[107] Sagan, Carl *Cosmos*. New York: Random House, 1980; Chaisson, Eric *Cosmic Dawn*. New York: Berkley, 1981; Chaisson, Eric *The Life Era: Cosmic Selection and Conscious Evolution*. Lincoln, Nebraska: iUniverse, 1987; Ferris, Timothy *Coming of Age in the Milky Way*. New York: William Morrow and Company, 1988; Ferris, Timothy *The Mind's Sky: Human Intelligence in a Cosmic Context*. New York: Bantam, 1992; Hubbard, Barbara Marx, 1998.
[108] Bradbury, Ray *The Martian Chronicles*. New York: Bantam Books, 1950; Bester, Alfred *The Stars, My Destination*. New York: Berkley Publishing Corporation, 1956; Niven, Larry *Ringworld*. New York: Ballantine Books, 1970; Pohl, Frederick *Gateway*. New York: Ballantine Books,1977; Brin, David *Startide Rising*. New York: Bantam Books, 1983; Simmons, Dan *Hyperion*. New York: Bantam Books, 1989; Vinge, Vernor *A Fire Upon the Deep*. New York: Tom Doherty Associates, 1992; Vinge, Vernor *A Deepness in the Sky*. New York: Tom Doherty Associates, 1999.
[109] Prantzos, Nikos *Our Cosmic Future: Humanity's Fate in the Universe*. Cambridge: Cambridge University Press, 2000.
[110] Berry, Adrian *The Next 500 Years: Life in the Coming Millennium*. New York: W. H. Freeman and Co., 1996; Sagan, Carl *Pale Blue Dot*. New York: Random House, 1994; Savage, Marshall *The Millennial Project: Colonizing the Galaxy in Eight Easy Steps*. Boston: Little, Brown, and Company, 1992; Wachhorst, Wyn *The Dream of Spaceflight: Essays on the Near Edge of Infinity*. New York: Basic Books, 2000.
[111] Sagan, Dorian *Biospheres: Metamorphosis of Planet Earth*. New York: McGraw - Hill, 1990.
[112] The Mars Society - http://www.marssociety.org/; Mars Direct - http://www.rps.psu.edu/0305/direct.html, http://en.wikipedia.org/wiki/Mars_Direct, http://www.astronautix.com/craft/marirect.htm; Zubrin, Robert *The Case for Mars: The Plan to Settle the Red Planet and Why We Must*. New York: The Free Press, 1996; Robinson, Kim Stanley *Red Mars*. New York: Bantam, 1991; Robinson, Kim Stanley *Green Mars*. New York: Bantam, 1994; Robinson, Kim Stanley *Blue Mars*. New York: Bantam, 1996; The Artemis Project - Moon Settlement - http://www.asi.org/.

[113] Tipler, Frank, 1994.
[114] Wachhorst, Wyn, 2000.
[115] Space Future - http://www.spacefuture.com/; The Planetary Society - http://www.planetary.org/; The NASA Homepage - http://www.nasa.gov/; The National Space Society - http://www.nss.org/
[116] Anderson, Walter Truett, 1990; Easterbrook, Gregg, 1995.
[117] Ray, Paul, 1998; Ray, Paul and Anderson, Sherry, 2000.
[118] Lovelock, James *Gaia*. Oxford: Oxford University Press, 1979; Lovelock, James *The Ages of Gaia*. New York: W. W. Norton, 1988.
[119] Hines, Andy "Population Growth: Two Warring Paradigms" *The Futurist*. January-February, 1998.
[120] Sahtouris, Elisabet, 2000.
[121] DeGraaf, John, Wann, David, and Naylor, Thomas *Affluenza: The All-Consuming Epidemic*. San Francisco: Berret-Koehler Publishers, Inc., 2001.
[122] World Watch Institute, 2001; World Watch Institute, 2003; World Watch Institute - http://www.worldwatch.org/.
[123] Carson, Rachel *Silent Spring*. New York: Fawcett, 1962; Ehrlich, Paul R. *The Population Bomb*. New York: Sierra Club/Ballantine, 1971; Meadows, Dennis, Meadows, Donella, et al., 1972; Meadows, Dennis, Meadows, Donella, and Randers, Jorgen, 1992.
[124] Easterbrook, Gregg, 1995; Anderson, Walter Truett, 1996; Bailey, Ronald (Ed.) *Earth Report 2000: Revisiting the True State of the Planet*. McGraw-Hill, 2000; Moore, Stephen and Simon, Julian, 2000; Lomberg, Bjørn *The Skeptical Environmentalist: Measuring the Real State of the World*. Cambridge, UK: Cambridge University Press, 2001.
[125] Capra, Fritjof *The Tao of Physics*. New York: Bantam, 1975; Capra, Fritjof *The Web of Life*. New York: Doubleday, 1996; Ackoff, Russell "From Mechanistic to Social Systematic Thinking" *Systems Thinking in Action Conference*, Pegasus Communications, Inc., 1993.
[126] Morowitz, Harold, 2002.
[127] Goerner, Sally, 1994.
[128] Lombardo, Thomas "The Scientific Revolution" in Lombardo, Thomas *The Evolution of Future Consciousness*. Bloomington, Indiana: Author House, 2006.
[129] Smolin, Lee, 1997.
[130] Gibson, James J. *The Ecological Approach to Visual Perception*. Boston: Houghton Mifflin, 1979.
[131] Goerner, Sally, 1999.
[132] Capra, Fritjof, 1975; Capra, Fritjof, 1983; Berman, Morris *The Reinchantment of the World*. New York: Bantam, 1981.
[133] Foundation for Global Community - http://www.globalcommunity.org/.
[134] Henderson, Hazel, 1991.

[135] Anderson, Walter Truett *The Future of the Self*. New York: Putnam, 1997.
[136] Wright, Robert, 2000.
[137] Calvin, William *A Brief History of the Mind: From Apes to Intellect and Beyond*. New York: Oxford University Press, 2004; Diamond, Jared *The Third Chimpanzee: The Evolution and Future of the Human Animal*. New York: HarperPernnial, 1992; Shlain, Leonard *Sex, Time, and Power: How Women's Sexuality Shaped Human Evolution*. New York: Viking, 2003; Jaynes, Julian *The Origin of Consciousness in the Breakdown of the Bicameral Mind*. Boston: Houghton Mifflin, 1976; White, Randall *Prehistoric Art: The Symbolic Journey of Humankind*. New York: Harry N. Abrams, 2003.
[138] Anderson, Walter Truett, 1997; Anderson, Walter Truett *The Next Enlightenment: Integrating East and West in a New Vision of Human Evolution*. New York: St. Martin's Press, 2003.
[139] Garreau, Joel, 2005.
[140] Wheatley, Margaret *Leadership and the New Science*. San Francisco: Berrett-Koehler, 1992.
[141] Hubbard, Barbara Marx, 1998; Hubbard, Barbara Marx *Emergence: The Shift from Ego to Essence*. Charlottesville, VA: Hampton Roads Publishing, 2001; Russell, Peter, 1992.
[142] Anderson, Walter Truett, 1997.
[143] Csikszentmihalyi, Mihalyi, 1993.
[144] O'Hara, Maureen "Future Mind: Three Scenarios for a Psychological Future" World Future Society, Washington, D.C., 1999.
[145] Paepke, C. Owen *The Evolution of Progress: The End of Economic Growth and the Beginning of Human Transformation*. New York: Random House, 1993.
[146] Henderson, Hazel "Economics and Evolution: An Ethos for an Action Researcher" in Loye, David (Ed.) *The Evolutionary Outrider: The Impact of the Human Agent on Evolution*. Westport, Connecticut: Praeger, 1998.
[147] Russell, Peter, 1992.
[148] Wishard, William Van Dusen "Future View: Humanity as a Single Entity" *The Futurist*, March-April, 1996; Wishard, William Van Dusen "Cultural Change" in Kurian, George Thomas, and Molitor, Graham T.T. (Ed.) *Encyclopedia of the Future*. New York: Simon and Schuster Macmillan, 1996b; Wishard, William Van Dusen "Epochal Change" in Kurian, George Thomas, and Molitor, Graham T.T. (Ed.) *Encyclopedia of the Future*. New York: Simon and Schuster Macmillan, 1996c; Molitor, Graham T.T. and Wishard, William Van Dusen "Values Change" in Kurian, George Thomas, and Molitor, Graham T.T. (Ed.) *Encyclopedia of the Future*. New York: Simon and Schuster Macmillan, 1996.
[149] Ray, Paul and Anderson, Sherry, 2000.

[150] Bolt, Martin *Pursuing Human Strengths: A Positive Psychology Guide*. New York: Worth Publishers, 2004; Martin Seligman - Authentic Happiness -http://www.authentichappiness.org/.

[151] Maslow, Abraham *Toward a Psychology of Being*. New York: D. Van Nostrand Co., 1968; Maslow, Abraham *The Farther Reaches of Human Nature*. Middlesex, England: Penguin Books, 1972; Rogers, Carl *On Becoming a Person*. Boston: Houghton-Mifflin, 1961; Rogers, Carl *A Way of Being*. Boston: Houghton-Mifflin, 1980.

[152] Seligman, Martin *Learned Optimism: How to Change Your Mind and Your Life*. New York: Pocket Books, 1998; Seligman, Martin *Authentic Happiness: Using the New Positive Psychology to Realize Your Potential for Lasting Fulfillment*. New York: The Free Press, 2002.

[153] Lombardo, Thomas "The Psychology and Value of Future Consciousness" in Lombardo, Thomas *The Evolution of Future Consciousness*. Bloomington, Indiana: Author House, 2006.

[154] Eisler, Riane, 1987; Tarnas, Richard *The Passion of the Western Mind: Understanding the Ideas that have Shaped Our World View*. New York: Ballantine, 1991; Shlain, Leonard *The Alphabet Versus the Goddess: The Conflict Between Word and Image*. New York: Penguin Arkana, 1998.

[155] O'Hara, Maureen, "Constructing Emancipatory Realities: Toward a Feminist Voice in Humanistic Psychology", *AHP Perspective*, August/September, 1989; Pinker, Steven *The Blank Slate: The Modern Denial of Human Nature*. New York: Penguin Books, 2002, Chapter Eighteen.

[156] Wagner, Cynthia (Ed.) "Women's Preferred Futures" *The Futurist*, May-June, 1997.

[157] Eisler, Riane, 1987; Eisler, Riane, 1995.

[158] Eisler, Riane "Conscious Evolution: Cultural Transformation and Human Agency" in Loye, David (Ed.) *The Evolutionary Outrider: The Impact of the Human Agent on Evolution*. Westport, Connecticut: Praeger, 1998; Eisler, Riane "A Multilinear Theory of Cultural Evolution: Genes, Culture, and Technology" in Loye, David (Ed.) *The Great Adventure: Toward a Fully Human Theory of Evolution*. Albany, New York: State University of New York Press, 2004.

[159] Tarnas, Richard, 1991.

[160] Belenky, Mary Field, Clinchy, Blythe McVicker, Goldberger, Nancy Rule, and Tarule, Jill Mattuck *Women's Ways of Knowing: The Development of Self, Voice, and Mind*. New York: Basic Books, 1986; Gray, Elizabeth Dodson "Women and Religion" in Kurian, George Thomas, and Molitor, Graham T.T. (Ed.) *Encyclopedia of the Future*. New York: Simon and Schuster Macmillan, 1996.

[161] Women's Resources on the Internet - http://www.ibiblio.org/cheryb/women/wresources.html; The Feminist Majority Foundation Online - http://www.feminist.org/.

[162] Eisler, Riane, 1987; Eisler, Riane, 1995; The Center for Partnership Studies - http://www.partnershipway.org/; Tarnas, Richard, 1991.

[163] Toffler, Alvin, 1980.
[164] Wheatley, Margaret, 1992.
[165] Senge, Peter *The Fifth Discipline: The Art and Practice of the Learning Organization*. New York: Doubleday, 1990.
[166] Theobald, Robert,1992.
[167] Henderson, Hazel, 1991.
[168] Shlain, Leonard, 1998.
[169] Wright , Robert, 2000.
[170] Bloom, Howard, 2000.
[171] Naisbitt, John, 1982; Naisbitt, John *Global Paradox* . New York: Avon Books, 1994; Naisbitt, John and Aburdene, 1990.
[172] Wishard, William Van Dusen "Future View: Humanity as a Single Entity" *The Futurist*, March-April, 1996.
[173] Friedman, Thomas, 1999.
[174] Anderson, Walter Truett *All Connected Now: Life in the First Global Civilization*. Boulder; Westview Press, 2001.
[175] Friedman, Thomas, 1999.
[176] Friedman, Thomas *The World is Flat: A Brief History of the Twenty-first Century*. New York: Farrar, Straus, and Giroux, 2005.
[177] Friedman, Thomas, 2005, Chapter One.
[178] "The Great Leveling" by Warren Bass in Washington Post.Com - http://www.washingtonpost.com/wp-dyn/articles/A17314-2005Mar31.html.
[179] "The Great Leveling" by Warren Bass in Washington Post.Com - http://www.washingtonpost.com/wp-dyn/articles/A17314-2005Mar31.html; The World is Flat - http://en.wikipedia.org/wiki/The_World_Is_Flat.
[180] Anderson, Walter Truett, 2001.
[181] Stock, Gregory, 1993.
[182] Bugliarello, George "Hyperintelligence: The Next Evolutionary Step" *The Futurist*, December, 1984; Glenn, Jerome *Future Mind: Artificial Intelligence*. Washington, D.C.: Acropolis Books, 1989; Glenn, Jerome "Post-Information Age" in Kurian, George Thomas, and Molitor, Graham T.T. (Ed.) *Encyclopedia of the Future*. New York: Simon and Schuster Macmillan, 1996.
[183] Naisbitt, John, 1994.
[184] Bloom, Howard, 2000.
[185] Toffler, Alvin, 1980; Capra, Fritjof, 1983; Fukuyama, Francis, 1992.
[186] Ray, Paul and Anderson, Sherry, 2000.
[187] Templeton, John "Assessing Human Progress: A Worldwide Rise in Living Standards", *The Futurist*, January, 1999; Henderson, Hazel, 1991; Moore, Stephen and Simon, Julian, 2000.
[188] Lombardo, Thomas "Enlightenment and the Theory of Secular Progress" in Lombardo, Thomas *The Evolution of Future Consciousness*. Bloomington, Indiana: Author House, 2006.
[189] Fukuyama, Francis, 1992.
[190] Barber, Benjamin, 1995, 2001.

[191] Toffler, Alvin, 1990; Drucker, Peter, 1993; Wheatley, Margaret, 1992.
[192] Schwartz, Peter, Leyden, Peter, and Hyatt, Joel *The Long Boom: A Vision for the Coming Age of Prosperity*. Cambridge, MA: Perseus, 1999.
[193] D'Souza Dinesh, 2000.
[194] Bell, Daniel *The Cultural Contradictions of Capitalism*. New York: Basic Books, 1976; Fukuyama, Francis *The Great Disruption: Human Nature and the Reconstitution of Social Order*. New York: The Free Press, 1999; Putnam, Robert *Bowling Alone: The Collapse and Revival of American Community*. New York: Touchstone, 2000; Easterbrook, Gregg, 2003.
[195] Barber, Benjamin, 1995, 2001.
[196] DeGraaf, John, Wann, David, and Naylor, Thomas, 2001.
[197] Sardar, Ziauddin *Rescuing All Our Futures: The Future of Future Studies*. Westport, Connecticut: Praeger, 1999; Sardar, Ziauddin "The Problem of Future Studies" in Sardar, Ziauddin (Ed.) *Rescuing All Our Futures: The Future of Future Studies*. Westport, Connecticut: Praeger, 1999b.
[198] Triandis, Harry *Individualism and Collectivism*. Boulder, CO: Westview Press, 1995; Triandis, Harry "The Psychological Measurement of Cultural Syndromes" *American Psychologist*, Vol. 51, 1996.
[199] Inglehart, Ronald and Baker, Wayne "Modernization's Challenge to Traditional Values: Who's Afraid of Ronald McDonald?" *The Futurist*, March-April, 2001.
[200] Nisbett, Richard *The Geography of Thought: How Asians and Westerners Think Differently ...and Why*. New York: The Free Press, 2003.
[201] Toffler, Alvin, 1980.
[202] Razak, Victoria and Cole, Sam "Culture and Society" in Kurian, George Thomas, and Molitor, Graham T.T. (Ed.) *Encyclopedia of the Future*. New York: Simon and Schuster Macmillan, 1996.
[203] Huntington, Samuel *The Clash of Civilizations and the Remaking of World Order*. New York: Touchtone, 1996; Bell, Wendell "The Clash of Civilizations and Universal Human Values", *Journal of Futures Studies*, Vol. 6, No. 3, February, 2002.
[204] Best, Steven and Kellner, Douglas *Postmodern Theory: Critical Interrogations*. New York: The Guilford Press, 1991; Best, Steven and Kellner, Douglas,1997.
[205] Lombardo, Thomas "Romanticism" in Lombardo, Thomas *The Evolution of Future Consciousness*. Bloomington, Indiana: Author House, 2006.
[206] Nisbet, Robert, 1994.
[207] Lombardo, Thomas "Enlightenment and the Theory of Secular Progress" in Lombardo, Thomas *The Evolution of Future Consciousness*. Bloomington, Indiana: Author House, 2006.
[208] Gleick, James *Chaos*. New York: Viking, 1987; Prigogine, Ilya and Stengers, Isabelle, 1984; Goerner, Sally, 1994.
[209] Wheatley, Margaret, 1992.
[210] Rucker, Rudy, Sirius, R.U., and Queen Mu, 1992.

[211] Wired News - http://www.hotwired.com/
[212] Fukuyama, Francis, 1992.
[213] Wright, Robert, 2000.
[214] Lombardo, Thomas "Hegel and Marx" in Lombardo, Thomas *The Evolution of Future Consciousness*. Bloomington, Indiana: Author House, 2006.
[215] Naisbitt, John and Aburdene, Patricia, 1990.
[216] Naisbitt, John, 1994.
[217] Putnam, Robert, 2000.
[218] Putnam, Robert and Feldstein, Lewis *Better Together: Restoring the American Community*. New York: Simon and Schuster, 2003; Robert Putnam - Bowling Alone; http://www.bowlingalone.com/; Better Together - http://www.bettertogether.org/; "Bowling Together" in The American Prospect, Feb. 2/11/2002 - http://www.prospect.org/print/V13/3/putnam-r.html.
[219] Florida, Richard *The Rise of the Creative Class*. New York: Basic Books, 2002.
[220] Florida, Richard "The World is Spiky" *Atlantic Monthly*, October, 2005.
[221] Florida, Richard *The Flight of the Creative Class: The New Global Competition for Talent*. New York: HarperCollins, 2005.
[222] Quinn, Daniel *Beyond Civilization: Humanity's Next Great Adventure*. New York: Three Rivers Press, 1999.
[223] Lombardo, Thomas "Ancient Myth, Religion, and Philosophy" in Lombardo, Thomas *The Evolution of Future Consciousness*. Bloomington, Indiana: Author House, 2006.
[224] Noss, David *A History of the World's Religions*. 10th Ed. Upper Saddle River, N.J.: Prentice Hall, 1999.
[225] Armstrong, Karen *A History of God: The Four Thousand Year Quest of Judaism, Christianity, and Islam*. New York: Alfred Knopf, 1994.
[226] Nisbett, Richard, 2003.
[227] Molé, Phil "Deepak's Dangerous Dogmas" *Skeptic*, Vol.6, No.2, 1998.
[228] Walsch, Neal Donald *Conversations with God: Book One*. New York: G.P. Putnam's Sons, 1995; Walsch, Neal Donald *Conversations with God: Book Two*. Charlottesville, VA: Hampton Roads, 1997; Walsch, Neal Donald *Conversations with God: Book Three*. Charlottesville, VA: Hampton Roads, 1998; Conversations with God Foundation – Neal Walsch - http://www.cwg.org/.
[229] Creedon, Jeremiah "God with a Million Faces", *Utne Reader*, July-August, 1998.
[230] World Transformation - http://newciv.org/worldtrans/; New Age Web Works - http://www.newageinfo.com/; Earth Portals - http://www.earthportals.com/; Enlightenment.com - http://www.enlightenment.com/; The Genesis of Eden - http://www.dhushara.com/book/genesis.

htm; IKosmos - The Portal for Cultural Creativity - http://www.ikosmos.com/; New Age Journal - http://www.newagejournal.com/.
[231] Wilber, Ken *A Brief History of Everything*. Boston: Shambhala, 1996; Wilber, Ken *The Marriage of Sense and Soul: Integrating Science and Religion*. New York: Random House, 1998; Hubbard, Barbara Marx, 1998.
[232] Fraser, J. T., 1978; Fraser, J. T., 1982; Fraser, J. T., 1987.
[233] Chardin, Teilhard de *The Phenomenon of Man*. New York: Harper, 1959.
[234] Omega Point Web page of Transhumanism - http://www.aleph.se/Trans/Global/Omega/.
[235] Dyson, Freeman *Infinite in All Directions*. New York: Harper and Row, 1988.
[236] Tipler, Frank, 1994.
[237] Barrow, John, and Tipler, Frank *The Anthropic Cosmological Principle*. Oxford: Oxford University Press, 1986.
[238] Gardner, James *Biocosm: The New Scientific Theory of Evolution: Intelligent Life is the Architect of the Universe*. Makawao, Maui, Hawaii: Inner Ocean, 2003.
[239] Integral Naced - http://integralnaked.org/.
[240] Goerner, Sally, 1999.
[241] Foundation for Global Community - http://www.globalcommunity.org/.
[242] Spayde, Jon "The New Renaissance" *Utne Reader*, February, 1998.
[243] Elgin, Duane *Awakening Earth: Exploring the Evolution of Human Culture and Consciousness*. William Morrow and Company, 1993; Awakening Earth - Duane Elgin - http://www.awakeningearth.org/; Eisler, Riane, 1987; Eisler, Riane, 1995; Henderson, Hazel, 1991; Henderson, Hazel *Building a Win-Win World: Life Beyond Global Economic Warfare*. San Francisco: Berrett-Koehler Publishers, 1996; Hubbard, Barbara Marx, 1998.
[244] Wilber, Ken, 1996; Wilber, Ken, 1998; Wilber, Ken *Sex, Ecology, and Spirit: The Spirit of Evolution*. Boston: Shambhala, 1995; Wilber, Ken *The Essential Ken Wilber: An Introductory Reader*. Boston and London: Shambhala, 1998b; The World of Ken Wilber - Integral World- http://www.worldofkenwilber.com/; Integral World - http://www.integralworld.net/.
[245] What is Enlightenment? - http://www.wie.org/; Integral Future Studies - http://www.swin.edu.au/afi/research/integral_futures.htm.
[246] Koestler, Arthur *Janus: A Summing Up*. New York: Random House, 1987.
[247] Beck, Don Edward and Cowan, Christopher *Spiral Dynamics: Mastering Values, Leadership, and Change*. Oxford, UK: Blackwell Publishers, 1996; Spiral Dynamics Home Page - http://www.spiraldynamics.com/; Spiral Dynamics Integral - http://www.spiraldynamics.net/ - http://www.spiraldynamicsgroup.com/.

[248] Summary of Spiral Dynamics by Don Beck and Christopher Cowan by Steve Dinan - http://www.spiraldynamics.com/book/SDreview_Dinan.htm; Colors of Thinking in Spiral Dynamics - http://www.spiraldynamics.org/Graves/colors.htm.
[249] Colors of Thinking in Spiral Dynamics - http://www.spiraldynamics.org/Graves/colors.htm.
[250] Spiral Dynamics Integral - http://www.spiraldynamics.net/.
[251] Zey, Michael G. *The Future Factor: The Five Forces Transforming Our Lives and Shaping Human Destiny*. New York: McGraw-Hill, 2000.
[252] Expansionary Institute – Michael Zey - http://www.zey.com/.
[253] Kelly, Eamon, 2006.
[254] Anderson, Walter Truett, 1990; Anderson, Walter Truett, 1996; Anderson, Walter Truett, 1997; Anderson, Walter 2001; Anderson, Walter Truett, 2003.
[255] Anderson, Walter Truett, 1995.
[256] Communities of the Future - http://communitiesofthefuture.org/; Futures Generative Dialogue for 2nd Enlightenment Clubs - http://communitiesofthefuture.org/articles/2nd%20enlightenment%20clubs.html.
[257] Goerner, Sally, 1999.
[258] The Wisdom Page - Wisdom Resources - http://www.isn.net/info/wisdompg.html; Macdonald, Copthorne *Matters of Consequence: Creating a Meaningful Life and a World that Works*. Charlottetown, Prince Edward Island, Canada: Big Ideas Press, 2004.
[259] Macdonald, Copthorne *Toward Wisdom: Finding Our Way Toward Inner Peace, Love, and Happiness*. Charlottesville, Virginia: Hampton Roads Publishing Company, 1996, Preface and Chapter One; Maslow, Abraham, 1968; Maslow, Abraham, 1972.
[260] Smolin, Lee, 1997.
[261] Bloom, Howard *The Lucifer Principle: A Scientific Expedition into the Forces of History*. New York: The Atlantic Monthly Press, 1995.
[262] Lombardo, Thomas "The Nature and Value of Future Consciousness" in Lombardo, Thomas *The Evolution of Future Consciousness*. Bloomington, Indiana: Author House, 2006.
[263] Bloom, Howard, 1995; Ghiglieri, Michael *The Dark Side of Man: Tracing the Origins of Male Violence*. New York: Helix Books/Basic Books, 1999.
[264] Lombardo, Thomas "Ancient Myth, Religion, and Philosophy" in Lombardo, Thomas *The Evolution of Future Consciousness*. Bloomington, Indiana: Author House, 2006.
[265] Davis, Don C. "Faith for the Future: Updating Religious Paradigms for the Infotech Age" in *The Futurist*, September-October, 2005.
[266] Sternberg, Robert (Ed.) *Wisdom: Its Nature, Origins, and Development*. New York: Cambridge University Press, 1990; The Wisdom Page - http://www.isn.net/info/wisdompg.html; Lombardo, Thomas "The Pursuit of Wisdom and the Future of Education" *Creating Global Strategies*

for Humanity's Future. Mack, Timothy C. (Ed.) World Future Society, Bethesda, Maryland, 2006.

Printed in the United States
61420LVS00003B/29

9 781425 945770